DIFFERENTIAL SCANNING CALORIMETRY

CALORIMETRY

Applications
in Fat and
Oil Technology

DIFFERENTIAL SCANNING CALORIMETRY

Applications in Fat and Oil Technology

EDITED BY

EMMA CHIAVARO

CRC Press
Taylor & Francis Group
Boca Raton London New York

CRC Press is an imprint of the
Taylor & Francis Group, an **informa** business

CRC Press
Taylor & Francis Group
6000 Broken Sound Parkway NW, Suite 300
Boca Raton, FL 33487-2742

First issued in paperback 2019

© 2015 by Taylor & Francis Group, LLC
CRC Press is an imprint of Taylor & Francis Group, an Informa business

No claim to original U.S. Government works

ISBN-13: 978-1-4665-9152-3 (hbk)
ISBN-13: 978-0-367-37806-6 (pbk)

Visit the Taylor & Francis Web site at
http://www.taylorandfrancis.com

and the CRC Press Web site at
http://www.crcpress.com

Dedication

To my mother, dear and unique.

Contents

SECTION I Recent and New Perspectives from DSC Application on Vegetable Oils and Fats

SECTION II Application of DSC in Oil and Fat Technology: Coupling with Other Thermal and Physical Approaches

SECTION III   DSC in Food Technology: Palm Products, Lipid Modification, Emulsion Stability

Preface

La dernière démarche de la raison est de reconnaître qu'il y a une infinité de choses qui la surpassent. Elle n'est que faible si elle ne va jusqu'à connaître cela.

Reason's last step is the recognition that there are an infinite number of things which are beyond it. It is merely feeble if it does not go as far as to realize that.

Blaise Pascal, *Pensées*

This book focuses on the application of thermal techniques in the field of oil and fat technology, almost all differential scanning calorimetry (DSC), but also other types of calorimetric apparatus and techniques. It is divided into three sections.

In the first section, comprising Chapters 1 through 4, the application of DSC to evaluate cooling and heating profiles obtained on fats and vegetable oils in relation to chemical composition is presented, taking into consideration the advantages and disadvantages of its use and discussing other possibilities in the determination of authenticity, adulteration, and oil quality.

In Chapter 1, the application of DSC to the main fats and vegetable oils is presented, relating cooling and heating profiles to their chemical composition. Chapter 2 deals with the evaluation of vegetable oil and fat adulteration by means of DSC and the main advantages that could be obtained in this field by applying the technique. Chapter 3 addresses the best-known application of the technique, to vegetable oils, summarizing the recent developments in DSC analysis to evaluate thermooxidation and efficacy of antioxidants in vegetable oils. Chapter 4 presents an overview of the literature on DSC use for the main quality aspects of olive oils.

The second section (Chapters 5 through 7) is more theoretical and deals with the application of DSC to oil and fat technology in relation to other thermal techniques and physical approaches.

Chapter 5 focuses on the application of thermogravimetry and differential thermal analysis in the field of oils and fats, also taking into consideration bio-oils used as fuels. In Chapter 6, the application of coupled DSC–x-ray diffraction (XRD) analysis is described, also taking into account its potential in the field of provisional stability of food preparation. Chapter 7 analyzes the coupling potential offered by DSC and other techniques such as low-resolution nuclear magnetic resonance (p-NMR), low-intensity ultrasound, and polarized light microscopy in the study of crystallization kinetics of lipid models.

The last section (Chapters 8 through 10) is more application related. Beginning with DSC use in the field of quality evaluation of the main current sources of oils and fats, such as palm, palm kernel, and coconut oils and their fractions (Chapter 8), some other important aspects of lipid technology in which DSC is currently employed are described.

Chapter 9 covers the use of DSC in the field of lipid physicochemical modification, such as interesterification, fractionation, and the addition of crystallization agents (seeding) or structural compounds to lipid systems. Chapter 10 provides an overview of the main literature data on the thermal behavior of emulsion ingredients and emulsion properties.

The book is planned as a summary of the scientific literature in the field, with relevant experts presenting data published in the last 20 years. It is structured so as to provide readers (academicians, food industry sector operators, and food engineers and technologists) with knowledge of applications of calorimetry in the field of oils and fats. With the information gained they will discover new ideas on the development of food products in which fats, oils, or both are usually present as ingredients, using the information reported in the text to lead innovation in their research areas and exploring new possibilities of DSC and related thermogravimetric techniques in the field, either alone or in combination with other chemical and physical approaches. The book is also novel, as there is currently no comprehensive text (including review articles) covering this topic.

All my kindly and friendly acknowledgment to the contributors, as this book could not have come into existence without their hard work, and to the CRC editors, Steve Zollo and Marsha Pronin, for their precious and enthusiastic support.

This is the book I would have liked to find on a bookshelf when I began my research activity in this field. Sometimes dreams come true.

Happy reading to everyone.

Emma Chiavaro

Editor

Emma Chiavaro was born in Rome, Italy, in 1968. She graduated in pharmaceutical chemistry and technology cum laude from the Sapienza University of Rome. She received a specialization degree in food chemistry and technology from the University of Parma in 1997 and a PhD in food of animal origin inspection from the University of Naples Federico II in 2000.

In 2005, she was promoted to assistant professor in food science and technology (SDS AGR/15) at the University of Parma, Department of Food Science. In 2014, she became associate professor in food science and technology at the same university. She teaches several food technology courses and acts as cosupervisor or supervisor for experimental work for more than 60 theses (first and advanced degree courses, degree in food science and technology, University of Parma).

Professor Chiavaro is associate editor of the *Journal of the Science of Food and Agriculture* (Wiley), a member of the editorial board for *Lipid Insights* (Libertas Academica), *Advances in Chemistry* (Hindawi Publishing Corporation), and *Journal of Food Research* (Canadian Center of Science and Education), and acts as reviewer for more than 30 scientific journals. She is the author of 70 publications in international peer-reviewed journals, 21 in national journals, and 58 abstracts or communications to national and international congresses and conferences. She is a participant in national and international research projects.

Her research activity has included the use of differential scanning calorimetry for the evaluation of the thermal properties of food, mainly taking into consideration the applicability of this technique to the identification of the quality and genuineness of vegetable oils, in particular extra virgin olive oil. The technology of oils and fats; the processing aspects of food composition, formulation, shelf life, and stability; and the effect of cooking and other preservation techniques on the physicochemical and nutritional properties of vegetables and meat are also topics of her investigations.

Contributors

Gianmichele Arrighetti
Institute of Crystallography
Italian National Research Council
Trieste, Italy

Ana Paula Badan Ribeiro
School of Food Engineering
University of Campinas
Campinas, Brazil

Luisa Barba
Institute of Crystallography
Italian National Research Council
Trieste, Italy

Rodrigo Corrêa Basso
Institute of Chemistry
Federal University of Alfenas
Alfenas, Brasil

Alessandra Bendini
Department of Agricultural and Food
    Sciences
University of Bologna
Bologna, Italy

and

Interdepartmental Centre for Agri-
    Food Industrial Research (CIRI
    Agroalimentare)
University of Bologna
Bologna, Italy

Sonia Calligaris
Department of Food Science
University of Udine
Udine, Italy

Lorenzo Cerretani
Pizzoli S.p.A
Budrio, Italy

Emma Chiavaro
Department of Food Science
University of Parma
Parma, Italy

Stefano Vecchio Ciprioti
Department of Basic and Applied
    Science for Engineering
Sapienza University of Rome
Rome, Italy

Glazieli Marangoni de Oliveira
School of Chemical Engineering
University of Campinas
Campinas, São Paulo, Brazil

Katarzyna Jodko-Piórecka
Faculty of Chemistry
University of Warsaw
Warsaw, Poland

Theo Guenter Kieckbusch
School of Chemical Engineering
University of Campinas
Campinas, São Paulo, Brazil

Hong Kwong Lim
Department of Food Technology
Putra University Malaysia
Selangor, Malaysia

Grzegorz Litwinienko
Faculty of Chemistry
University of Warsaw
Warsaw, Poland

Like Mao
School of Food and Nutritional Sciences
University College Cork
Cork, Ireland

Mohammed Nazrim Marikkar
Department of Biochemistry and Halal
    Products Research Institute
Putra University Malaysia
Selangor, Malaysia

Silvana Martini
Department of Nutrition, Dietetics, and
    Food Sciences
Utah State University
Logan, Utah

Monise Helen Masuchi
School of Chemical Engineering
University of Campinas
Campinas, São Paulo, Brazil

Song Miao
Department of Food Chemistry and
    Technology
Teagasc Food Research Centre
Fermoy, Ireland

Imededdine Arbi Nehdi
Department of Chemistry
King Saud University
Riyadh, Saudi Arabia

Siou Pei Ng
Department of Food Technology
Putra University Malaysia
Selangor, Malaysia

Maria Cristina Nicoli
Department of Food Science
University of Udine
Udine, Italy

Maria Teresa Rodriguez-Estrada
Department of Agricultural and Food
    Sciences
University of Bologna
Bologna, Italy
and
Interdepartmental Centre for Agri-
    Food Industrial Research (CIRI
    Agroalimentare)
University of Bologna
Bologna, Italy

Valter Luís Zuliani Stroppa
School of Chemical Engineering
University of Campinas
Campinas, São Paulo, Brazil

Chin Ping Tan
Department of Food Technology
Putra University Malaysia
Selangor, Malaysia

Introduction

Stefano Vecchio Ciprioti and Emma Chiavaro

INTRODUCTION TO THE TECHNIQUE

Calorimetry is a universal method to measure the heat flow into or out of a sample, or, better, to investigate the generation or consumption of heat connected with chemical reactions and physical transitions that encompass endothermic or exothermic processes or changes in heat capacity.

Differential scanning calorimetry (DSC) measures the temperatures and heat flows associated with transitions in materials as a function of time and temperature in a controlled atmosphere: endothermic transitions (e.g., melting, transition helix-coil transitions in DNA, protein denaturation, dehydrations, reduction reactions) in which heat flows into the sample or exothermic ones (e.g., crystallization, some cross-linking processes, oxidation reactions) in which heat flows out of the sample. The well-known and well-written book of Höhne et al. (2003) summarizes the principles and fundamentals of DSC theory, giving detailed information on calibration modes, curve evaluation, and general applications.

In power compensation-type DSC, the temperature of the sample is constantly adjusted to match that of an inert reference material as the temperature range is scanned. In a heat flux instrument, heat flow is determined from the differential temperature across a fixed thermal path between the sample and the reference. Generally, a flow of inert gas (i.e., dry helium or nitrogen) is maintained over the samples to create a reproducible and dry atmosphere that also eliminates air oxidation of the samples at high temperatures. A reactive atmosphere can be used to study a particular process occurring in a given sample. A very interesting topic that has recently attracted interest is the absorption of CO_2 from a nanostructured material, which can be followed properly using a DSC (or a simultaneous thermogravimetry [TG]/DSC) device under isothermal or nonisothermal conditions. Another common application of a reactive atmosphere is for an oxidation study, which is carried out under a flowing pure oxygen or air atmosphere instead of nitrogen.

The sample is sealed, sometimes hermetically, into a small aluminum or stainless steel pan that holds up to about 1–10 mg of material. The reference is usually made up of an empty pan and cover.

Each peak of a DSC curve, in which heat flow is reported as a function of time or temperature, is due to an enthalpy change associated with a specific physical or chemical process. From each typical DSC peak, the most relevant quantitative and qualitative parameters that can be determined and associated with the process occurring are the onset and the peak temperatures (corresponding to the temperatures of the flex and the peak), the temperature range, and the enthalpy change of the process (from the area of the DSC peak).

Finally, a special case in which the temperature of a phase transformation is of great importance in food analysis is the glass transition temperature T_g. This is the temperature at which samples are converted from a brittle, glasslike form to a rubbery, flexible form. Glass transition theory derived from the study of polymer science may help to understand textural properties of food systems and explain changes that occur during food processing and storage, such as stickiness, caking, softening, and hardening. In drying of food products, T_g is one of the crucial factors that need to be considered seriously (Ted-Labuza et al., 2004; Ratti, 2001; Stefan Kasapis, 2006). In a DSC experiment, T_g is manifested by a drastic change in the baseline due to a significant change in the heat capacity of the sample, with no transition enthalpy associated with such a transition. For this reason, since it involves no discontinuity in thermodynamic properties such as molar volume and enthalpy, according to the Ehrenfest classification it is called a second-order transition.

When a second-order glass transition occurs in a sample, the change in the heat capacity produces a sigmoidal shape with a very slight effect. A careful calibration of heat flow is needed, along with a good sensitivity of the instrument, to determine the T_g value at the flex of this curve, as Tolstorebrov et al. (2014) did recently for 18 pure triacylglycerides and four fish oils.

FOOD APPLICATION: A BRIEF OVERVIEW BESIDES FAT AND OIL TECHNOLOGY

The application of DSC to food and biological samples present several advantages, as this reliable calorimetric technique does not require time-consuming manipulation practices and chemical treatment of the sample, avoiding the use of toxic chemicals that could be hazardous for the analyst and the environment, and uses a well-automated analysis protocol. In addition, for shelf-life and stability tests, DSC is efficient, as it measures in a day what might otherwise take months to discover. But some parameters must be carefully taken into consideration during analysis, such as moisture loss of a sample, which could lead to an overestimation of the transition temperature and a change of its enthalpy, and the interpretation of overlapping peaks, which can be resolved experimentally or by means of appropriate software. Some DSC limitations are now overcome using modulated differential scanning calorimetry (MDSC), in which the heating rate is no longer constant but varies in a periodic (modulated) fashion, such that a sinusoidal modulation is overlaid on the linear ramp.

Besides its use in fat and oil technology, which is discussed in depth in this book, DSC is applied in food research and analysis to study the effects of components and ingredients on thermal denaturation and aggregation of food protein and protein gels; to evaluate starch gelatinization and retrogradation in the field of cereal and cereal-based product technology; to evaluate authentication or potential fraud or both; and to estimate the effect of well-known (e.g., extrusion, spray drying) or

TABLE I.1
DSC Food Applications: Main Recently Published Literature (2011–2014)

Application	Food	Reference(s)
Authenticity, adulteration	Almond, baby formulas, pig derivatives	García et al. (2013), Ostrowska-Ligeza et al. (2012), and Rohman and Che Man (2012)
Food packaging, innovative biofilm, edible film		Valderrama Solano and Rojas de Gante (2014), Santos et al. (2014), Fakhreddin Hosseini et al. (2013), Xia et al. (2011), and Nair et al. (2011)
Effects of high pressure, ozone treatments, osmotic dehydration	Starch, peanut protein, meat product, vegetables and fruits	Ahmed et al. (2014), Dong et al. (2011), Khan et al. (2014), Acero-Lopez et al. (2012), and Panarese et al. (2012)
Effect of spray drying, extrusion	Aloe, green tea, guava	Cervantes-Martínez et al. (2014), Zhang et al. (2013), and Osorio et al. (2011)
Technology of cereal and cereal-based product	Gluten-free bread, spaghetti, dough	Demirkesen et al. (2013), Sim et al. (2012), and Rahman et al. (2011)
Characterization and denaturation of protein and protein gels	Cooked meat, liquid egg products, pea protein, bovine plasma, *Phaseolus vulgaris*, cod fillet	Ishiwatari et al. (2013), de Souza and Fernández (2013), Sirtori et al. (2012), Rodriguez Furlán et al. (2012), Yin et al. (2011), and Thorarinsdottir et al. (2011)
Physical properties of starch and other polysaccharides		Teng et al. (2013), Tran et al. (2013), and Du et al. (2011)

innovative technological treatments (e.g., osmotic dehydration, high-pressure ozonization) of macrocomponents in order to produce foods with improved characteristics and storage stability. A recent book revisited in depth specific applications for characterization of food systems (Gönül Kaletunç, 2009).

A summary of the recent scientific literature published in the area in the three-year period 2011–2014 is shown in Table I.1. In addition to the research activities cited above, new perspectives may be found for the application of DSC in the study of thermal properties and transitions of new biofilms, edible or otherwise, for innovative food packaging solutions.

This increasing interest in DSC and thermal techniques suggests that its application in several areas is still underexplored and opens new perspectives in food technology, not only for academicians but also for researchers of different sectors of food manufacture.

REFERENCES

Acero-Lopez, A., Ullah, A., Offengenden, M., Jung, S., and Wu, J. Effect of high pressure treatment on ovotransferrin. *Food Chemistry* 135 (2012): 2245–2252.

Ahmed, J., Singh, A., Ramaswamy, H. S., Pandey, P. K., and Raghavan, G. S. V. Effect of high-pressure on calorimetric, rheological and dielectric properties of selected starch dispersions. *Carbohydrate Polymers* 103 (2014): 12–21.

Cervantes-Martínez, C. V., Medina-Torres, L., González-Laredo, R. F., et al. Study of spray drying of the Aloe vera mucilage (*Aloe vera barbadensis* Miller) as a function of its rheological properties. *LWT—Food Science and Technology* 55 (2014): 426–435.

Demirkesen, I., Sumnu, G., and Sahin, S. Quality of gluten-free bread formulations baked in different ovens. *Food and Bioprocess Technology* 6 (2013): 746–753.

de Souza, P. M. and Fernández, A. Rheological properties and protein quality of UV-C processed liquid egg products. *Food Hydrocolloids* 31 (2013): 127–134.

Dong, X., Zhao, M., Yang, B., Yang, X., Shi, J., and Jiang, Y. Effect of high-pressure homogenization on the functional property of peanut protein. *Journal of Food Process Engineering* 34 (2011): 2191–2204.

Du, X., MacNaughtan, B., and Mitchell, J. R. Quantification of amorphous content in starch granules. *Food Chemistry* 127 (2011): 188–191.

Fakhreddin Hosseini, S., Rezaei, M., Zandi, M., and Ghavi, F. F. Preparation and functional properties of fish gelatin-chitosan blend edible films. *Food Chemistry* 136 (2013): 1490–1495.

García, A. V., Beltrán Sanahuja, A., and Garrigós Selva, M. C. Characterization and classification of almond cultivars by using spectroscopic and thermal techniques. *Journal of Food Science* 78 (2013): C138–C144.

Höhne, G., Hemminger, W., and Flammersheim, H.-J. *Differential Scanning Calorimetry.* Berlin: Springer (2003).

Ishiwatari, N., Fukuoka, M., Tamego, A., and Sakai, N. Validation of the quality and microbiological risk of meat cooked with the vacuum-pack cooking method (sous-vide). *Japan Journal of Food Engineering* 14 (2013): 19–28.

Kaletunç, G. *Calorimetry in Food Processing: Analysis and Design of Food Systems.* Ames, IA: Wiley-Blackwell (2009).

Kasapis, S. Definition and applications of the network glass transition temperature. *Food Hydrocolloids* 20 (2006): 218–228.

Khan, M. A., Ali, S., Abid, M., et al. Enhanced texture, yield and safety of a ready-to-eat salted duck meat product using a high pressure-heat process. *Innovative Food Science and Emerging Technologies* 21 (2014): 50–57.

Labuza, T., Roe, K., and Payne, C. Storage stability of dry food systems: Influence of state changes during drying and storage. In *Proceedings of the 14th International Drying Symposium (IDS 2004). Drying 2004,* vol. A (2004), Saõ Paulo, Brazil, 22–25 August, pp. 48–68.

Nair, S. B., Jyothi, A. N., Sajeev, M. S., and Misra, R. Rheological, mechanical and moisture sorption characteristics of cassava starch-konjac glucomannan blend films. *Starch/ Staerke* 63 (2011): 728–739.

Osorio, C., Forero, D. P., and Carriazo, J. G. Characterisation and performance assessment of guava (*Psidium guajava* L.) microencapsulates obtained by spray-drying. *Food Research International* 44 (2011): 1174–1181.

Ostrowska-Ligeza, E., Górska, A., Wirkowska, M., and Koczoń, P. An assessment of various powdered baby formulas by conventional methods (DSC) or FT-IR spectroscopy. *Journal of Thermal Analysis and Calorimetry* 110 (2012): 465–471.

Panarese, V., Tylewicz, U., Santagapita, P., Rocculi, P., and Dalla Rosa, M. Isothermal and differential scanning calorimetries to evaluate structural and metabolic alterations of osmo-dehydrated kiwifruit as a function of ripening stage. *Innovative Food Science and Emerging Technologies* 15 (2012): 66–71.

Rahman, M. S., Senadeera, W., Al-Alawi, A., Truong, T., Bhandari, B., and Al-Saidi, G. Thermal transition properties of spaghetti measured by differential scanning calorimetry (DSC) and thermal mechanical compression test (TMCT). *Food and Bioprocess Technology* 4 (2011): 1422–1431.

Ratti, C. Hot air and freeze-drying of high-value foods: A review. *Journal of Food Engineering* 49 (2001): 311–319.

Rodriguez Furlán, L. T., Lecot, J., Pérez Padilla, A., Campderrós, M. E., and Zaritzky, N. Stabilizing effect of saccharides on bovine plasma protein: A calorimetric study. *Meat Science* 91 (2012): 478–485.

Rohman, A. and Che Man, Y. B. Analysis of pig derivatives for Halal authentication studies. *Food Reviews International* 28 (2012): 97–112.

Santos, T. M., Pinto, A. M. B., de Oliveira, A. V., et al. Physical properties of cassava starch-carnauba wax emulsion films as affected by component proportions. *International Journal of Food Science and Technology* (2014) (forthcoming).

Sim, S. Y., Noor Aziah, A. A., Teng, T. T., and Cheng, L. H. Thermal and dynamic mechanical properties of frozen wheat flour dough added with selected food gums. *International Food Research Journal* 19 (2012): 333–340.

Sirtori, E., Isak, I., Resta, D., Boschin, G., and Arnoldi, A. Mechanical and thermal processing effects on protein integrity and peptide fingerprint of pea protein isolate. *Food Chemistry* 134 (2012): 113–121.

Teng, L. Y., Chin, N. L., and Yusof, Y. A. Rheological and textural studies of fresh and freeze-thawed native sago starch-sugar gels. II. Comparisons with other starch sources and reheating effects. *Food Hydrocolloids* 31 (2013): 156–165.

Thorarinsdottir, K. A., Arason, S., Sigurgisladottir, S., Valsdottir, T., and Tornberg, E. Effects of different pre-salting methods on protein aggregation during heavy salting of cod fillets. *Food Chemistry* 124 (2011): 7–14.

Tolstorebrov, I., Eikevik, T. M., and Bantle, M. A DSC determination of phase transitions and liquid fraction in fish oils and mixtures of triacylglycerides. *Food Research International* 58 (2014): 132–140.

Tran, P. L., Lee, J.-S., and Park, K.-H. Molecular structure and rheological character of high-amylose water caltrop (*Trapa bispinosa* Roxb.) starch. *Food Science and Biotechnology* 22 (2013): 979–985.

Valderrama Solano, A. C. and Rojas de Gante, C. Development of biodegradable films based on blue corn flour with potential applications in food packaging. Effects of plasticizers on mechanical, thermal, and microstructural properties of flour films. *Journal of Cereal Science* 60 (2014): 60–66.

Xia, Y., Wang, Y., and Chen, L. Molecular structure, physicochemical characterization, and in vitro degradation of barley protein films. *Journal of Agricultural and Food Chemistry* 59 (2011): 13221–13229.

Yin, S.-W., Tang, C.-H., Wen, Q.-B., and Yang, X.-Q. Conformational and thermal properties of phaseolin, the major storage protein of red kidney bean (*Phaseolus vulgaris* L.). *Journal of the Science of Food and Agriculture* 91 (2011): 94–99.

Zhang, X., Chen, H., Zhang, N., Chen, S., Tian, J., Zhang, Y., and Wang, Z. Extrusion treatment for improved physicochemical and antioxidant properties of high-molecular weight polysaccharides isolated from coarse tea. *Food Research International* 53 (2013): 726–731.

Section I

Recent and New Perspectives from DSC Application on Vegetable Oils and Fats

1 DSC Analysis of Vegetable Oils

Relationship between Thermal Profiles and Chemical Composition

Chin Ping Tan and Imededdine Arbi Nehdi

CONTENTS

1.1 DIFFERENTIAL SCANNING CALORIMETRY

Differential scanning calorimetry (DSC) belongs to the family of thermal analysis. The Nomenclature Committee of the International Confederation for Thermal Analysis and Calorimetry (ICTAC) has defined DSC as "A technique in which the difference in energy inputs into a substance and a reference material is measured as a function of temperature whilst the substance and reference materials are subjected to a controlled temperature program" (Wendlandt 1986). Much scientific literature has indicated the importance of DSC in food analysis (Biliaderis 1983; Wright 1986; Harwalkar and Ma 1990; Tocci and Mascheroni 1998; Brake and Fennema 1999; Tan and Che Man 2002a; Gill et al. 2010).

For the last 50 years, it has been known that DSC is useful in applications of practical importance in the field of fats and oils. Today, DSC is widely used for the evaluation of various physicochemical transformations occurring during quality control of fat and oil products and development of new fat and oil products (Herrera and Anon 1991; Cebula and Smith 1992; Lee and Foglia 2001; Oomah et al. 2002).

Almost all DSC studies have been conducted using commercial instruments. Most commercial DSC instruments currently used in measurement of fat and oil systems are of either the power-compensation design or the heat-flux design (Ford and Timmins 1989). There have been many reports in monographs and papers indicating differences in response between the two types of design (Wendlandt 1986; Brown 1988; Griffin and Laye 1992; Noble 1995). Although the two designs provide the same information, the instrumentation for the two is fundamentally different.

In power-compensation DSC, heat is added to the sample or to the reference material as needed so that the two substances remain at the identical temperature. Figure 1.1 is a schematic diagram showing the basic holder design of a power-compensation DSC. The sample and reference material are supplied with separate heaters or micro ovens and are maintained at nominally the same temperature by means of a system operated through platinum resistance thermometers and resulting in different amounts of heat being supplied to each specimen as appropriate.

Thermal events in the sample appear as deviations from the DSC baseline, in either an endothermic or exothermic direction, depending on whether more or less energy is supplied to the sample relative to the reference material (Brown 1988). In power-compensation DSC, endothermic responses are usually represented as being positive, that is, above the baseline, corresponding to an increased transfer of heat to the sample compared with the reference. The opposite is the case in heat-flux DSC, in which endothermic responses are represented as negative differences in heat flow, below the baseline.

Since the introduction of commercial DSC instruments, great strides have been made in their instrumentation through advances in microcomputers and microelectronics. Today, commercial DSC instruments can operate from −200°C to about 750°C, which covers most applications for the field of fats and oils. The temperature can be maintained isothermally or programmed at scanning rates from 0.1°C/min up to unrealistically high values such as 500°C/min.

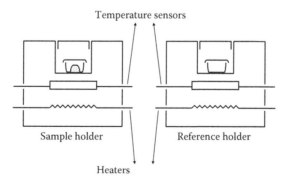

FIGURE 1.1   Basic holder system for power-compensation DSC.

1.2 APPLICATIONS OF DSC IN VEGETABLE OILS

Vegetable oils form a part of almost all food products, and the properties of the fat or oil often play an integral part in the production and, in many cases, the consumption of the food (Talbot 1995). In order to use vegetable oil materials efficiently, it is important to understand the complex structures and properties of fats and oils. Heat-related phenomena in vegetable oils are fundamental to elucidating their physical and chemical properties.

1.2.1 MELTING AND CRYSTALLIZATION BEHAVIORS OF FAT AND OIL PRODUCTS BY DSC

In the field of fats and oils, one major area of application that is eminently suitable for study by DSC is the various melting or crystallization profiles of vegetable oils. The thermal properties of a large number of fat and oil compounds have been studied by DSC. In most of the studies, DSC data have been used to complement the results obtained from other analytical instruments such as nuclear magnetic resonance (NMR), x-ray diffraction (XRD), high-pressure liquid chromatography (HPLC), and gas chromatography (GC). For example, Tan and Che Man (2000) studied the chemical composition and melting and crystallization behavior of various vegetable oils by HPLC, GC, and DSC; Noor Lida et al. (2006) used DSC to evaluate the melting properties of palm oil, sunflower oil, and palm kernel olein blends before and after chemical interesterification; and Fredrick et al. (2008) used DSC and XRD to evaluate the crystallization behavior of palm oil. XRD, DSC, and polarized light microscopy were used to monitor the thermal and structural properties of fully hydrogenated and interesterified fat and vegetable oil (Zhang et al. 2011).

It is well known that the properties of fats and oils are profoundly influenced by physicochemical interactions, particularly among triacylglycerols (TAG). TAG are the main chemical species in fats and oils. Although a species-specific TAG compositional profile is often observed in natural oil and fat systems, the possible mixtures of individual TAG appear to be almost infinite and, as a result, are not simple to resolve (Fouad et al. 1990). Their physicochemical interactions are often complex, and a full understanding of their thermal properties requires examination of these interactions. Recently, Nehdi (2011a,b, 2013) and Nehdi et al. (2010) evaluated the thermal properties of various plant oils from selected seeds using DSC. Tan and Che Man (2000) studied the thermal properties of 17 different vegetable oils by DSC. This work characterizes the DSC melting and crystallization profiles of 17 edible oils. Figures 1.2 and 1.3 show the melting and crystallization profiles of six different edible oils. The DSC melting and crystallization profiles of edible oils are discussed in relation to each other and also to their fatty acid (FA) and TAG composition. The authors concluded that DSC does not provide any direct information about the chemical composition of edible oils under a given set of experimental conditions. However, it provides useful information regarding the nature of the thermodynamic changes that are associated with the edible oils transforming from one physical state to another. These thermodynamic characteristics are sensitive to the general chemical composition of edible oils and fats and thus can be used in qualitative and quantitative ways for identification of edible oils.

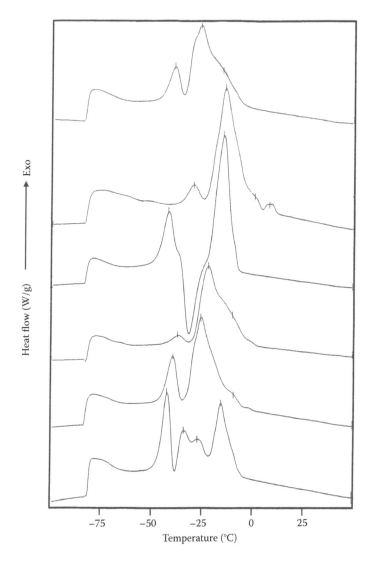

FIGURE 1.2 Differential scanning calorimetry melting curves of corn oil, peanut oil, sesame oil, safflower oil, soybean oil, and sunflower oil, from top to bottom. (Reprinted with permission from Tan, C.P., Che Man, Y.B., *J Am Oil Chem Soc*, 77, 143–155, 2000.)

DSC is particularly suitable for studying the physicochemical interactions within TAG, because these techniques readily afford phase equilibrium diagrams, which provide a wealth of physicochemical information. The classifications of the types of interactions that occur in phase diagrams are usually determined by DSC heating-curve data (Minato et al. 1996). Naturally occurring fats and oils are complex mixtures of TAG and melt over a wide range of temperatures (Barbano and Sherbon 1978). Therefore, intensive study on the physicochemical interactions within TAG

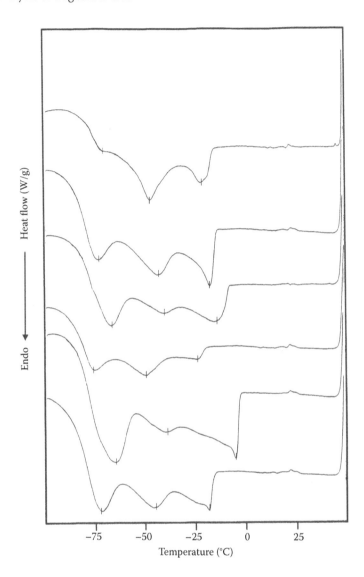

FIGURE 1.3 Differential scanning calorimetry crystallization curves of corn oil, peanut oil, sesame oil, safflower oil, soybean oil, and sunflower oil, from top to bottom. (Reprinted with permission from Tan, C.P., Che Man, Y.B., *J Am Oil Chem Soc*, 77, 143–155, 2000.)

in edible oils is carried out by the use of a simple TAG system (Minato et al. 1996, 1997; Barbano and Sherbon 1978).

Beyond the compositional variation and their physicochemical interactions, TAG in edible oils also show temperature-dependent polymorphic behavior, complicating their thermal properties. Polymorphism describes phase changes and structural modification of the solid-fat phase (Herrera and Marquez Rocha 1996). The scientific literature abounds with indications of the importance of polymorphism in the field of fats and oils (Barbano and Sherbon 1978; Herrera and Marquez Rocha 1996;

Gray et al. 1976; Gray and Lovegren 1978; Lovegren and Gray 1978; Reddy and Prabhakar 1986; Guth et al. 1989; Yap et al. 1989; Desmedt et al. 1990; Arishima et al. 1991; Elisabettini et al. 1996). Attention has been directed largely toward many physically important characteristics, such as ease of handling, flow properties, processing properties, and physical stability. DSC has been employed in the study of polymorphism of fats and oils. Nevertheless, polymorphism obtained by DSC does not yield absolute results and requires supportive evidence from other techniques, such as infrared spectroscopy and x-ray diffraction studies (deMan and deMan 1994).

The unique sensory characteristics of cocoa butter, which are mainly due to its sharp melting point, make it the preferred fat in confectionery products. For many years DSC has been used in the characterization of confectionery fats, cocoa butter, and pure TAG (Cebula and Smith 1992; Chapman 1971; Lovegren et al. 1976a,b; Cebula and Smith 1991; Reddy et al. 1994; Md. Ali and Smith 1994a,b; Chaiseri and Dimick 1995a,b). Among fats, the best-known example is cocoa butter, which exists in six different polymorphic forms (Talbot 1995). Polymorphism of cocoa butter has been the subject of extensive research. Reddy et al. (1994) evaluated the degree of tempering of chocolate by using DSC. The melting points and the heat of fusion of the β'-seed crystals of cocoa butter were the deciding parameters in assessing the degree of temper of chocolate. Md. Ali and Dimick (1994a) reported that the melting curves of palm mid fraction, anhydrous milk fat, and cocoa butter show distinct differences among the three fats.

The crystallization of fats and oils is a complex phenomenon that has interested researchers for many years (Smith et al. 1994). The process is complicated by the slow rate of crystal growth, caused by the polymorphic behavior of fats and the complex shape. Crystal lattice imperfections develop during crystallization and exert major effects in fat and oil processing. The effect of lauric-based molecules on trilaurin crystallization was described by Smith et al. (1994). In this study, DSC showed that chilled liquid trilaurin crystallizes at 21°C, and the addition of 2% monolaurin leads to a slight increase in this temperature, while the addition of 2% dilaurin decreases the crystallization temperature. Factors influencing the crystallization of palm oil are the presence of free fatty acids (FFA), partial glycerides, and oxidation products (Jacobsberg and Ho 1976). Ng (1990) studied nucleation from a supercooled melt of palm oil by optical microscopy and DSC. The author confirmed that palm oil exhibits a rather simple cooling curve with its high- and low-temperature exotherms, which is exclusively related to the "hard" and "soft" components of the oil. On the other hand, Che Man and Swe (1995) determined the thermal behavior of failed-batch palm oil. They concluded that a rapid and sudden surge of heat demand is observed for samples from failed crystallizers. The presence of polar compounds in fats has a major impact on the crystallization behavior of fat products. Recently, Ray et al. (2013) conducted a study to evaluate the crystallization and polymorphic behavior of shea stearin and the effect of removal of polar components.

DSC provides an opportunity not only for thermodynamic analysis but also for isothermal analysis. In isothermal analysis, the samples are held in the calorimeter at a given temperature but for varying lengths of time. DSC isothermal analysis was used to investigate the crystallization behavior of high-erucic-acid rapeseed in conjunction with the usual cooling and heating methods (Kawamura 1981). The iso-thermal crystallization of palm oil was studied by Ng and Oh (1994). Results from DSC experiments showed interesting crystallization curves from each temperature of crystallization (from 0°C to 30°C). In contrast, Dibildox-Alvarado and Toro-Vazquez (1997) investigated the isothermal crystallization of tripalmitin in sesame oil. The results obtained for the thermal behavior of sesame oil indicated that, at tripalmitin concentrations <0.98%, TAG of sesame oil with tripalmitin developed into mixed crystals.

1.2.2 DSC AS A TOOL TO MONITOR CHANGES IN CHEMICAL COMPOSITION DURING PHYSICAL AND CHEMICAL PROCESSES OF VEGETABLE OILS

Fractionation and hydrogenation are two widely used procedures that help to alter the melting profile, physical properties, and chemical composition of the feed oil or fat. These two important industrial processes produced new products suitable for use in applications in which the original oil or fat could never have been used or would have performed poorly (Allen 1982; Hastert 1996; Krishnamurthy and Kellens 1996). DSC has been widely used to monitor these two processes (Lee and Foglia 2001; deMan et al. 1989; Busfield and Proschogo 1990; D'Souza et al. 1991; Herrera et al. 1992; Herrera 1994; Dimick et al. 1996; Che Man et al. 1999). For example, Che Man et al. (1999) described the thermal profiles of palm oil and its products. They outlined the importance of thermal behavior of various palm oil products and concluded that the DSC thermal profiles can be used as guidelines for fractionation of crude palm oil or refined–bleached–deodorized (RBD) palm oil. Lee and Foglia (2001) observed the crystallization pattern of fractionated menhaden oil and partially hydroge-nated menhaden oil by DSC. Danthine and Deroanne (2003) determined the efficacy of blending between vegetable oil and palm-based oil for the production of shortenings using DSC.

Besides fractionation and hydrogenation processes, chemical/enzymatic inter-esterification is another useful process to produce new products from fats and oils. Thermal behavior is one of the most important characteristics of chemically or enzy-matically transesterified fat and oil products. DSC has been widely used to evaluate the thermal behavior of these products (Lai et al. 1999; Sellappan and Akoh 2000; Chu et al. 2001, 2002a). DSC has been used to monitor the transesterification process and to compare the transesterified products with the raw materials. Noor Lida et al. (2006) evaluated the melting properties of palm oil, sunflower oil, and palm kernel olein blends before and after chemical interesterification using DSC, while Siddique et al. (2010) evaluated the physicochemical parameters and DSC thermal curves of blends of palm olein with other vegetable oils. Recently, Adhikari and Hu (2012) monitored changes in blends of rice bran oil, shea olein, and palm stearin during chemical and enzymatic interesterification by DSC. DSC is also frequently used to monitor phase transitions of various oil blends. For example, Saberi et al. (2011) monitored the phase behavior of palm oil in blends with palm-based diacylglycerol (Figure 1.4).

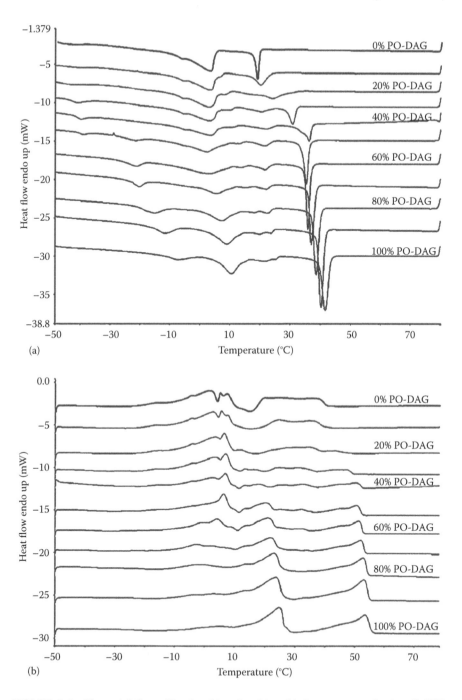

FIGURE 1.4 Figure 1.4 Crystallization (a) and melting (b) thermograms of palm oil (PO) and PO with 10%–100% of palm-based diacylglycerol oil (PO-DAG) at 10% intervals. (Reprinted with permission from Saberi, A. H., Tan, C.-P., and Lai, O.-M. *J Am Oil Chem Soc* 88, 1857–1865, 2011.)

1.2.3 USE OF MODULATED DSC AND EFFECT OF SCANNING RATES ON THERMAL PROFILES OF VEGETABLE OILS

One of the significant developments in DSC in recent years is the development of modulated DSC (MDSC). This is a temperature programming method in which a sine wave is superimposed on the temperature oscillating around a regular programmed gradient (Gmelin 1997; Verdonck et al. 1999). In recent years, research has commenced into the use of MDSC in the field of fats and oils (Satish et al. 1999). The ability to disentangle reversing and nonreversing components of a thermal event is the most important advantage of MDSC over conventional DSC. Satish et al. (1999) concluded that MDSC enables overlapping thermal events of tristearin to be separated, thus increasing the information obtained compared with conventional DSC. They also observed that reversible thermal processes are strongly influenced by the underlying heating rate. They recommended the use of low to moderate heating rates. In another paper (Satish et al. 1999), they observed the ability of MDSC to separate the melting of tristearin β′-form from the simultaneous crystallization of the β-form. Oomah et al. (2002) used MDSC at various cooling rates to characterize commercial hempseed oil. They also concluded that the thermal structural transitions of hempseed oil are highly sensitive to the cooling rate. Recently, Samyn et al. (2012) monitored the quality of Brazilian vegetable oils using MDSC. In this work, a first or second heating scan was used to study the thermal behavior of palm, soy, sunflower, corn, castor, and rapeseed oils in relation to their composition.

Tan and Che Man (2000) showed that, if edible oils give rise to identical DSC scans at the same scan rate, the DSC technique promises to offer a sensitive, rapid, and reproducible fingerprint method for quality-control purposes. However, results in most of the scientific literature show that one critical limitation of using DSC is the dependence of the thermal transition on the scanning rate. Tan and Che Man (2002b,c) and Che Man and Tan (2002) have focused their study on how the thermal properties of edible oils are influenced by variations in DSC scanning rate. These works also characterize the DSC melting and cooling curves of 17 edible oils. In general, edible oil samples behave differently depending on the heating/cooling rate of the DSC. Although it could be seen that the number of endothermic or exothermic peaks was dependent on scanning rate, the melting/cooling curves of oil samples were not straightforward, in that there was no correlation between the number of endothermic or exothermic peaks and the scanning rate. These studies concluded that accurate comparisons of the calorimetric experiments in the vegetable oils could only be made when these DSC experiments were done at the same scanning rate. The use of slow scan rates is advisable in that it minimizes instrumental lag in output response and, at a given temperature, the reaction being examined is closer to chemical equilibrium as well as giving the true line shape of the transition, which will be important if identification of the data is to be undertaken. Consequently, they also concluded that a relatively low scanning rate (e.g., 1°C or 5°C/min) is required to ensure thermal equilibrium of the sample at the time of data collection. In addition, Abdulkarim and Ghazali (2007) also conducted a thorough evaluation on the use of slow and fast scanning rates on the melting behavior of canola, sunflower, palm olein, rice bran oils, and cocoa butter. This study concluded that increasing the scan rate resulted in an increase in the peak temperature and the elimination of shoulder peaks.

1.2.4 QUANTITATIVE CHEMICAL ANALYSIS BY DSC

Application of statistical techniques to DSC data as an indirect method for the determination of various quality parameters of fat and oil products has been widely applied. Some are listed in Table 1.1. Perhaps the earliest quantitative analysis by DSC in the field of fats and oils was the application of DSC thermal curve data to calculate solid fat index (SFI) (deMan and deMan 1994; Bentz and Breidenbach 1969; Miller et al. 1969; Walker and Bosin 1971; Menard et al. 1994). This is classically determined by dilatometry and essentially measures the amount of melted solid in the sample at various temperatures (Firestone 1993). Currently, SFI is measured by NMR. Walker and Bosin (1971) showed that DSC compares favorably with NMR and dilatometric methods for determining SFI. While the rheological properties of fat products have frequently been compared with DSC data, Toro-Vazquez et al. (2004) studied the crystallization behavior of cocoa butter by comparing the rheological properties of fat under static and stirring conditions with thermal properties obtained from DSC.

TABLE 1.1
Quantitative Analysis of DSC Thermal Data to Evaluate Various Quality Parameters in Fat and Oil Products

Quality Parameter	Reference(s)
Solid fat index	Bentz and Breidenbach (1969) and Márquez et al. (2013)
Melting point of vegetable oils	Nassu and Gonçalves (1999)
Detection of animal body fats in ghee	Lambelet and Ganguli (1983)
Recombined butter	Tunick et al. (1997)
Detection of butter adulteration with water	Tomaszewska-Gras (2012)
Imitation Mozzarella cheese	Tunick et al. (1989)
Differentiation of genuine and randomized lard	Marikkar et al. (2001)
Degree of saturation in transesterified blends of jojoba wax ester	Sessa et al. (1996) and Sessa (1996)
Iodine value	Haryati (1997) and Chu et al. (2002)
Determination of thermally oxidized olive oil	Vittadini et al. (2003)
Evaluation of oxidative stability of interesterified fats	Bryś et al. (2013)
Country of origin for oil-bearing nuts	Dyszel (1990, 1993) and Dyszel and Pettit (1990)
Concurrent determination of total polar compounds, free fatty acid content, and iodine value in heated corn oil, RBD palm olein, and soybean oil	Tan and Che Man (1999) and Tan (2001)
Determination of quality of frying oil	Abdulkarim et al. (2008)
Detection of lard in oils and foods	Marikkar et al. (2002, 2003)

Through the use of DSC thermal profiles, many researchers have addressed the issue of authentication in fat and oil products. For example, Chen et al. (2004) used DSC to determine the presence of wax in various crude oils. Encouraging results have been obtained for detecting animal body fats in ghee using DSC (Lambelet and Ganguli 1983). The authors concluded that detecting adulteration of ghee using the DSC technique is more sensitive with cooling curves than with melting curves. Lambelet and Ganguli (1983) further detected cow or buffalo ghee adulteration with pig or buffalo body fat. The authors' results showed that ghee adulteration with these animal fats at levels down to 5% was clearly seen in the cooling curves. DSC can also be used to quickly determine whether a product labeled as butter is, in fact, recombined butter made without milk (Tunick et al. 1997). Previously, Tunick et al. (1989) developed a DSC method for examining imitation Mozzarella cheese containing calcium caseinate and found that the emulsifying properties of the caseinate affected fat crystallization in untempered samples. Tunick and Malin (1997) also distinguished the DSC curves of Mozzarella cheese from those of water buffalo and cow milk fat. Al-Rashood et al. (1996) described the use of DSC to differentiate genuine and randomized lard. The authors concluded that the DSC curves and thermodynamics of phase transitions of both samples were quite different but did not reveal any common characteristic that could be used for immediate detection of lard substances in fat admixtures. Marikkar et al. (2001) conducted a study to detect the adulteration of RBD palm oil with enzymatically randomized lard by DSC. They concluded that DSC is a suitable method for detection of lard in RBD palm oil with a detection limit of 1%.

The degree of unsaturation or iodine value (IV) of fat and oil products is one of the most frequently used quality parameters. This quality parameter is also closely related to the SFI. Therefore, DSC has been used to quantify this important quality parameter. Sessa et al. (1996) used mathematical and statistical techniques to devise a DSC index to estimate the amount of saturation present in transesterified blends of jojoba wax esters based on heats of fusion enthalpies. In this study, a series of jojoba liquid wax esters was constructed by transesterifying native jojoba oil with 5%–50% completely hydrogenated jojoba wax esters. The authors found that, when this series was subjected to a standardized DSC tempering method with heating or cooling cycles, an excellent correlation was exhibited between level of saturation based on area and changes in endothermic enthalpies. Furthermore, Sessa (1996) also devised mathematical indices based on heats of crystallization enthalpies as well as heats of fusion enthalpies to define the level of saturation in transesterified wax ester blends and used them to select the optimum level of saturation needed for obtaining a cocoa butter equivalent.

Haryati et al. (1997) also quantified IV in palm oil products. In the preliminary study, regression analysis showed that the peak characteristics in the heating and cooling curves can predict the IV of palm oil with coefficient of determination higher than 0.99. Besides, Haryati (1999) also showed that the onset temperature of the cooling curve and the offset temperature of the heating curve can predict the cloud point and melting point of palm oil products. Chu et al. (2002b) also determined IV in blends of palm olein with six different edible oils by DSC.

DSC has been shown to be a sensitive and useful technique for assignment of country of origin for oil-bearing nuts (Dyszel 1990, 1993; Dyszel and Pettit 1990). This method has been adopted by the Research Division of the Office of Laboratories and Scientific Services, U.S. Customs Service as the choice of method for screening pistachio (Dyszel and Pettit 1990) and macadamia (Dyszel 1990) nuts for country of origin. The U.S. Customs Service also characterized peanuts for country of origin (Dyszel 1993). This study used DSC to profile oils extracted from peanuts grown in the United States, Argentina, and China. The melting behavior of TAG and other components in the oil matrix is obtained by controlled heating from the subambient. A series of variables for each oil is assigned to the observed DSC thermal curve for the temperature region between 240 and 340 K. A graphical presentation of the canonical discriminant functions scores grouped the samples into three areas by country of origin.

A DSC technique has been developed by Tan and Che Man (1999a) and Tan (2001) for concurrently determining three important quality parameters in the deep-fat frying industry—total polar compounds (TPC), FFA, and IV—of heated corn oil (CO), RBD palm olein, and soybean oil (SO), using the crystallization curves of DSC. The DSC variables were used as independent variables while values from standard chemical methods were used as dependent variables. In heated oils, the polar compounds and FFA increased during the heating process. As their level increased, these compounds would contribute to the changes in DSC crystallization peak parameters. From the calibration and validation analyses, this study revealed that a single DSC cooling curve could predict the TPC, FFA, and IV of heated oils using stepwise regression analysis. Most recently, Cuvelier et al. (2012) also determined TPC in thermooxidized oil samples by using DSC. Teles Dos Santos et al. (2012) also compared the experimental and predicted DSC curves for palm oil, peanut oil, and grapeseed oil. The predicted curves are generated from the solid–liquid equilibrium modeling and direct minimization of the Gibbs free energy.

1.2.5 Monitoring Oxidation in Heated Vegetable Oils by DSC

Deep-fat frying is a popular way of cooking (Lawson 1995). During the past 30 years, scientists have extensively reported on the physical, chemical, and sensory changes that occur during frying and on the wide variety of decomposition products formed in frying oils (Paul and Mittal 1997; Warner 1998). Microwave heating is the most versatile method worldwide, mainly because consumers appreciate the advantages of convenience, economy, and time saving. It is a novel method of heating in that the primary mechanism of heat transfer to a food product is neither conduction nor convection. During these heating processes, a variety of reactions, such as oxidative, hydrolytic, and thermolytic degradation, occur in the fat, and numerous decomposition products are formed. Therefore, numerous analytical methods have been described for the measurement of changes that occur in heated oils (Blumenthal 1991). In recent years, DSC has been used as one of the analytical instruments to measure changes in heated oils.

Gloria and Aguilera (1998) studied the quality changes in three different heated oils (rapeseed oil, sunflower oil, and peanut oil) by DSC. These commercial frying oils were subjected to heating at 180°C for up to 10 h. DSC cooling curves of oils scanned at 1°C/min were characterized by a single crystallization peak. The DSC thermal characteristics of oils correlated well with TPC, viscosity, and color changes. They concluded that DSC is a rapid method for analyzing the quality of oils. In a similar paper by Aguilera and Gloria (1997), the authors also studied the uptake of oil in commercial frozen parfried potatoes after frying at 180°C by DSC.

Tan and Che Man (1999) have developed a simple and reliable DSC method for monitoring the oxidation in three different heated oils (corn oil, RBD palm olein, and soybean oil). In this study, heated oils exhibited a simple curve after cooling in the DSC with a well-defined single crystallization peak. Two DSC parameters of this single crystallization peak, namely peak temperature and enthalpy, were determined. Figure 1.5 shows the changes in peak enthalpy for heated oils. In addition to DSC, deterioration of heated oils was also quantified by means of seven chemical methods. The high correlation found between the DSC curve parameters and changes in chemical parameters suggests that DSC can be recommended as an appropriate objective method for assessing the extent of oxidation in edible oils.

Tan et al. (2002a,b) also studied the effects of microwave heating on the changes in DSC cooling and melting profiles, or both, of edible oils in two separate papers. The influence of microwave power (low-, medium-, and high-power settings) and heating time on lipid deterioration produced during the microwave heating of vegetable oils was evaluated. The DSC method was based on the cooling or melting curve, or both, of oils at a scanning rate of 5°C/min. The DSC results were explained on the basis of the endothermic or exothermic peak temperatures. A statistical comparative study was carried out on the DSC and chemical parameters. In general, there were good correlations between these parameters. These papers concluded that DSC could be employed as a time–microwave power indicator during microwave heating.

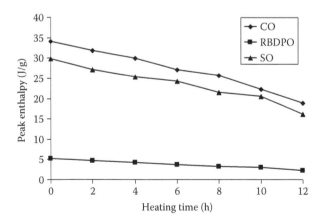

FIGURE 1.5 Scatter plot showing changes of peak enthalpy versus heating time for corn oil (CO), RBD palm olein (RBDPO), and soybean oil (SO).

1.2.6 LIPID OXIDATION OF VEGETABLE OILS BY DSC

Lipid oxidation is of great concern to the vegetable fats and oils industry because it is directly related to economic, nutritional, flavor, safety, and storage problems (Halliwell et al. 1995; Min 1998). Oxidative stability is one of the most important indicators of the keeping quality of edible oils. In general, the time before a dramatic increase in the rate of lipid oxidation is a measure of oxidative stability and is referred to as the induction time (Coppin and Pike 2001). A different approach is to measure the induction period before the rapid oxidation phase that occurs in a lipid matrix exposed to conditions of accelerated oxidation (Frankel 1993). Some of the conventional stability tests and their limitations have been reviewed by Rossell (1975) and Frankel (1993). Currently, oxidative stability of fats and oils can be determined automatically by two commercially available pieces of equipment, the Rancimat from Brinkmann Instruments (Westbury, NY) and the Oxidative Stability Instrument from Omnion, Inc. (Rockland, MA) (Akoh 1994).

The transfer of an oxygen molecule to an unsaturated fatty acid requires energy (it is an exothermic process). Therefore, the oxidative stability of edible oils can also be established by DSC. The application of DSC as an accelerated oil stability test has been studied by several researchers. Cross (1970) and Hassel (1976) used DSC in isothermal mode with an oxygen purge to measure the stability of oils. The end point of DSC was taken at the time when a rapid exothermic reaction of oil and oxygen occurred. Hassel's results showed that oil samples that required 14 days by the active oxygen method (AOM), for example, were evaluated in <4 h by DSC (Hassel 1976). Raemy et al. (1987) evaluated the oxidative stability of saturated C_{18} fatty acid methyl esters, vegetable oils, and chicken fat by normal-pressure isothermal DSC. Kowalski (1989, 1991) and Kowalski et al. (1997) have extensively monitored the oxidative stability of vegetable oils by pressure DSC.

In general, isothermal or dynamic DSC techniques can be applied to obtain the kinetic data of lipid oxidation in vegetable oils. As compared with the dynamic condition, the isothermal condition is rather time-consuming and is usually coupled with a high-pressure cell (pressure DSC) (Litwinienko and Kasprzycka-Guttman 1998). However, many of the complex oxidation phenomena during dynamic DSC have not been well explained by many researchers. These phenomena include variations in reaction rate, differences in oxygen solubility in vegetable oils during linear programmed heating, and differences in oxidation pathways.

Litwinienko et al. (1995) used both isothermal and dynamic DSC to investigate the kinetics of thermooxidative decomposition of edible oils. Litwinienko and Kasprzycka-Guttman (1998) calculated the activation energies and Arrhenius kinetic parameters for thermal oxidation of mustard oil by dynamic DSC. In another paper, Litwinienko (2001) also observed oxidation of unsaturated fatty acids and their esters by dynamic DSC. Two main exothermic peaks, partially overlapped, were observed. The first peak is caused by hydroperoxide formation and the second peak by further oxidation of peroxides. They also compared the activation energies of unsaturated fatty acids and their esters by both isothermal and nonisothermal DSC. In an earlier publication, Litwinienko et al. (1999) also studied the oxidation of C_{12}–C_{18} saturated fatty acids and their esters. They found that the activation energies of these fatty acids are similar (106–134 kJ/mol) and do not depend on length of the carbon chain. Furthermore, they also observed that the kinetic parameters are similar for each investigated fatty acid and ester.

Simon et al. (2000) studied the oxidation of rapeseed and sunflower oils by DSC under dynamic conditions. Simon and Kolman (2001) also presented a theory based on dynamic DSC measurements to evaluate the kinetic parameters of edible oils and polyolefins. Kowalski et al. (2000) studied the kinetics of oxidation in rapeseed oil by pressure DSC under isothermal conditions, and compared the results obtained by DSC with Oxidograph measurements. The discrepancies in the results obtained by these two instruments are accounted for by oxygen diffusion within the samples.

Tan et al. (2002c) have conducted a comparative study to determine the oxidative stability of 12 different edible oils by DSC and the oxidative stability instrument (OSI). The DSC technique was based on an isothermal condition with purified oxygen as purge gas. A dramatic increase in evolved heat was observed, with the appearance of a sharp exothermic curve during initiation of the oxidation reaction. The results indicated a good correlation ($p < .01$) between the DSC and OSI values. In another paper, Tan et al. (2001a) applied the isothermal DSC method to obtain the kinetic data for lipid oxidation of 10 different edible oils. The temperature dependence of the rates of lipid oxidation gave highly significant correlations when analyzed by the DSC method. In addition, based on the Arrhenius equation and activated complex theory, reaction rate constants, activation energies, activation enthalpies, and activation entropies were calculated for the oxidative stability of vegetable oils. However, fat-containing foods are difficult to test directly using DSC. The feasibility of using DSC to determine the oxidative stability of complex food products is restricted. This is mainly because of the difficulty of obtaining a representative sample due to the use of small sample sizes (5–15 mg) in the DSC analysis.

1.2.7 EFFICACY OF ANTIOXIDANTS IN VEGETABLE OILS BY DSC

Free radical oxidation of the lipid components in foods is a major problem for food manufacturers; thus, the early attempts to measure antioxidative activity were mainly focused on lipid protection (Frankel 1980). Today, the effectiveness of synthetic and natural antioxidants is measured by monitoring the oxidative stability of the food lipids. After the sample is oxidized under standard conditions, the extent of oxidation after the addition of the antioxidant is measured by chemical or instrumental means or as induction time (Pratt 1996). The extension in induction time is also expressed as an antioxidant index or protection factor.

Kowalski (1993) evaluated the activities of various antioxidants in rapeseed oil samples by pressure DSC. An evaluation of the efficacy of antioxidants in soap by DSC was conducted by Gupta and Jaworski (1990). The procedure involved forced oxidation of a sample in an oxygen-pressurized DSC cell. The DSC method described in the study may be a useful tool in the development and optimization of an antioxidant system for bar soap products and for fatty materials in general. A simple DSC method for measuring the antioxidant activity in RBD palm olein was developed by Tan et al. (2001b). The oxidation temperature was 150°C and the oxygen flowed at a rate of 50 ml/min. In this method, the thermal changes occurring

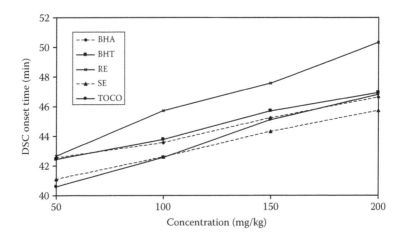

FIGURE 1.6 Scatter plot showing DSC induction time versus antioxidant concentration. BHT, butylated hydroxytoluene; BHA, butylated hydroxyanisole; RE, rosemary extract; SE, sage extract; TOCO, tocopherol.

during oxidation of the oil are recorded. Figure 1.6 illustrates the changes in induction time values for five added antioxidants with different concentrations in RBD palm olein. Generally, the results show that the antioxidants act mainly by increasing the induction time of lipid oxidation. Irwandi et al. (2000) also evaluated the synergistic effects between various mixtures of rosemary, sage, and citric acid in RBD palm olein by isothermal DSC. This work could contribute to the selection of an appropriate antioxidant (or combination of antioxidants) at optimum level in RBD palm olein.

1.3 CONCLUSION AND FUTURE RESEARCH

As shown in this chapter, DSC is one of the frequently used instrumental techniques for characterizing the physicochemical properties of vegetable oils. It has been applied in the field of fats and oils to observe a variety of complex reactions such as phase transitions, melting and crystallization processes, and lipid oxidation in vegetable oils.

Almost all physicochemical changes in fats and oils involve endothermic or exothermic reactions. In principle, these thermal processes can always be measured calorimetrically, without interference with the processes. However, DSC is a nonspecific analytical technique. Therefore, the use of DSC to measure various physicochemical changes in fats and oils normally generates nonspecific calorimetric results. The nonspecific calorimetric results from a complex physical or chemical reaction in fats and oils are usually difficult to interpret on a micro scale, such as at molecular level. Therefore, the presence of more specific analytical information obtained by the use of another instrument is required. Nevertheless, the calorimetric results generated by DSC do give an overall account of the complex process, which a specific analytical instrument will rarely offer.

Today, calorimetric analytical problems continue to arise in new forms. The variation of the DSC results obtained by two different designs, namely the power-compensation and heat-flux systems, are not well studied. Therefore, further study comparing thermal profiles obtained by these two different designs is required. As pointed out in this chapter, a significant number of systematic methodologies have been reported, but no application of practical importance has yet been developed. The nonspecific nature of the calorimetric data is believed to be the main cause of this lack of success. Therefore, it is felt that a combination of DSC with different specific analytical techniques could be very useful in many research applications. Moreover, the DSC method also presents an alternative for the analysis of vegetable oils due to its minimal sample preparation and no need for toxic chemicals.

REFERENCES

Abdulkarim, S. M., Frage, A., Tan, C. P., and Ghazali, H. M. Determination of the extent of frying fat deterioration using differential scanning calorimetry. *Journal of Food, Agriculture and Environment* 6(3–4) (2008): 54–59.

Abdulkarim, S. M. and Ghazali, H. M. Comparison of melting behaviors of edible oils using conventional and hyper differential scanning calorimetric scan rates. *International Food Research Journal* 14(1) (2007): 25–35.

Adhikari, P. and Hu, P. Enzymatic and chemical interesterification of rice bran oil, sheaolein, and palm stearin and comparative study of their physicochemical properties. *Journal of Food Science* 77(12) (2012): C1285–C1292.

Aguilera, J. M. and Gloria, H. Determination of oil in fried potato products by differential scanning calorimetry. *Journal of Agricultural and Food Chemistry* 45(3) (1997): 782–785.

Akoh, C. C. Oxidative stability of fat substitutes and vegetable oils by the oxidative stability index method. *Journal of the American Oil Chemists' Society* 71(2) (1994): 211–216.

Al-Rashood, K. A., Abou-Shaaban, R. R. A., Abdel-Moety, E. M., and Rauf, A. Compositional and thermal characterization of genuine and randomized lard: A comparative study. *Journal of the American Oil Chemists' Society* 73(3) (1996): 303–309.

Allen, R. R. Hydrogenation. In D. Swern (ed.), *Bailey's Industrial Oil and Fat Products*, vol. 2, 4th edn., pp. 1–96. New York: Wiley (1982).

Arishima, T., Sagi, N., Mori, H., and Sato, K. Polymorphism of POS. I. Occurrence and polymorphic transformation. *Journal of the American Oil Chemists' Society* 68(10) (1991): 710–715.

Barbano, P. and Sherbon, J. W. Phase behavior of tristearin/trioctanoin mixtures. *Journal of the American Oil Chemists' Society* 55(5) (1978): 478–481.

Bentz, A. P. and Breidenbach, B. G. Evaluation of the differential scanning calorimetry method for fat solids. *Journal of the American Oil Chemists' Society* 46(2) (1969): 60–63.

Biliaderis, C. G. Differential scanning calorimetry in food research—A review. *Food Chemistry* 10(4) (1983): 239–265.

Blumenthal, M. M. A new look at the chemistry and physics of deep-fat frying. *Food Technology* 45(2) (1991): 68–72.

Brake, N. C. and Fennema, O. R. Glass transition values of muscle tissue. *Journal of Food Science* 64(1) (1999): 10–15.

Brown, M. E. *Introduction to Thermal Analysis: Techniques and Applications.* London: Chapman and Hall (1988).

Bryś, J., Wirkowska, M., Górska, A., Ostrowska-Ligęza, E., Bryś, A., and Koczoń, P. The use of DSC and FT-IR spectroscopy for evaluation of oxidative stability of interesterified fats. *Journal of Thermal Analysis and Calorimetry* 112(1) (2013): 481–487.

Busfield, W. K. and Proschogo, P. N. Hydrogenation of palm stearine: Changes in chemical composition and thermal properties. *Journal of the American Oil Chemists' Society* 67(3) (1990): 176–181.

Cebula, D. J. and Smith, K. W. Differential scanning calorimetry of confectionery fats. Pure triglycerides: Effects of cooling and heating rate variation. *Journal of the American Oil Chemists' Society* 68(8) (1991): 591–595.

Cebula, D. J. and Smith, K. W. Differential scanning calorimetry of confectionery fats: Part II—Effects of blends and minor components. *Journal of the American Oil Chemists' Society* 69(10) (1992): 992–998.

Chaiseri, S. and Dimick, P. S. Dynamic crystallization of cocoa butter. I. Characterization of simple lipids in rapid- and slow-nucleating cocoa butters and their seed crystals. *Journal of the American Oil Chemists' Society* 72(12) (1995a): 1491–1496.

Chaiseri, S. and Dimick, P. S. Dynamic crystallization of cocoa butter. II. Morphological, thermal, and chemical characteristics during crystal growth. *Journal of the American Oil Chemists' Society* 72(12) (1995b): 1497–1504.

Chapman, G. M. Cocoa butter and confectionery fats. Studies using programmed temperature x-ray diffraction and differential scanning calorimetry. *Journal of the American Oil Chemists' Society* 48(12) (1971): 824–830.

Che Man, Y. B., Haryati, T., Ghazali, H. M., and Asbi, B. A. Composition and thermal profile of crude palm oil and its products. *Journal of the American Oil Chemists' Society* 76(2) (1999): 237–242.

Che Man, Y. B. and Swe, P. Z. Thermal analysis of failed-batch palm oil by differential scanning calorimetry. *Journal of the American Oil Chemists' Society* 72(12) (1995): 1529–1532.

Che Man, Y. B. and Tan, C. P. Comparative differential scanning calorimetric analysis of vegetable oils: II. Effects of cooling rate variation. *Phytochemical Analysis* 13(3) (2002): 142–151.

Chen, J., Zhang, J., and Li, H. Determining the wax content of crude oils by using differential scanning calorimetry. *Thermochimica Acta* 410(1–2) (2004): 23–26.

Chu, B. S., Ghazali, H. M., Lai, O. M., Che Man, Y. B., and Yusof, S. Physical and chemical properties of a lipase-transesterified palm stearin/palm kernel olein blend and its isopropanol-solid and high melting triacylglycerol fractions. *Food Chemistry* 76(2) (2002a): 155–164.

Chu, B. S., Ghazali, H. M., Lai, O. M., Che Man, Y. B., Yusof, S., Tee, S. B., and Yusoff, M. S. A. Comparison of lipase-transesterified blend with some commercial solid frying shortenings in Malaysia. *Journal of the American Oil Chemists' Society* 78(12) (2001): 1213–1219.

Chu, B. S., Tan, C. P., Ghazali, H. M., and Lai, O. M. Determination of iodine value of palm olein mixtures using differential scanning calorimetry. *European Journal of Lipid Science and Technology* 104(8) (2002b): 472–482.

Coppin, E. A. and Pike, O. A. Oil stability index correlated with sensory determination of oxidative stability in light-exposed soybean oil. *Journal of the American Oil Chemists' Society* 78(1) (2001): 13–18.

Cross, C. K. Oil stability: A DSC alternative for the active oxygen method. *Journal of the American Oil Chemists' Society* 47(6) (1970): 229–230.

Cuvelier, M.-E., Lacoste, F., and Courtois, F. Application of a DSC model for the evaluation of TPC in thermo-oxidized oils. *Food Control* 28(2) (2012): 441–444.

Danthine, S. and Deroanne, C. Blending of hydrogenated low-erucic acid rapeseed oil, low-erucic acid rapeseed oil, and hydrogenated palm oil or palm oil in the preparation of shortenings. *Journal of the American Oil Chemists' Society* 80(11) (2003): 1069–1075.

deMan, J. M. and deMan, L. Differential scanning calorimetry techniques in the evaluation of fats for the manufacture of margarine and shortening. *INFORM* 5 (1994): 522.

deMan, L., deMan, J. M., and Blackman, B. Polymorphic behavior of some fully hydroge-nated oils and their mixtures with liquid oil. *Journal of the American Oil Chemists' Society* 66(12) (1989): 1777–1780.

Desmedt, A., Culot, C., Deroanne, C., Durant, F., and Gibon, V. Influence of *cis* and *trans* dou-ble bonds on the thermal and structural properties of monoacid triglycerides. *Journal of the American Oil Chemists' Society* 67(10) (1990): 653–660.

Dibildox-Alvarado, E. and Toro-Vazquez, J. F. Isothermal crystallization of tripalmitin in ses-ame oil. *Journal of the American Oil Chemists' Society* 74(2) (1997): 69–76.

Dimick, P. S., Reddy, S. Y., and Ziegler, G. R. Chemical and thermal characteristics of milk-fat fractions isolated by a melt crystallization. *Journal of the American Oil Chemists' Society* 73(12) (1996): 1647–1652.

D'Souza, V., de Man, L., and de Man, J. M. Polymorphic behavior of high-melting glycerides from hydrogenated canola oil. *Journal of the American Oil Chemists' Society* 68(12) (1991): 907–911.

Dyszel, S. M. Characterization of macadamia nuts by differential scanning calorimetry for country of origin. *Thermochimica Acta* 166 (1990): 291–300.

Dyszel, S. M. Characterization of peanut by DSC for country of origin. *Thermochimica Acta* 226 (1993): 265–274.

Dyszel, S. M. and Pettit, B. C. Determination of the country of origin of pistachio nuts by DSC and HPLC. *Journal of the American Oil Chemists' Society* 67(12) (1990): 947–951.

Elisabettini, P., Desmedt, A., and Durant, F. Polymorphism of stabilized and nonstabilized tristearin, pure and in the presence of food emulsifiers. *Journal of the American Oil Chemists' Society* 73(2) (1996): 187–192.

Firestone, D. *Official Methods and Recommended Practices of the American Oil Chemists' Society*, 4th edn. Champaign, IL: AOCS (1993).

Ford, J. L. and Timmins, P. *Pharmaceutical Thermal Analysis: Techniques and Applications*, pp. 9–24, 136–149. Chichester: Ellis Horwood Ltd. (1989).

Fouad, F. M., van de Voort, F. R., Marshall, W. D., and Farrell, P. G. A critical evaluation of thermal fractionation of butter oil. *Journal of the American Oil Chemists' Society* 67(12) (1990): 981–988.

Frankel, E. N. Lipid oxidation. *Progress in Lipid Research* 19(1) (1980): 1–22.

Frankel, E. N. In search of better methods to evaluate natural antioxidants and oxidative stabil-ity in food lipids. *Trends in Food Science and Technology* 4(7) (1993): 220–225.

Fredrick, E., Foubert, I., Van Sype, J. D., and Dewettinck, K. Influence of monoglycerides on the crystallization behavior of palm oil. *Crystal Growth and Design* 8(6) (2008): 1833–1839.

Gill, P., Moghadam, T. T., and Ranjbar, B. Differential scanning calorimetry techniques: Applications in biology and nanoscience. *Journal of Biomolecular Techniques* 21(4) (2010): 167–169.

Gloria, H. and Aguilera, J. M. Assessment of the quality of heated oils by differential scanning calorimetry. *Journal of Agricultural and Food Chemistry* 46(4) (1998): 1363–1368.

Gmelin, E. Classical temperature-modulated calorimetry: A review. *Thermochimica Acta* 304–305 (1997): 1–26.

Gray, M. S. and Lovegren, N. V. Polymorphism of saturated triglycerides: II. 1, 3-dipalmito triglycerides. *Journal of the American Oil Chemists' Society* 55(8) (1978): 601–606.

Gray, M. S., Lovegren, N. V., and Feuge, R. O. Effect of 2-oleodipalmitin and 2-elaidodipal-mitin on polymorphic behavior of cocoa butter. *Journal of the American Oil Chemists' Society* 53(12) (1976): 727–731.

Griffin, V. J. and Laye, P. G. Differential thermal analysis and differential scanning calorimetry. In E. L. Charsley and S. B. Warrington (eds), *Thermal Analysis*, pp. 17–30. Cambridge: Royal Society of Chemistry (1992).

Gupta, S. K. and Jaworski, R. J. An evaluation of the efficacy of antioxidants in soap by differential scanning calorimetry (DSC). *Journal of the American Oil Chemists' Society* 68(4) (1990): 278–279.

Guth, O. J., Aronhime, J., and Garti, N. Polymorphic transitions of mixed triglycerides, SOS, in the presence of sorbitan monostearate. *Journal of the American Oil Chemists' Society* 66(11) (1989): 1606–1613.

Halliwell, B., Murcia, M. A., Chirico, S., and Aruoma, O. I. Free radicals and antioxidants in food and *in vivo*: What they do and how they work. *Critical Reviews in Food Science Nutrition* 35(1–2) (1995): 7–20.

Harwalkar, V. R. and Ma, C. Y. *Thermal Analysis of Foods*. London: Elsevier Applied Science (1990).

Haryati, T. Development and application of differential scanning calorimetric methods for physical and chemical analysis of palm oil. PhD Dissertation, Universiti Putra Malaysia, Serdang, Malaysia (1999).

Haryati, T., Che Man, Y. B., Asbi, A., Ghazali, H. M., and Buana, L. Determination of iodine value of palm oil by differential scanning calorimetry. *Journal of the American Oil Chemists' Society* 74(8) (1997): 939–942.

Hassel, R. L. Thermal analysis: An alternative method of measuring oil stability. *Journal of the American Oil Chemists' Society* 53(5) (1976): 179–181.

Hastert, R. C. Hydrogenation. In Y. H. Hui (ed.), *Bailey's Industrial Oil and Fat Products, Volume 4: Edible Oil and Fat Products: Processing Technology*, 5th edn., pp. 213–300. New York: Wiley (1996).

Herrera, M. L. Crystallization behavior of hydrogenated sunflowerseed oil: Kinetics and polymorphism. *Journal of the American Oil Chemists' Society* 71(11) (1994): 1255–1260.

Herrera, M. L. and Añón, M. C. Crystalline fractionation of hydrogenated sunflowerseed oil. II. Differential scanning calorimetry (DSC). *Journal of the American Oil Chemists' Society* 68(11) (1991): 799–803.

Herrera, M. L. and Marquez Rocha, F. J. Effects of sucrose ester on the kinetics of polymorphic transition in hydrogenated sunflower oil. *Journal of the American Oil Chemists' Society* 73(3) (1996): 321–326.

Herrera, M. L., Segura, J. A., Rivarola, G. J., and Añón, M. C. Relationship between cooling rate and crystallization behavior of hydrogenated sunflowerseed oil. *Journal of the American Oil Chemists' Society* 69(9) (1992): 898–905.

Irwandi, J., Che Man, Y. B., Kitts, D. D., Bakar, J., and Jinap, S. Synergies between plant antioxidant blends in preventing peroxidation reactions in model and food oil systems. *Journal of the American Oil Chemists' Society* 77(9) (2000): 945–950.

Jacobsberg, B. and Ho, O. C. Studies in palm oil crystallization. *Journal of the American Oil Chemists' Society* 53(10) (1976): 609–617.

Kawamura, K. The DSC thermal analysis of crystallization behavior in high erucic acid rapeseed oil. *Journal of the American Oil Chemists' Society* 58(8) (1981): 826–829.

Kowalski, B. Determination of oxidative stability of edible vegetable oils by pressure differential scanning calorimetry. *Thermochimica Acta* 156(2) (1989): 347–358.

Kowalski, B. Thermal-oxidative decomposition of edible fats and oils. DSC studies. *Thermochimica Acta* 184(1) (1991): 49–57.

Kowalski, B. Evaluation of activities of antioxidants in rapeseed oil matrix by pressure differential scanning calorimetry. *Thermochimica Acta* 213(14) (1993): 135–46.

Kowalski, B., Gruczynska, E., and Maciaszek, K. Kinetics of rapeseed oil oxidation by pressure differential scanning calorimetry measurements. *European Journal of Lipid Science and Technology* 102(5) (2000): 337–341.

Kowalski, B., Ratusz, K., Miciula, A., and Krygier, K. Monitoring of rapeseed oil autoxidation with a pressure differential scanning calorimeter. *Thermochimica Acta* 307(2) (1997): 117–121.

Krishnamurthy, R. and Kellens, M. Fractionation and winterization. In Y. H. Hui (ed.), *Bailey's Industrial Oil and Fat Products, Volume 4, Edible Oil and Fat Products: Processing Technology*, 5th edn., pp. 301–338. New York: Wiley (1996).

Lai, O. M., Ghazali, H. M., and Chong, C. L. Use of enzymatic transesterified palm stearin-sunflower oil blends in the preparation of table margarine formulation. *Food Chemistry* 64(1) (1999): 83–88.

Lambelet, P. and Ganguli, N. C. Detection of pig and buffalo body fat in cow and buffalo ghees by differential scanning calorimetry. *Journal of the American Oil Chemists' Society* 60(5) (1983): 1005–1008.

Lawson, H. Deep fat frying. In H. Lawson (ed.), *Food Fats and Oils: Technology, Utilization, and Nutrition*, pp. 66–115. New York: Chapman and Hall (1995).

Lee, K.-T. and Foglia, T. A. Fractionation of menhaden oil and partially hydrogenated menhaden oil: Characterization of triacylglycerol fractions. *Journal of the American Oil Chemists' Society* 78(3) (2001): 297–303.

Litwinienko, G. Autooxidation of unsaturated fatty acids and their esters. *Journal of Thermal Analysis and Calorimetry* 65(2) (2001): 639–646.

Litwinienko, G., Daniluk, A., and Kasprzycka-Guttman, T. Differential scanning calorimetry study on the oxidation of C_{12}-C_{18} saturated fatty acids and their esters. *Journal of the American Oil Chemists' Society* 76(6) (1999): 655–657.

Litwinienko, G. and Kasprzycka-Guttman, T. A DSC study on thermoxidation kinetics of mustard oil. *Thermochimica Acta* 319(1–2) (1998): 185–191.

Litwinienko, G., Kasprzycka-Guttman, T., and Jarosz-Jarszewska, M. Dynamic and isothermal DSC investigation of the kinetics of thermooxidative decomposition of some edible oils. *Journal of Thermal Analysis* 45(4) (1995): 741–750.

Lovegren, N. V. and Gray, M. S. Polymorphism of saturated triglycerides: I. 1,3-disteapo triglycerides. *Journal of the American Oil Chemists' Society* 55(3) (1978): 310–316.

Lovegren, N. V., Gray, M. S., and Feuge, R. O. Polymorphic changes in mixtures of confectionery fats. *Journal of the American Oil Chemists' Society* 53(2) (1976a): 83–88.

Lovegren, N. V., Gray, M. S., and Feuge, R. O. Effects of liquid fat on melting point and polymorphic behavior of cocoa butter and a cocoa butter fraction. *Journal of the American Oil Chemists' Society* 53(3) (1976b): 108–112.

Marikkar, J. M. N., Ghazali, H. M., Long, K., and Lai, O. M. Lard uptake and its detection in selected food products deep-fried in lard. *Food Research International* 36(9–10) (2003): 1047–1060.

Marikkar, J. M. N., Lai, O. M., Ghazali, H. M., and Che Man, Y. B. Detection of lard and randomized lard as adulterants in refined-bleached-deodorized palm oil by differential scanning calorimetry. *Journal of the American Oil Chemists' Society* 78(11) (2001): 1113–1119.

Marikkar, J. M. N., Lai, O. M., Ghazali, H. M., and Che Man, Y. B. Compositional and thermal analysis of RBD palm oil adulterated with lipase-catalyzed interesterified lard. *Food Chemistry* 76(2) (2002): 249–258.

Márquez, A. L., Pérez, M. P., and Wagner, J. R. Solid fat content estimation by differential scanning calorimetry: Prior treatment and proposed correction. *Journal of the American Oil Chemists' Society* 90(4) (2013): 467–473.

Md. Ali, A. R. and Dimick, P. S. Thermal analysis of palm mid-fraction, cocoa butter and milk fat blends by differential scanning calorimetry. *Journal of the American Oil Chemists' Society* 71(3) (1994a): 299–302.

Md. Ali, A. R. and Dimick, P. S. Melting and solidification characteristics of confectionery fats: Anhydrous milk fat, cocoa butter and palm kernel stearin blends. *Journal of the American Oil Chemists' Society* 71(8) (1994b): 803–806.

Menard, K. P., Rogers, R., and Huff, K. Prediction of SFI values by differential scanning calorimetry. *INFORM* 5 (1994): 523.

Miller, W. J., Koester, W. H., and Freeberg, F. E. The measurement of fatty solids by differential scanning calorimetry. *Journal of the American Oil Chemists' Society* 46(7) (1969): 341–343.

Min, D. B. Lipid oxidation of edible oil. In C. C. Akoh and D. B. Min (eds), *Food Lipids: Chemistry, Nutritional, and Biotechnology*, pp. 283–296. New York: Marcel Dekker (1998).

Minato, A., Ueno, S., Yano, J., Smith, K., Seto, H., Amemiya, Y., and Sato, K. Thermal and structural properties of *sn*-1,3-dioleoyl-2-palmitoylglycerol binary mixtures examined with synchrotron radiation X-ray diffraction. *Journal of the American Oil Chemists' Society* 74(10) (1997): 1213–1220.

Minato, A., Ueno, S., Yano, J., Wang, Z. H., Seto, H., Amemiya, Y., and Sato, K. Synchrotron radiation X-ray diffraction study on phase behavior of PPP-POP binary mixtures. *Journal of the American Oil Chemists' Society* 73(11) (1996): 1567–1572.

Nassu, R. T. and Gonçalves, L. A. G. Determination of melting point of vegetable oils and fats by differential scanning calorimetry (DSC) technique. *Grasas y Aceites* 50(1) (1999): 16–22.

Nehdi, I. A. Characteristics, chemical composition and utilisation of *Albizia julibrissin* seed oil. *Industrial Crops and Products* 33(1) (2011a): 30–34.

Nehdi, I. A. Characteristics and composition of *Washingtonia filifera* (Linden Ex André) H. Wendl. seed and seed oil. *Food Chemistry* 126(1) (2011b): 197–202.

Nehdi, I. A. *Cupressus sempervirens* var. *horizentalis* seed oil: Chemical composition, physicochemical characteristics, and utilizations. *Industrial Crops and Products* 41(1) (2013): 381–385.

Nehdi, I., Omri, S., Khalil, M. I., and Al-Resayes, S. I. Characteristics and chemical composition of date palm (*Phoenix canariensis*) seeds and seed oil. *Industrial Crops and Products* 32(3) (2010): 360–365.

Ng, W. L. A study of the kinetics of nucleation in a palm oil melt. *Journal of the American Oil Chemists' Society* 67(11) (1990): 879–882.

Ng, W. L. and Oh, C. H. A kinetic study on isothermal crystallization of palm oil by solid fat content measurements. *Journal of the American Oil Chemists' Society* 71(10) (1994): 1135–1139.

Noble, D. DSC balances out. *Analytical Chemistry* 67 (1995): 323A–327A.

Noor Lida, H. M. D., Sundram, K., and Idris, N. A. DSC study on the melting properties of palm oil, sunflower oil, and palm kernel olein blends before and after chemical interesterification. *Journal of the American Oil Chemists' Society* 83(8) (2006): 739–745.

Oomah, B. D., Busson, M., Godfrey, D. V., and Drover, J. C. G. Characteristics of hemp (*Cannabis sativa* L.) seed oil. *Food Chemistry* 76(1) (2002): 33–43.

Paul, S. and Mittal, G. S. Regulating the use of degraded oil/fat in deep-fat/oil food frying. *Critical Review in Food Science and Nutrition* 37(7) (1997): 635–662.

Pratt, D. E. Antioxidants: Technical and regulatory considerations. In Y. H. Hui (ed.), *Bailey's Industrial Oil and Fat Products, Volume 3: Edible Oil and Fat Products: Products and Application Technology*, 5th edn., pp. 523–546. New York: Wiley (1996).

Raemy, A., Froelicher, I., and Loelinger, J. Oxidation of lipids studied by isothermal heat flux calorimetry. *Thermochimica Acta* 114(1) (1987): 159–164.

Ray, J., Smith, K. W., Bhaggan, K., Nagy, Z. K., and Stapley, A. G. F. Crystallization and polymorphic behavior of shea stearin and the effect of removal of polar components. *European Journal of Lipid Science and Technology* 115(10) (2013): 1094–1106.

Reddy, S. Y., Full, N., Dimick, P. S., and Ziegler, G. R. Degree of temper in chocolate by differential scanning calorimetry. *INFORM* 5 (1994): 522.

Reddy, S. Y. and Prabhakar, J. V. Study on the polymorphism of normal triglycerides of Sal (*Shorea robusta*) fat by DSC. I. Effect of diglycerides. *Journal of the American Oil Chemists' Society* 63(5) (1986): 672–676.

Rossell, J. B. Differential scanning calorimetry of palm kernel oil products. *Journal of the American Oil Chemists' Society* 52(12) (1975): 505–511.

Saberi, A. H., Tan, C.-P., and Lai, O.-M. Phase behavior of palm oil in blends with palm-based diacylglycerol. *Journal of the American Oil Chemists' Society* 88(12) (2011): 1857–1865.

Samyn, P., Schoukens, G., Vonck, L., Stanssens, D., and Van Den Abbeele, H. Quality of Brazilian vegetable oils evaluated by (modulated) differential scanning calorimetry. *Journal of Thermal Analysis and Calorimetry* 110(3) (2012): 1353–1365.

Satish, S., Fathe J. A., and Maggie, A. Modulated temperature of differential scanning calorimetry for examination of tristearin polymorphism: II. Isothermal crystallization of metastable forms. *Journal of the American Oil Chemists' Society* 76(4) (1999): 507–510.

Sellappan, S. and Akoh, C. C. Enzymatic acidolysis of tristearin with lauric and oleic acids to produce coating lipids. *Journal of the American Oil Chemists' Society* 77(11) (2000): 1127–1133.

Sessa, D. J. Derivation of a cocoa butter equivalent from jojoba transesterified ester via a differential scanning calorimetry index. *Journal of the Science of Food and Agriculture* 72(3) (1996): 295–298.

Sessa, D. J., Nelsen, T. C., Kleiman, R., and Arquette, J. D. Differential scanning calorimetry index for estimating level of saturation in transesterified wax esters. *Journal of the American Oil Chemists' Society* 73(2) (1996): 271–273.

Siddique, B. M., Ahmad, A., Ibrahim, M. H., Hena, S., Rafatullahb, M., and Mohd, O. A. K. Physico-chemical properties of blends of palm olein with other vegetable oils. *Grasas y Aceites* 61(4) (2010): 423–429.

Simon, P. and Kolman, L. DSC study of oxidation induction periods. *Journal of Thermal Analysis and Calorimetry* 64(2) (2001): 813–820.

Simon, P., Kolman, L., Niklova, I., and Schmidt, S. Analysis of the induction period of oxidation of edible oils by differential scanning calorimetry. *Journal of the American Oil Chemists' Society* 77(6) (2000): 639–642.

Smith, P. R., Cebula, D. J., and Povey, M. J. W. The effect of lauric-based molecules on trilaurin crystallization. *Journal of the American Oil Chemists' Society* 71(12) (1994): 1367–1372.

Talbot, G. Fat eutectics and crystallization. In S. T. Beckett (ed.), *Physicochemical Aspects of Food Processing*, pp. 142–166. London: Chapman and Hall (1995).

Tan, C. P. Application of differential scanning calorimetric method for assessing and monitoring various physical and oxidative properties of vegetable oils. PhD Dissertation, University Putra Malaysia, Serdang, Malaysia (2001).

Tan, C. P. and Che Man, Y. B. Differential scanning calorimetric analysis for monitoring oxidation of heated oils. *Food Chemistry* 67(2) (1999a): 177–184.

Tan, C. P. and Che Man, Y. B. Quantitative differential scanning calorimetric analysis for determining total polar compounds in heated oils. *Journal of the American Oil Chemists' Society* 76(9) (1999b): 1047–1057.

Tan, C. P. and Che Man, Y. B. Differential scanning calorimetric analysis of edible fats and oils: Comparison of thermal properties and chemical composition. *Journal of the American Oil Chemists' Society* 77(2) (2000): 143–155.

Tan, C. P. and Che Man, Y. B. Comparative differential scanning calorimetric analysis of vegetable oils: I. Effects of heating rate variation. *Phytochemical Analysis* 13(3) (2002a): 129–141.

Tan, C. P. and Che Man, Y. B. Differential scanning calorimetric analysis of palm oil, palm based products and coconut oil: Effects of scanning rate variation. *Food Chemistry* 76(1) (2002b): 89–102.

Tan, C. P. and Che Man, Y. B. Recent developments in differential scanning calorimetry for assessing oxidative deterioration of vegetable oils. *Trends in Food Science Technology* 13(9–10) (2002c): 312–318.

Tan, C. P., Che Man, Y. B., Jinap, S., and Yusoff, M. S. A. Effects of microwave heating on changes in chemical and thermal properties of vegetable oils. *Journal of the American Oil Chemists' Society* 78(12) (2002a): 1227–1232.

Tan, C. P., Che Man, Y. B., Jinap, S., and Yusoff, M. S. A. Effects of microwave heating on changes in thermal properties of RBD palm olein by differential scanning calorimetry. *Innovative Food Science and Emerging Technologies* 3(2) (2002b): 157–163.

Tan, C. P., Che Man, Y. B., Selamat, J., and Yusoff, M. S. A. Comparative studies of oxidative stability of edible oils by differential scanning calorimetry and oxidative stability index methods. *Food Chemistry* 76(3) (2002c): 385–389.

Tan, C. P., Che Man, Y. B., Selamat, J., and Yusoff, M. S. A. Application of Arrhenius kinetics to evaluate oxidative stability of vegetable oils by isothermal different scanning calorimetry. *Journal of the American Oil Chemists' Society* 78(11) (2001a): 1133–1138.

Tan, C. P., Che Man, Y. B., Selamat, J., and Yusoff, M. S. A. Efficacy of natural and synthetic antioxidants in RBD palm olein by differential scanning calorimetry. In K. Nesaretnam and L. Packer (eds.), *Micronutrients and Health: Molecular Biological Mechanisms*, pp. 108–118. Champaign, IL: AOCS Press (2001b).

Teles Dos Santos, M., Gerbaud, V., and Le Roux, G. A. C. Comparison of predicted and experimental DSC curves for vegetable oils. *Thermochimica Acta* 545 (2012): 96–102.

Tocci, A. M. and Mascheroni, R. H. Characteristics of differential scanning calorimetry determination of thermophysical properties of meats. *Food Science Technology-LWT* 31(5) (1998): 418–426.

Tomaszewska-Gras, J. Detection of butter adulteration with water using differential scanning calorimetry. *Journal of Thermal Analysis and Calorimetry* 108(2) (2012): 433–438.

Toro-Vazquez, J. F., Pérez-Martínez, D., Dibildox-Alvarado, E., Charó-Alonso, M., and Reyes-Hernández, J. Rheometry and polymorphism of cocoa butter during crystallization under static and stirring conditions. *Journal of the American Oil Chemists' Society* 81(2) (2004): 195–202.

Tunick, M. H., Basch, J. J., Maleeff, B. E., Flanagan, J. F., and Holsinger, V. H. Characterization of natural and imitation Mozzarella cheeses by differential scanning calorimetry. *Journal of Dairy Science* 72(8) (1989): 1976–1980.

Tunick, M. H. and Malin, E. L. Differential scanning calorimetry of water buffalo and cow milk fat in mozzarella cheese. *Journal of the American Oil Chemists' Society* 74(12) (1997): 1565–1568.

Tunick, M. H., Malin, E. L., Smith, P. W., and Holsinger, V. H. Reorganization of casein submicelles in Mozarella cheese during storage. *International Dairy Journal* 7(2–3) (1997): 149–155.

Verdonck, E., Schaap, K., and Thomas, L. C. A discussion of the principles and applications of modulated temperature DSC (MTDSC). *International Journal of Pharmaceuticals* 192(1) (1999): 3–20.

Vittadini, E., Lee, J. H., Frega, N. G., Min, D. B., and Vodovotz, Y. DSC determination of thermally oxidized olive oil. *Journal of the American Oil Chemists' Society* 80(6) (2003): 533–537.

Walker, R. C. and Bosin, W. A. Comparison of SFI, DSC and NMR methods for determining solid-liquid ratios of fats. *INFORM* 5 (1971): 50–53.

Warner, K. Chemistry of frying fats. In C. C. Akoh and D. B. Min (eds), *Food Lipids: Chemistry, Nutritional, and Biotechnology*, pp. 167–180. New York: Marcel Dekker (1998).

Wendlandt, W. W. *Thermal Analysis*, 3rd edn., pp. 299–358, 799–810. New York: Wiley (1986).

Wright, D. J. Application of DSC to the study of food behavior. *Analytical Proceedings* 23 (1986): 389–390.

Yap, P. H., de Man, J. M., and de Man, L. Polymorphism of palm oil and palm oil products. *Journal of the American Oil Chemists' Society* 66(5) (1989): 693–697.

Zhang, L., Muramoto, H., Ueno, S., and Sato, K. Crystallization of fully hydrogenated and interesterified fat and vegetable oil. *Journal of Oleo Science* 60(6) (2011): 287–292.

2 DSC as a Valuable Tool for the Evaluation of Adulteration of Oils and Fats

Mohammed Nazrim Marikkar

CONTENTS

2.1 INTRODUCTION

Food authentication has become an important aspect of food quality control in recent times. As there are numerous practices of a fraudulent nature taking place, it has become essential to establish procedures to monitor food quality at different stages of production. The fat and oil processing industry is no exception to fraudulent practices. Over the years, many cases of adulteration have been reported for highly priced vegetable oils and fats such as virgin olive oil (Lai et al., 1995), cocoa butter (Lipp and Anklam, 1998), and dietary supplement oils such as cod liver oil (Rohman and Che Man, 2011), evening primrose oil, flaxseed oil (Ozen et al., 2003), borage oil (Aparicio and Aparicio-Ruiz, 2000), grape seed oil, and pumpkin seed oil (Butinar et al., 2010). Adulteration of virgin olive oil has been common in many parts of the world. For instance, extra-virgin olive oil is a premium product, which is produced through a cold-pressed extraction process. It is in short supply and high demand, which leads to the temptation to perform adulteration practices

(Mavromoustakos et al., 2000; Lai et al., 1995). It is subjected to adulteration either by blending with olive oils of lower grade (e.g., refined olive oil or olive-pomace oil) or by subtle mixing with other types of seed oils. Adulteration of coconut oil is another issue, which is seriously affecting consumers in many Asian countries. The coconut oil industry shows a lack of competitiveness with other major vegetable oils because of the ever-increasing cost of production. Traders in coconut-producing countries are forced to adulterate coconut oil with palm olein, which is cheaper to import. Mixing edible-grade coconut oil with inferior-quality coconut paring oil is yet another fraud carried out by unscrupulous traders to meet the increasing demand for coconut oil (Marikkar and Nasyrah, 2012). There has also been speculation concerning the mixing of edible-grade coconut oil with oil extracted from the waste sludge of desiccated coconut manufacture. If this is truly happening, it will have serious health implications in the long run.

Animal body fats such as lard, chicken, beef, and mutton fats are known for their uses in food applications. Apart from using lard as a component in food, deliberate mixing of lard or pork with other food materials has been highlighted several times in the past (Sosa et al., 2000; Jowder et al., 1997; Saeed et al., 1986, 1989). As lard shows some similarity to palm oil in fatty acid and triacylglycerol (TAG) composition (Marikkar et al., 2001), it could easily be mixed with palm oil. Seriburi and Akoh (1998) demonstrated that mixing lard with sunflower oil in different ratios could produce a variety of plastic shortenings. According to Farag et al. (1983), lard was used as a substitute for butter fat because it was known to show a good spreadability like butter fat. There have also been efforts to blend lard with plant fats to produce substitutes for butter and other dairy products, as pointed out by Lambelet and Ganguli (1983). However, mixing of animal fats (lard, beef, and mutton fats) with plant oils may not be desirable due to religious restrictions. As swine are considered to be unclean in Muslim and Jewish beliefs, eating swine flesh or any product derived from the animal is prohibited (Regenstein et al., 2003). Aside from religious reasons, consumption of animal fats, including lard, is not recommended because of negative perceptions about the implications of animal fats for human health. Hence, there has been much effort to develop methods to authenticate food against the adulteration of animal fats.

2.2   THERMAL EVENTS MEASURED BY DIFFERENTIAL SCANNING CALORIMETRY

In the study of thermal behavior of lipids, differential scanning calorimetry (DSC) is generally operated under two different conditions: cooling process (cooling curve) and heating process (melting curve). While the cooling process provides details of the thermal events associated with crystallization, the heating process could provide information about the thermal events associated with the melting of the lipid constituents. Generally, DSC cooling/heating curves of most edible oils and fats are rather complex, and are closely related to the characteristics of the component TAG molecules in these substances (Dyszel and Pettit, 1990). For this reason, lipid compositional changes caused by adulteration might lead to changes in both cooling and heating curves. Adulteration can change the DSC cooling or heating curves of a lipid sample in a number of ways: new peaks may appear, or existing peaks may undergo changes in their associated DSC parameters. According to many past studies, both of these could provide a basis for qualitative detection of adulterations in oils and fats.

DSC analysis could provide an ample amount of information. Each and every thermal transition in the DSC curve is associated with parameters such as onset temperature (ON), offset temperature (OF), enthalpy change (EN), peak temperature (PT), and peak height (PH). According to Dyszel and Pettit (1990), the relative areas under different thermal transitions in the DSC curve reflect the amount of energy it takes, or that is released, to melt or crystallize a particular portion of the sample. Assuming that the molar heat of fusion for TAG is equivalent, the areas under the DSC curve represent the relative amount of the TAG fraction present. Thus, if adulteration causes changes in the peak areas of the DSC curve of a particular oil or fat, this can be related to TAG compositional changes due to adulteration. Apart from peak area, other DSC parameters, namely PT, PH, ON, and OF, are also shown to be sensitive to TAG compositional changes due to adulteration. Hence, these five DSC parameters could be employed effectively for quantitative estimation of adulteration in lipids.

2.3 DETECTION OF ADULTERATIONS IN PLANT OILS AND FATS

2.3.1 PALM OIL AND PALM OLEIN

Palm oil is a semisolid fat having multiple uses in the food and nonfood sectors. As reported by different researchers (Yanty et al., 2012; Marikkar and Ghazali, 2011; Tan and Che Man, 2000; Haryati, 1999), the TAG profile of palm oil is generally found to include MMM, PLL, MPL, OOL, PLO, PPL, OOO, OOP, OOS, POS, PPO, PPP, and PPS (Table 2.1). Some of these TAG molecules tend to crystallize at relatively lower temperatures (<0°C), while others crystallize in the higher-temperature region (>0°C) (De Man, 1999). The nature of the profile of the cooling curve of palm oil is believed to be due to its content of TAG molecular species with wide-ranging melting points (Yanty et al., 2012; Tan and Che Man, 2000; Che Man et al., 1999). According to several previous studies, the DSC cooling curve of palm oil usually has a major sharp peak at 17.8°C and a broad peak at 1.3°C, with two small shoulder peaks at −6.8°C and −43.9°C (Marikkar et al., 2001; Tan and Che Man, 2000; Che Man et al., 1999) (Figure 2.1). Palm olein, being the liquid fraction of palm oil, usually displays its major exothermic thermal transition at 1.35°C with a minor shoulder peak at −44.57°C (Marikkar and Ghazali, 2011; Tan and Che Man, 2000; Che Man et al., 1999) (Figure 2.2). In the cooling curves of palm oil and palm olein, the shoulder peak appearing around −43.9°C is of considerable interest, since it has been found to be sensitive to lard adulteration. As this peak usually appears in the low-temperature region, it was attributed to the presence of di- and triunsaturated TAG molecules in palm oil and palm olein (Haryati, 1999). In both these oils, this particular peak was found to increase in size with increasing proportion of lard (Marikkar et al., 2002). This was due to the fact that adulteration of either palm oil or palm olein with lard was found to cause an increase in oleic-rich TAGs with a concurrent decrease in palmitic-rich TAGs (Marikkar et al., 2003a, 2005).

TABLE 2.1

TAG Compositional Changes in Palm Oil (PO) after Adulteration (20% Level) with Different Animal Fats

TAG	PO	PO+20%LD	PO+20%BF
MMM	0.54	1.17	0.33
PLL	2.74	3.53	2.20
MPL	0.71	0.73	0.72
OOL	1.81	2.71	1.60
PLO	10.54	12.83	9.75
PPL	10.41	9.05	10.25
OOO	4.19	4.50	4.43
OOP	23.17	23.49	22.29
PPO	31.31	27.22	29.57
PPP	5.38	4.66	5.66
OOS	2.30	2.69	3.18
POS	5.29	6.20	6.99
PPS	0.97	1.02	1.76
SOS	0.44	0.20	0.65
UK	—	—	0.62

Source: Marikkar, J.M.N., Detection of lard as adulterant in some vegetable oils and selected fried food products. PhD Thesis, Universiti Putra Malaysia, Selangor D.E., Malaysia, 2004.

UK, unknown.

As mentioned before, detection of animal fat, such as beef fat, in plant oils is also of considerable interest for food regulatory purposes. Therefore, adulteration of palm oil and palm olein with beef fat (BF) was also investigated by these authors (Marikkar et al., 2001, 2003). According to Marikkar et al. (2001), adulteration of palm oil with BF was not found to give any adulteration peak in the negative-temperature region of the DSC cooling curve. This was due to the fact that adulteration of palm oil with BF caused significant increases in the disaturated and trisaturated TAG molecular species (Table 2.1). Hence, deviations were mostly noticed in the major sharp thermal transition at 17.8°C of the higher-temperature region. With increasing BF adulteration, this transition not only increased in size but also shifted to the higher-temperature region. This finding spurred further interest among researchers to investigate the effect of BF adulteration on the cooling curve of pure palm olein. According to several previous reports, pure palm olein was not found to show any peaks in the higher-temperature region beyond its major exothermic peak appearing at 1.5°C. Once it was adulterated with BF, a clear adulteration peak was found to emerge in the higher-temperature region (from 8.5°C to 19.0°C), indicating the presence of BF (Figure 2.2). Interestingly, this approach could detect

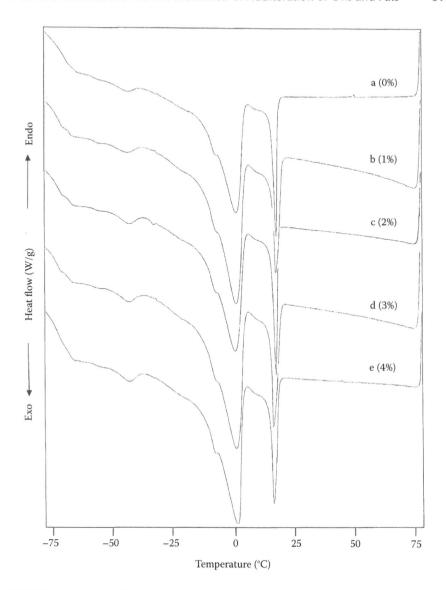

FIGURE 2.1   DSC cooling curves of (a) palm oil, and palm oil adulterated with (b) 1% LD, (c) 2% LD, (d) 3% LD, and (e) 4% LD. (With kind permission from Springer: *Journal of the American Oil Chemists' Society*, Detection of lard and randomized lard as adulterants in RBD palm oil by differential scanning calorimetry, 78, 2001, 1113–1119, Marikkar, J.M.N., Lai, O.M., Ghazali, H.M., and Che Man, Y.B.)

palm olein adulterated with as little as 3% BF. According to TAG compositional analysis, there was a steady increase in the proportions of di- and trisaturated TAG molecular species after adulteration (Table 2.2).

The DSC heating curve of unadulterated palm oil displayed two distinct endo-thermic regions called HMG and LMG, representing high- and low-melting TAG

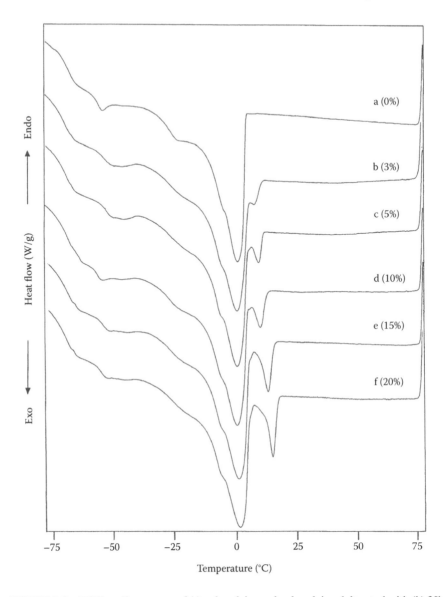

FIGURE 2.2 DSC cooling curves of (a) palm olein, and palm olein adulterated with (b) 3% BF, (c) 5% BF, (d) 10% BF, (e) 15% BF, and (f) 20% BF. (From Marikkar, J.M.N., Ghazali, H.M., Che Man, Y.B., and Lai, O.M.: Differential scanning calorimetric analysis for determination of some animal fats as adulterants in palm olein. *Journal of Food Lipids*. 2003. 10. 63–79. Copyright Wiley-VCH Verlag GmbH & Co. KGaA. Reproduced with permission.)

components, respectively (Figure 2.3). The HMG peaks appeared to emerge in the temperature range from 16.5°C to 47.3°C, while those belonging to LMG appeared in the temperature region from −30.4°C to 16.5°C (Marikkar, 2004). These were in accordance with the pattern of the profile of the heating curve of palm oil reported from other sources (Yanty et al., 2012; Tan and Che Man, 2000; Che Man et al.,

TABLE 2.2

TAG Compositional Changes in Palm Olein (POle) after Adulteration (20% Level) with Different Animal Fats

TAG	POle	POle + 20%LD	POle + 20%BF
MMM	0.70	1.24	0.68
PLL	3.37	4.03	2.97
MPL	0.86	1.08	0.76
OOL	2.10	3.17	1.95
PLO	12.43	13.68	10.63
PPL	10.97	9.59	9.79
OOO	4.54	4.94	4.46
OOP	27.86	26.98	24.57
PPO	29.09	24.76	25.86
PPP	—	—	1.12
OOS	3.05	3.83	4.59
POS	5.04	6.72	7.25
PPS	—	—	1.40
SOS	—	—	2.09
UK	—	—	1.91

Source: Marikkar, J.M.N., Ghazali, H.M., Che Man, Y.B., and Lai, O.M., *Journal of Food Lipids*, 10, 63–79, 2003a.

UK, unknown.

1999). In contrast to the cooling curve, the nature of the profile of the DSC heating curve was rather complex. Because of this, in detection of animal fats using DSC heating curves of palm oil, there is some difficulty in tracing the adulteration peak corresponding to lard. According to the thermal profiles shown in Figure 2.3, lard (LD) adulteration of palm oil, even at 20%, did not produce any additional peaks; instead, major changes were noticed only in the existing peaks. This was due to the fact that there is a complete overlap of thermal transitions occurring in both LMG and HMG regions. Hence, any peak resulting from animal fat adulteration could possibly be masked within the existing peaks of the heating curve.

2.3.1.1   Quantification of Adulteration in Palm Oil and Palm Olein

In addition to qualitative detection of adulteration in palm oil and palm olein, the method of quantifying the level of adulteration in these oils has also been investigated. For palm oil, the particular adulteration peak mentioned earlier was used for quantification of adulteration, as there were good correlations between the DSC parameters of the adulteration peak ($r = 0.9967$, $p < .0001$) and adulteration levels ranging from 1% to 20%. However, oil samples containing <1% LD did not show good correlation with any of the peak parameters. From the adulteration peak, three DSC parameters (peak area, A; peak height, HT; and ON) were derived to serve as independent variables in

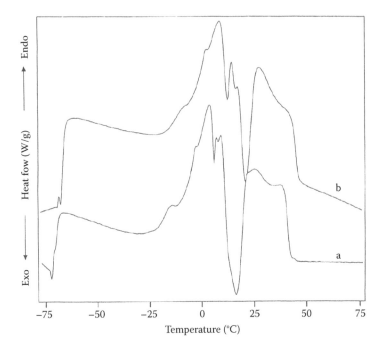

FIGURE 2.3 DSC heating curves of (a) palm oil and (b) palm oil adulterated with 20% LD. (From Marikkar, J.M.N., Detection of lard as adulterant in some vegetable oils and selected fried food products. PhD Thesis, Universiti Putra Malaysia, Selangor D.E., Malaysia, 2004.)

a stepwise multiple linear regression (SMLR) analysis, with percentage of LD (added into palm oil) as the dependent variable. The SMLR analysis showed that only A and HT were necessary to predict the LD adulteration level in palm oil based on DSC cooling thermograms. In a subsequent study, the same approach was also found to be useful for estimation of animal fat adulteration (with LD and BF) of palm olein (Marikkar et al., 2003a). The regression models obtained to predict percentage of LD adulteration in palm oil and palm olein are shown in Table 2.3.

2.3.2 CANOLA AND SUNFLOWER OILS

Canola oil (CaO) comes from an improved version of rapeseed plant that gives an oil that is high in oleic acid, resembling olive oil. Similar to canola oil, sunflower oil (SFO) is also among the most popular vegetable oils used in the world. SFO usually differs from CaO by its high content of linoleic acids (>65%) (Anwar et al., 2007; Davidson et al., 1996). Both of these oils are characterized by possessing high proportions of di- and triunsaturated TAG molecular species (Tables 2.4 and 2.5). Owing to the differences in TAG composition, their cooling and heating curves have slight differences despite some common characteristic features. As reported in the literature (Figure 2.4), the DSC cooling curve of CaO usually displays a major exothermic peak at −54.66°C with two minor exothermic peaks at −40.5°C and −17.0°C (Marikkar et al., 2002). This is in accordance with the pattern of the DSC profiles

TABLE 2.3

Summary of the Regression Analysis to Predict Quantitative Levels of Lard in Selected Vegetable Fats and Oils

Oil Type	Type of Thermal Curve	Regression Equation	R^2 with p Value
Palm oil	Cooling	LD (%) = 14.2675 A +479.9473 HT − 10.159	.9967 ($p < .0001$)
Palm olein	Cooling	LD (%) = −4.6798 A − 0.9577	.9941($p < .0001$)
Canola oil	Heating	LD (%) = 8.2792 A + 9.2597	.9777 ($p < .0001$)
Sunflower oil	Heating	LD (%) = 1.30534 ON + 6.81126 A − 8.53653	.9815 ($p < .0001$)
Palm kernel oil	Cooling	LD (%) = 2.41073 ON + 3.60231 A − 9.33421	.9745 ($p < .0001$)

Source: Marikkar, J.M.N., Detection of lard as adulterant in some vegetable oils and selected fried food products. PhD Thesis, Universiti Putra Malaysia, Selangor D.E., Malaysia, 2004.

R^2, coefficient of determination.

of CaO reported from other sources (Tan and Che Man, 2000). For an unadulterated sample of CaO, there was hardly any exothermic thermal transition beyond −10°C of the DSC cooling curve (Marikkar et al., 2002). This is generally attributed to the presence of predominantly low-melting triunsaturated TAG groups in canola oil. Past studies (Nur Illiyin et al., 2013; Tan and Che Man, 2000) showed that CaO usually contains about 81% monounsaturated TAG and 17.5% diunsaturated (SUU) TAG, while the total amount of trisaturated (SSS) TAG present in the oil was only 1% (Table 2.4). The changes taking place in the DSC cooling profile of canola oil due to adulteration with LD (Figure 2.4) indicated the influence of adulteration on the existing peaks at −17.0°C, −40.5°C, and −60.8°C. The major peak of CaO at −60.8°C was most affected by the increasing level of LD adulteration. While the peak at −40.5°C seems to have gradually disappeared, the peak at −17.0°C gradually increased in size, giving rise to a broad peak shifting slightly toward the higher-temperature region. Apart from this, changes were also noticed in other DSC parameters, namely peak height and enthalpy. Hence, addition of animal fats into a liquid oil like CaO caused peak shifts in its original thermal transitions in the upward direction, primarily due to the fact that the oil sample tends to behave as a binary mixture after having been adulterated with animal fats, which contain more saturated TAG molecular species.

Detection of BF as adulterant in CaO was also investigated using DSC. In contrast to LD, BF is a substance composed of higher proportions of SSS and disaturated (SSU) TAG molecular species (Marikkar et al., 2002). For this reason, adulteration of CaO with BF tended to give rise to HMG peaks in the cooling curve, as shown in Figure 2.5. As the adulteration level increased from 2% to 20%, the major exothermic peak of CaO at −54.66°C was also found to have gradually reduced in size, with

TABLE 2.4

TAG Compositional Changes in Canola Oil (CaO) after Adulteration (20% Level) with Different Animal Fats

TAG	CaO	CaO + 20% LD	CaO + 20% BF
LnLnL	1.43	1.30	1.89
LLLn	6.14	4.95	4.40
LLL	12.02	10.20	9.55
PLLn	1.85	1.70	1.20
OLLn	9.58	8.90	8.50
OOLn	15.84	14.25	12.95
OOL	19.00	16.90	15.90
POL + SLL	6.48	8.70	5.70
PSLn	0.71	0.95	0.60
OOO	19.04	17.00	18.20
POO + SOL	5.43	8.00	7.65
PPO	0.28	2.30	2.75
OOGa	0.52	0.65	1.70
SOO	1.68	2.00	3.35
UK-1	—	2.10	2.60
UK-2	—	—	1.10
UK-3	—	—	0.85
UK-4	—	—	1.11

Source: Marikkar, J.M.N., Ghazali, H.M., Che Man, Y.B., Peiris, T.S.G., and Lai, O.M. *Food Chemistry*, 91, 5–14, 2005.

UK, unknown.

the shifting of its peak position and onset temperatures toward a higher-temperature region. This was primarily due to the fact that the adulterated samples tended to behave as binary mixtures after being adulterated with animal fats, which contain higher-melting TAG molecules.

The nature of the profile of the heating curve of CaO is not as complicated as that of palm oil. Some interesting features were found to emerge in the heating curve of canola oil after adulteration with animal fats. As reported in the literature (Figure 2.6), the DSC heating curve of CaO is characterized by the appearance of two overlapping endotherms: a large, higher-temperature transition at −17.8°C and a small, lower-temperature transition at −28°C. As the completion of the endothermic thermal transitions was marked at −5.5°C, no further thermal transitions were observed beyond this point. According to the changes illustrated in Figure 2.6, the major peak of CaO, appearing at −17.8°C, was affected by the addition of animal fat. In addition, the minor

TABLE 2.5

TAG Compositional Changes in Sunflower Oil (SFO) after Adulteration (20% Level) with Different Animal Fats

TAG	SFO	SFO+20%LD	SFO+20%BF
LLLn	0.59	0.54	0.55
LLL	29.7	24.99	26.80
OLL	28.08	24.27	25.19
PLL	11.46	10.92	10.08
OOL	10.05	9.38	9.28
POL+SLL	12.13	5.81	12.20
UK-1	—	7.72	—
PPL	0.83	0.91	0.45
POO+SOL	5.04	5.51	5.99
PPO	1.22	2.61	2.08
PPP	—	3.52	3.36
UK-2	—	0.87	—
UK-3	—	—	0.77
SOO	0.32	0.37	0.57
POS	—	0.84	1.81
PPS	—	1.74	0.92

Source: Marikkar, J.M.N., Dzulkifly, M.H., Nor Nadiha, M.Z., and Che Man, Y.B., *International Journal of Food Properties*, 15, 683–690, 2012.

UK, unknown.

peak appearing at −28.0°C tended to overlap the major peak as the level of adulteration of animal fat (BT and LD) increased. The temperature region above −5.5°C of the thermal curve was of particular interest in detecting animal fat adulterations in CaO. When CaO was adulterated with LD, a sharp adulteration peak corresponding to LD appeared in this region (from −5.5°C to 50°C) of the heating curve. On adulteration of CaO with BF, this particular adulteration peak was found to appear as a broad peak, as shown in Figure 2.6, which could be visualized even at an adulteration level as low as 2%. According to statistical analysis, high correlations were observed between animal fat adulteration level and the DSC parameters associated with this particular adulteration peak. This clearly showed the influence of SSS and SSU TAG molecular species of BT (Table 2.4) on the original heating curve of CaO.

SFO is another highly unsaturated oil composed of TAG molecules that are mainly esterified with oleic and linoleic acids (Table 2.5). Since SFO shares some common characteristics with CaO, it was employed in a separate study to validate the findings of CaO adulteration with animal fats. As reported in the literature (Marikkar et al., 2012), curve A of Figure 2.7, representing the uncontaminated sample of SFO, was found to display two distinct endothermic transitions at −39.0°C and −25.1°C.

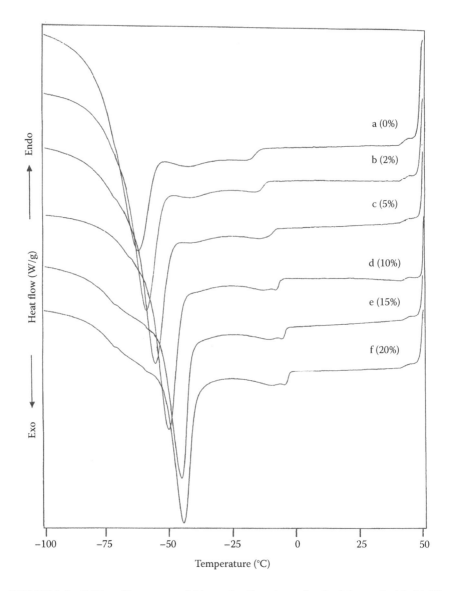

FIGURE 2.4 DSC cooling curves of (a) canola oil, and canola oil adulterated with (b) 2% LD, (c) 5% LD, (d) 10% LD, (e) 15% LD, and (f) 20% LD. (Reprinted from *Food Research International*, 35, Marikkar, J.M.N., Ghazali, H.M., Che Man, Y.B., and Lai, O.M. The use of cooling and heating thermograms for monitoring of tallow, lard and chicken fat adulterations in canola oil, 1007–1014, Copyright 2002, with permission from Elsevier.)

In common with CaO, its heating curve also consisted of a major transition at −25.1°C with a smaller shoulder peak appearing at −8°C, and there were hardly any significant thermal transitions in the temperature region from −5.5°C to 50°C of heating curve A (Marikkar et al., 2012). As mentioned previously, this could be due to the fact that SFO is largely composed of 68.8% triunsaturated (UUU) TAG and 28.9% SUU

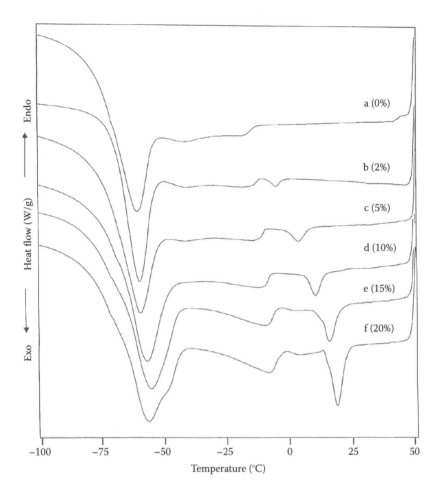

FIGURE 2.5   DSC cooling curves of (a) canola oil, and canola oil adulterated with (b) 2% BF, (c) 5% BF, (d) 10% BF, (e) 15% BF, and (f) 20% BF. (Reprinted from *Food Research International*, 35, Marikkar, J.M.N., Ghazali, H.M., Che Man, Y.B., and Lai, O.M. The use of cooling and heating thermograms for monitoring of tallow, lard and chicken fat adulterations in canola oil, 1007–1014, Copyright 2002, with permission from Elsevier.)

TAG, while the total amount of SSS TAG present in the oil was negligible (Table 2.5). Adulteration of SFO with LD (1%–20%) caused significant changes in the heating curve, as the TAG distribution of LD is considerably different from that of SFO. In this case, too, the adulteration peak corresponding to LD was found to appear within the range from 13°C to 17°C. The DSC parameters associated with the adulteration peaks were also found to show good correlation with the increasing concentration of LD. At this point, it is worth mentioning that the method of detection of lard in CaO and SFO can also be extended to soybean oil (SBO). According to a separate study conducted for SBO adulterated with LD (Marikkar et al., 2003b), the results obtained were similar to and in conformity with the results obtained for detection of LD in CaO and SFO.

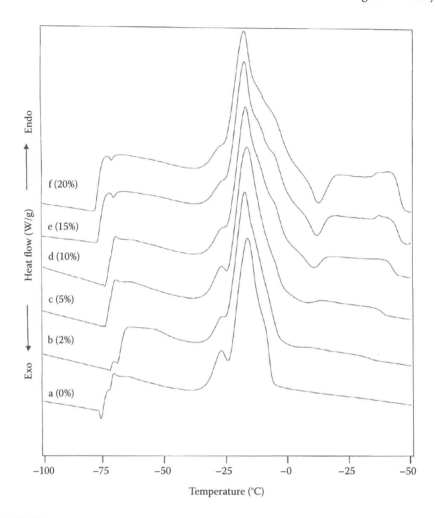

FIGURE 2.6   DSC heating curves of (a) canola oil, and canola oil adulterated with (b) 2% BF, (c) 5% BF, (d) 10% BF, (e) 15% BF, and (f) 20% BF. (Reprinted from *Food Research International*, 35, Marikkar, J.M.N., Ghazali, H.M., Che Man, Y.B., and Lai, O.M. The use of cooling and heating thermograms for monitoring of tallow, lard and chicken fat adulterations in canola oil, 1007–1014, Copyright 2002, with permission from Elsevier.)

2.3.2.1   Quantification of Adulteration in Canola and Sunflower Oils

The methods of quantifying the level of adulteration in these oils were similar to those described previously for palm oil and palm olein. In both of these oils, the DSC parameters A, HT, and ON were found to show good correlation with increasing levels of adulteration. The regression models obtained to predict percentage of LD adulteration in canola and sunflower oils are shown in Table 2.3. In the execution of SMLR analysis, percentage of LD (added to canola oil) was used as the dependent variable, while DSC parameters of the adulterant peak served as independent variables.

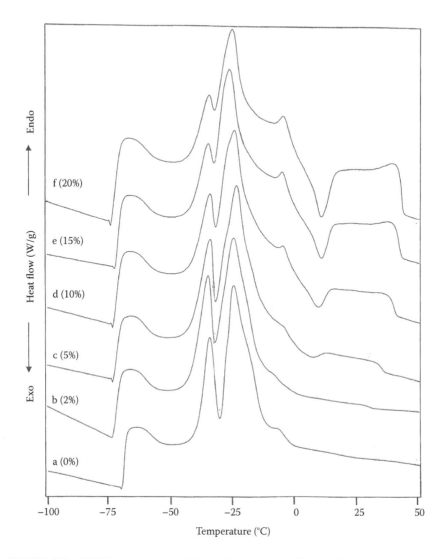

FIGURE 2.7 DSC heating curves of (a) sunflower, and sunflower adulterated with (b) 2% BF, (c) 5% BF, (d) 10% BF, (e) 15% BF, and (f) 20% BF. (From Marikkar, J.M.N., Dzulkifly, M.H., Nor Nadiha, M.Z., and Che Man, Y.B. *International Journal of Food Properties*, 15, 683–690, 2012. With kind permission from Taylor and Francis.)

2.3.3 COCONUT AND PALM KERNEL OILS

Coconut oil and palm kernel oil (PKO) are usually categorized as lauric oils because of the predominance of TAG molecules esterified with lauric acid. The two oils differ from each other only slightly with respect to their fatty acid and TAG compositions, which results in differences in degree of unsaturation and melting temperatures. While coconut oil was found to possess CCLa, CLaLa, LaLaLa, and LaLaM (where C stands for capric, La for lauric, and M for myristic) as major TAG molecular species present at >10% (Marikkar et al., 2013), PKO was found to contain only LaLaLa and LaLaM at >10% (Table 2.6) (Tan and Che Man, 2000). For this reason, there were some differences between the DSC thermal curves of coconut oil (Marikkar et al., 2013; Marina et al., 2009) and PKO (Tan and Che Man, 2000). The DSC cooling curve of PKO was found to display a sharp single transition at 3.1°C with a shoulder peak appearing at 6.4°C (Figure 2.8). It is believed that the sharpness of the exothermic thermal transition is an indication of PKO undergoing crystallization within a narrow temperature range because of the close similarity in thermal characteristics of constituent TAG molecules (Marikkar et al., 2013). The patterns of the DSC cooling curves of PKO reported from different sources are more or less similar (Marikkar, 2004; Tan and Che Man, 2000). According to Marikkar et al. (2005), adulteration of PKO with LD (1%–20%) causes significant changes in its TAG composition because the TAG distribution of LD is considerably different from that of PKO (Table 2.6). For PKO adulteration, as shown in Table 2.6, there was a decline in the proportions of lauric-based TAG molecular species with a concurrent increase in palmitic-based TAG molecular species. In the case of coconut oil adulteration with LD, Mansor et al. (2012) also noticed a similar trend in the proportions of TAG species such as SPO, PPO, and OPO. As a consequence of PKO adulteration with LD, the shoulder peak that originally appeared at 6.4°C of curve A of Figure 2.9 was found to increase in size to become a doublet with increasing level of adulteration (up to 20%). This alone is more than sufficient for preliminary detection of PKO adulteration with animal fats. In a separate study, Mansor et al. (2012) observed a similar trend for the thermal curves of virgin coconut oil adulterated with LD. However, caution must be exercised while making a decision about coconut oil adulteration, since pure coconut oil itself has a doublet thermal transition in the curve. The newly emerging exothermic peak of PKO not only increased in size but also displayed an upward shift in the position of transition as the adulteration level (% LD) gradually increased (0%–20%). However, with regard to limit of detection, the changes in the DSC parameters (A, HT, and ON) of the newly emerging adulteration peak were significant only at the 5% level of adulteration. These were also in conformity with the results reported for virgin coconut oil adulterated with LD by Mansor et al. (2012).

The usefulness of the DSC heating curve of PKO for detection of adulteration was also investigated. As reported in the literature, the DSC heating curve of PKO (curve A in Figure 2.9) had its major endothermic peak at 30.45°C with a shoulder appearing at 18.7°C. In addition, two minor peaks appeared at 5.9°C and −15.75°C. In most cases, the patterns of the DSC heating curve of PKO were almost the same as those reported from different sources (Marina et al., 2009; Marikkar, 2004; Tan and Che Man, 2000). Being lauric-based oils, coconut oil and PKO share some common characteristics in their thermal curves despite slight differences. As a common feature, the nature of thermal transitions occurring in the heating curves of these two oils could be considered to be rapid melting behavior within a narrow temperature range. As seen previously with cooling curves, the differences in the profile of the heating curves of coconut oil and PKO are based on the differences in fatty acid and TAG compositions. According to Marikkar et al. (2013), coconut oil displayed a major melting transition at 23.4°C with a small shoulder peak at 12.5°C. When

TABLE 2.6

TAG Compositional Changes in Palm Kernel Oil (PKO) after Adulteration (20% Level) with Different Animal Fats

TAG	PKO	PKO + 20%LD	PKO + 20%BF
UK-1	2.09	1.83	1.89
CCLa	6.63	5.29	5.08
CLaLa	9.57	7.64	7.33
LaLaLa	20.70	16.86	15.89
LaLaM	16.98	13.52	12.65
LaLaP	6.00	4.89	4.72
LaMO	8.50	6.77	6.85
LaPM	1.11	1.16	1.04
LaOO	4.98	3.88	4.18
LaPO	4.30	4.39	3.67
LaPP/MMO	3.70	4.05	3.30
MMP	4.29	3.72	3.76
MOO	2.30	3.41	2.50
MPO/POL	1.84	5.25	2.67
PPL	1.89	2.34	2.99
OOO	1.67	2.71	2.58
POO	1.54	6.25	4.08
PPO	0.79	2.57	2.12
PPP	0.29	—	1.63
SOO	0.50	1.16	1.70
PSO	0.33	2.32	3.30
UK-2	—	—	2.24
UK-3	—	—	1.73
UK-4	—	—	2.10

Source: Marikkar, J.M.N., Detection of lard as adulterant in some vegetable oils and selected fried food products. PhD Thesis, Universiti Putra Malaysia, Selangor D.E., Malaysia, 2004.

UK, unknown.

both these oils were adulterated with LD and BT (0%–20%), there was no additional peak indicative of the adulteration in the thermal curve (Figure 2.9); instead, the changes due to adulteration were mostly found to take place in the existing peaks. Particularly, in PKO, the shifting of the peak maximum of the major endothermic peak at 30.45°C was clearly seen (Figure 2.9). In a separate study, Mansor et al. (2012) observed a similar trend for the heating curves of virgin coconut oil adulterated with LD. A peak broadening phenomenon was also noticeable with the increasing concentration of animal fats. The peak broadening could be attributed to the higher-melting group of TAG present in LD (Table 2.6).

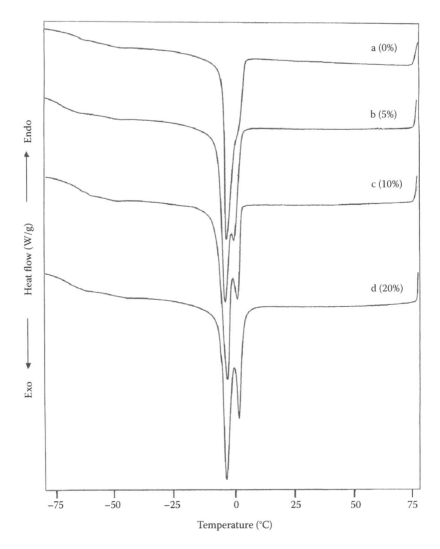

FIGURE 2.8 DSC cooling curves of (a) PKO, and PKO adulterated with (b) 5% LD, (c) 10% LD, and (d) 20% LD. (From Marikkar, J.M.N., Detection of lard as adulterant in some vegetable oils and selected fried food products. PhD Thesis, Universiti Putra Malaysia, Selangor D.E., Malaysia, 2004.)

2.3.3.1 Quantification of Adulterations in Palm Kernel Oil

The methods of quantifying the level of adulteration in these oils were similar to those described previously for other vegetable oils. For estimation of LD adulteration level in PKO using the cooling curve, DSC parameters of the adulteration peak, namely, A, HT, and ON, were useful, since they were found to show good correlation with increasing levels of adulteration. The regression models obtained to predict percentage LD adulteration in PKO are shown in Table 2.3. In the execution of SMLR analysis, percentage LD was used as the dependent variable while DSC parameters

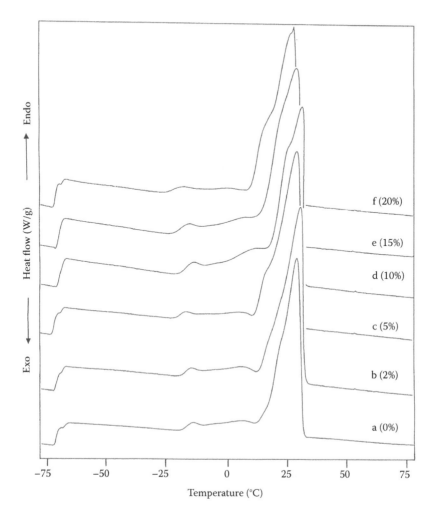

FIGURE 2.9   DSC heating curves of (a) PKO, and PKO adulterated with (b) 2% LD, (c) 5% LD, (d) 10% LD, (e) 15% LD, and (f) 20% LD. (From Marikkar, J.M.N., Detection of lard as adulterant in some vegetable oils and selected fried food products. PhD Thesis, Universiti Putra Malaysia, Selangor D.E., Malaysia, 2004.)

of the adulterant peak served as independent variables. As mentioned previously, caution must be exercised while applying this approach to quantify adulteration of coconut oil, since pure coconut oil itself has a doublet thermal transition in the curve.

2.4   CONCLUDING REMARKS

Authentication is part of quality assessment to maintain purity standards in oils and fats. It is highly useful for oils and fats, as it helps to combat adulteration practices taking place in trade and industries. Among the different analytical techniques, DSC is a valuable tool for establishing the identity of pure oils and fats for authentication purposes. It is a nondestructive technique and works with minimal sample preparation. DSC analyses of palmitic, lauric, and oleic oils can provide well-defined thermal curves, which can be compiled as a database to become a reference for investigations leading to detection of adulterations. As DSC curves of plant oils and fats differ considerably from those of animal fats, adulteration in plant oils and fats caused by animal fats can be detected easily. There is a basis for qualitative detection of adulteration if significant deviations are noticed in the DSC curve of a particular oil or fat with respect to that of the reference authentic sample. However, based on the characteristic composition of individual oils and fats, the detection approaches and limits of detection for either LD or BF differ considerably. DSC could also be used effectively for quantification of adulteration, as DSC variables associated with adulteration peaks were found to show high correlation with increasing levels of adulteration. For most oils, predictive models for estimation of LD adulteration can be obtained using SMLR analysis with percentage LD (adulteration level) as the dependent variable and DSC parameters (A, HT, and ON) of the adulteration peak as independent variables.

REFERENCES

Anwar, F., Hussain, A. I., Iqbal, S., and Bhanger, M. I. Enhancement of the oxidative stability of some vegetable oils by blending with *Moringa oleifera* oil. *Food Chemistry* 93 (2007): 1181–1191.

Aparicio, R. and Aparicio-Ruiz, R. Authentication of vegetable oils by chromatographic techniques. *Journal of Chromatography A* 881 (2000): 93–104.

Butinar, B., Bucar-Miklavcic, M., Valencic, V., and Raspor, P. Stereospecific analysis of triacylglycerols as a useful means to evaluate genuineness of pumpkin seed oils: Lesson from virgin olive oil analyses of minor components. *Journal of Agricultural and Food Chemistry* 58 (2010): 5227–5234.

Che Man, Y. B., Haryati, T., Ghazali, H. M., and Asbi, B. A. Composition and thermal profiles of crude palm oil and its products. *Journal of the American Oil Chemists' Society* 76 (1999): 215–220.

Davidson, H. F., Campbell, E. J., Bell, R. J., and Pritchard, R. A. Sunflower oil. In Y. H. Hui (ed.), *Bailey's Industrial Oil and Fat Products*, 5th edn., vol. 2, pp. 603–676. New York: Wiley (1996).

De Man, J. M. Lipids. In *Principles of Food Chemistry*, 3rd edn., pp. 33–110. New York: Springer Science+Business Media (1999).

Dyszel, S. M. and Pettit, B. C. Determination of the country of origin of pistachio nuts by DSC and HPLC. *Journal of the American Oil Chemists' Society* 67 (1990): 947–951.

Farag, R. S., Aboraya, S. H., Ahmed, F. A., Hewedi, F. M., and Khalifa, H. H. Fractional crystallization and gas chromatographic analysis of fatty acids as a means of detecting butter fat adulteration. *Journal of the American Oil Chemists' Society* 60 (1983): 1665–1669.

Haryati, T. Development and applications of differential scanning calorimetric methods for physical and chemical analysis of palm oil. PhD Thesis, Universiti Putra Malaysia (1999).

Jowder, O. A., Kemsley, E. K., and Wilson, R. H. Mid-infrared spectroscopy and authenticity problems in selected meats: A feasibility study. *Food Chemistry* 59 (1997): 195–201.

Lai, Y. W., Kemsley, E. K., and Wilson, R. H. Quantitative analysis of potential adulterants of extra virgin olive oil using infrared spectroscopy. *Food Chemistry* 53 (1995): 95–98.

Lambelet, P. and Ganguli, N. C. Detection of pig and buffalo body fat in cow and buffalo ghees by differential scanning calorimetry. *Journal of the American Oil Chemists' Society* 60 (1983): 1005–1008.

Lipp, M. and Anklam, E. Review of cocoa butter and alternative fats for use in chocolate— Part B. Analytical approaches for identification and determination. *Food Chemistry* 62 (1998): 99–108.

Mansor, T. S. T., Che Man, Y. B., and Shuhaimi, M. Employment of differential scanning calorimetry in detecting lard adulteration in virgin coconut oil. *Journal of the American Oil Chemists' Society* 89 (2012): 485–496.

Marikkar, J. M. N. Detection of lard as adulterant in some vegetable oils and selected fried food products. PhD Thesis, Universiti Putra Malaysia, Selangor D.E., Malaysia (2004).

Marikkar, J. M. N., Dzulkifly, M. H., Nor Nadiha, M. Z., and Che Man, Y. B. Detection of animal fat contaminations in sunflower oil by differential scanning calorimetry. *International Journal of Food Properties* 15 (2012): 683–690.

Marikkar, J. M. N. and Ghazali, H. M. Effect of *Moringa oleifera* oil blending on fractional crystallization behavior of palm oil. *International Journal of Food Properties* 14 (2011): 1049–1059.

Marikkar, J. M. N., Ghazali, H. M., Che Man, Y. B., and Lai, O. M. The use of cooling and heating thermograms for monitoring of tallow, lard and chicken fat adulterations in canola oil. *Food Research International* 35 (2002): 1007–1014.

Marikkar, J. M. N., Ghazali, H. M., Che Man, Y. B., and Lai, O. M. Differential scanning calorimetric analysis for determination of some animal fats as adulterants in palm olein. *Journal of Food Lipids* 10 (2003a): 63–79.

Marikkar, J. M. N., Ghazali, H. M., Che Man, Y. B., Peiris, T. S. G., and Lai, O. M. Distinguishing lard from other animal fats in admixtures of some vegetable oils using liquid chromatographic data coupled with multivariate data analysis. *Food Chemistry* 91 (2005): 5–14.

Marikkar, J. M. N., Lai, O. M., Ghazali, H. M., and Che Man, Y. B. Detection of lard and randomized lard as adulterants in RBD palm oil by differential scanning calorimetry. *Journal of the American Oil Chemists' Society* 78 (2001): 1113–1119.

Marikkar, J. M. N. and Nasyrah, A. R. Distinguishing coconut oil from coconut paring oil using principle component analysis of fatty acid data. *International Journal of Coconut Research and Development* 28 (2012): 9–13.

Marikkar, J. M. N., Saraf, D., and Dzulkifly, M. H. Effect of fractional crystallization on composition and thermal behavior of coconut oil. *International Journal of Food Properties* 16 (2013): 1284–1292.

Marina, A. M., Che Man, Y. B, Nazimah, S. A. H., and Amin, I. Monitoring adulteration of virgin coconut oil by selected vegetable oils using differential scanning calorimetry. *Journal of Food Lipids* 16 (2009): 50–61.

Mavromoustakos, T., Zervou, M., Bonas, G., Kolocouris, A., and Petrakis, P. A novel analytical method to detect adulteration of virgin olive oil by other oils. *Journal of the American Oil Chemists' Society* 77 (2000): 405–411.

Nur Illiyin, M. R., Marikkar, J. M. N., Shuhaimi, M., Mahiran, B., and Miskandar, M. S. A comparison of the thermo physical behavior of Engkabang (*Shorea macrophylla*) seed fat—Canola oil blends and lard. *Journal of the American Oil Chemists' Society* 90 (2013): 1485–1493.

Ozen, B. F., Weiss, I., and Mauer, L. J. Dietary supplement oil classification and detection of adulteration using Fourier transform infrared spectroscopy. *Journal of Agricultural and Food Chemistry* 51 (2003): 5871–5876.

Regenstein, J. M., Chaudry, M. M., and Regenstein, C. E. The kosher and halal food laws. *Comprehensive Review in Food Science and Food Safety* 2 (2003): 111–127.

Rohman, A. and Che Man, Y. B. FTIR spectroscopy combined with chemometrics for authentication of cod liver oil. *Vibrational Spectroscopy* 55 (2011): 141–145.

Saeed, T., Abu-Dagga, F., and Rahman, H. A. Detection of pork and lard as adulterants in beef and mutton mixtures. *Journal of the Association of Official Analytical Chemists* 69 (1986): 999–1002.

Saeed, T., Ali, S. G., Rahman, H. A. A., and Sawaya, W. N. Detection of pork and lard as adulterants in processed meat: Liquid chromatographic analysis of derivatized triglycerides. *Journal of the Association of Official Analytical Chemists* 72 (1989): 921–925.

Seriburi, V. and Akoh, C. C. Enzymatic interesterification of lard and high-oleic sunflower oil with *Candida antartica* lipase to produce plastic fats. *Journal of the American Oil Chemists' Society* 75 (1998): 1339–1345.

Sosa, J. F. M., Pesini, E. R., Montoya, J., Roncales, P., Perez, M. J. L., and Martos, A. P. Direct and highly species-specific detection of pork meat and fat in meat products by PCR amplification of mitochondrial DNA. *Journal of Agricultural and Food Chemistry* 48 (2000): 2829–2832.

Tan, C. P. and Che Man, Y. B. Differential scanning calorimetric analysis of edible oils: Comparison of thermal properties and chemical composition. *Journal of the American Oil Chemists' Society* 77 (2000): 143–155.

Yanty, N. A. M., Marikkar, J. M. N., and Miskandar, M. S. Comparing the thermo-physical properties of lard and selected plant fats. *Grasas y Aceites* 63 (2012): 328–334.

3 Recent Developments in DSC Analysis to Evaluate Thermooxidation and Efficacy of Antioxidants in Vegetable Oils

Grzegorz Litwinienko and
Katarzyna Jodko-Piórecka

CONTENTS

3.1 MECHANISM AND KINETICS OF AUTOXIDATION

Oxidative decomposition of fats and oils is a relatively well-understood phenomenon of crucial importance for quality of food products as well as quality of other organic materials and goods like, for example, polymers, cosmetics, and pharmaceuticals. Lipids, together with proteins and sugars, are building materials of all living organisms and are essential constituents of food. The importance of lipids in the food industry as well as in the pharmaceutical and cosmetic industries is the main reason for their large-scale production and refinement. Moreover, due to high caloric value and bioavailability of plant oils, during the last several years a substantial intensification of research on their applications as renewable sources of energy (biofuels) has been observed.

Rapid development in lipid science and technology, new advances in nutrition research, new methods of food production (such as oil crops producing high-oleic and low-linolenic oils), and new methods of processing have resulted in a number of scientific problems connected with the oxidation of lipids. The extent of oxidation is usually monitored periodically by detection of products of oxidation, such as conjugated dienes (by UV–vis spectroscopy), hydroperoxides (by titration to determine the peroxide number), or volatile products of oxidation: 4-hydroxy-2-nonenal, hexanal, and propanal (by gas chromatography). More advanced methods of detection include chemiluminescence and electron paramagnetic resonance (EPR).

The present state of knowledge on the mechanisms of lipid oxidation is mainly due to the work of Porter, Frankel and others (Porter 1986; Porter et al. 1995; Frankel 2005; Chan 1987). During the last 20 years their fundamental works have been frequently cited as a classical and undeniable contribution to the field. The understanding of the kinetics of free radical chain processes and mechanisms of autoxidation has led to an enormous increase in research in biology and medicine, as has been summarized in an excellent monograph (Halliwell and Gutteridge 2007).

Autoxidation of lipids and hydrocarbons is a multistep chain process that can be described by the following sequence of reactions:

$$\text{initiation: Initiator} \rightarrow \text{L}^{\bullet} \quad R_i; k_i \tag{3.1}$$

$$\text{propagation: } \text{L}^{\bullet} + O_2 \rightarrow \text{LOO}^{\bullet} \quad k_{p1} \tag{3.2}$$

$$\text{LOO}^{\bullet} + \text{LH} \rightarrow \text{L}^{\bullet} + \text{LOOH} \quad k_{p2} \tag{3.3}$$

$$\text{termination: } \text{LOO}^{\bullet} + \text{LOO}^{\bullet} \rightarrow \left(\text{LOO}\right)_2 \quad v_{t1}; k_{t1} \tag{3.4}$$

$$\text{L}^{\bullet} + \text{LOO}^{\bullet} \rightarrow \text{L-OO-L} \quad v_{t2}; k_{t2} \tag{3.5}$$

$$\text{L}^{\bullet} + \text{L}^{\bullet} \rightarrow \text{L-L} \quad v_{t3}; k_{t3} \tag{3.6}$$

where:
LH and LOOH are lipid and hydroperoxide, respectively
$\text{L}^{\bullet}$ and $\text{LOO}^{\bullet}$ are lipid radical and lipid peroxyl radical, respectively
R_i is the rate of initiation
symbols v and k with indexes denote the rate and the rate constant of a particular reaction (3.2 through 3.6), respectively

For relatively high oxygen pressure (>13 kPa), the rate of oxidation, R_{ox}, can be expressed by a simple rate law equation:

$$R_{ox} = k_{p2} \sqrt{\frac{R_i}{2k_{t1}}} \left[\text{LH}\right] \tag{3.7}$$

Abstraction of a hydrogen atom from the lipid molecule (Equation 3.3) is the rate-limiting step of oxidation; thus, autoxidation is a first-order process with respect to the lipid concentration.

Autoxidation leads to the formation of thermally unstable molecules: peroxides (LO_nL', where $n = 2-4$) and hydroperoxides (LOOH). In hydroperoxides, the LO–OH bond is the weakest (LO–OH bond energy is about 40 kcal/mol compared with 80 kcal/mol for a C–C bond and above 90 kcal/mol for a C–H bond). Thus, the decomposition of LOOH gives rise to alkoxyl ($LO^•$) and hydroxyl ($HO^•$) radicals, being a source of new kinetic chains of oxidation. Moreover, the alkoxyl radicals of unsaturated lipids (hydrocarbons) can undergo further decomposition. Scission of the C–C bond results in the formation of unsaturated oxo-compounds, alkyl radicals, and olefin radicals to generate short-chain volatile alcohols, aldehydes, and ketones.

Regardless of the well-established mechanisms, many problems are still a matter of lively debate. For example, the problem of the origin of the very first radicals initiating the chain processes (Equation 3.1), the stereochemistry of the process of hydroperoxide formation, the stability and fate of hydroperoxides, and mechanisms of formation of primary or secondary products of oxidation are still under discussion. Other important problems are the impact of lipid autoxidation (in biochemistry this process is usually called "peroxidation") on health and well-being. Currently, there are three emerging areas of lipid oxidation research:

1. Development of experimental methods that could give more precise and more reliable data on oxidation kinetics and oxidative stability of lipids in the bulk phase
2. Research on oxidation kinetics and oxidative stability in multiphase lipid systems
3. Development of new antioxidants and optimization of their activity and efficiency in suppression of lipid oxidation

In food chemistry, these three areas are overlapping; therefore, the multidisciplinary research involves physics, chemistry, biology, and toxicology. Physical and analytical chemistry are focused on the development of analytical methods to evaluate the status/quality of oxidation as well as methods for evaluation of the oxidative stability of foods/lipids, studies on natural or artificial antioxidants, and studies on the mechanisms and kinetics of lipid oxidation in various systems.

During the last 15 years, rapid development of advanced analytical techniques of sensing and imaging has made significant progress in biology, biophysics, material engineering, and physical chemistry. However, the majority of these new techniques have not been transferred to food chemistry to improve the understanding of the phenomenon of lipid oxidation in complex multiphase systems. Almost all of these sophisticated techniques are unable to follow complex chain reactions when hundreds of compounds are generated at the same moment at elevated temperature, including simultaneous generation of radicals, primary and secondary products of oxidation, degradation products, and so on. Thus, US and European norms (Firestone 2003; European Committee for Standards [CEN] 2003) recommend much simpler techniques, called accelerated tests. In these methods the oxidation status of a sample is monitored at elevated temperature (in a continuous or periodic manner) by a simple detection/determination of oxidized products or intermediates, such as hydroperoxides (the peroxide value, PV), thiobarbituric acid adducts (TBA), *para*-anisidine, conjugated dienes, or volatile oxidation products. The common feature of all accelerated tests is that the sample is subjected to accelerated oxidation (as a result of exposure to increased temperature, elevated oxygen pressure, or the presence of initiators such as transition metal ions). Common tests invented several decades ago (the Schaal Oven test, the active oxygen method [AOM], the Rancimat method, the oxidative stability index [OSI], and the oxygen bomb method) are traditionally used in both scientific and commercial laboratories. In the Shaal Oven test a sample is kept at 60°C and periodically analyzed (titration of PV or TBA test). This manual method is time-consuming (several days or weeks) with poor reproducibility of results. Other accelerated tests are based on detection of volatile products of oxidation. Usually a lipid sample is kept at 100°C–150°C, with oxygen purged through the lipid. Then, the oxygen stream is directed into ultrapure water. The volatile products of oxidation are detected on the basis of changes in water conductivity. This method is rather qualitative and cannot give information on the rate and kinetics of oxidation as well as reliable knowledge on the behavior of the system at lower temperatures because: (a) volatile compounds are the secondary products of oxidation and (b) decomposition of hydroperoxides (primary products) into volatile (secondary) products is strongly dependent on temperature with various mechanisms of the formation of volatile aldehydes.

A critical review of accelerated tests is included in the book edited by Kamal-Eldin and Pokorný (2005). Accelerated tests like Rancimat, AOM, and OSI are used in industry as standard tests for determination of oxidative stability of food, fats, and oils. However, they give results that are still far from the scientific quality expected in basic sciences. Thus, new, simpler, and more reliable methods are needed.

3.2 KINETIC MEASUREMENTS OF LIPID AUTOXIDATION BY CALORIMETRY

The rate of oxidation can be determined by monitoring any physical or chemical parameter dependent on R_{ox}. Therefore, the heat evolved during oxidation is a valuable tool to follow the oxidation course. During the initial stage of oxidation the rate of initiation R_i is effectively constant (there is no branching of the chain reaction), and, a large part of Equation 3.7 can be regarded as constant:

$$k_{p2}\sqrt{\frac{R_i}{2k_{tl}}} = k = \text{const} \tag{3.8}$$

and k is a global (overall) first-order rate constant describing the autoxidation kinetics.

During the progress of oxidation the hydroperoxides decompose to ketones, alcohols, and fatty acids:

$$LCH(OOH)L_1 \quad \rightarrow \quad L(C=O)L_1 + H_2O \quad k_{3.9} \tag{3.9}$$

$$LCH(OOH)CH_2L_1 + O_2 \quad \rightarrow \quad LCOOH + L_1COOH \quad k_{3.10} \tag{3.10}$$

$$LOOH \quad \rightarrow \quad LO^{\cdot} + {}^{\cdot}OH \quad k_{3.11} \tag{3.11}$$

$$LO^{\cdot} + LH \quad \rightarrow \quad LOH + L^{\cdot} \quad k_{3.12} \tag{3.12}$$

Since $k_{3.11} \ll k_{3.12}$, the rate of formation of alcohols (Reaction 3.12) is determined by the rate of Reaction 3.11, and the overall decomposition of hydroperoxides (Reactions 3.9 through 3.12) is assumed to be a first-order process:

$$R_{decomp} = (k_{3.9} + k_{3.10} + k_{3.11})[LOOH] \tag{3.13}$$

In conclusion, according to the kinetic equations 3.7, 3.8, and 3.13, both the formation of peroxides and their decomposition are first-order processes. This relatively simple description of complex kinetics has led to the development of new analytical techniques able to give satisfactory kinetic data on lipid autoxidation and thermooxidation.

In differential scanning calorimetry (DSC), the progress of reaction is monitored through measurements of the heat flow, which is recorded as a function of time (in isothermal mode) or as a function of linearly increasing temperature (when the nonisothermal mode is employed). Degree of conversion (α) is the ratio of the amount of reactant consumed in the reaction during time τ to the total amount of the reactant. In scanning calorimetry, the parameter α can be determined as a ratio of the amount of heat released from the beginning of the process to time τ (ΔH_τ) to the total amount of heat evolved during the process (ΔH_∞) (see Figure 3.1).

$$\alpha = \frac{\Delta H_\tau}{\Delta H_\infty} \tag{3.14}$$

Therefore, the rate of the process can be followed with respect to the amount of heat released or absorbed during the reaction course:

$$\frac{d\alpha}{d\tau} = \frac{1}{\Delta H_\infty} \times \frac{dH}{d\tau} \tag{3.15}$$

where $dH/d\tau$ is the heat flow at the time τ. The rate of the process can also be expressed by the general equation

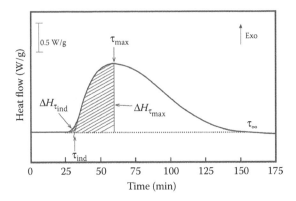

FIGURE 3.1 Example of a DSC curve of isothermal oxidation with definition of the oxidation induction time (τ_{ind}) and time of the maximal heat flow (τ_{max}). The integration of the area under the DSC curve from the start of thermal effect to time τ_α (τ_{ind}, ..., τ_{max}, ..., τ_∞) gives a partial heat of the process (ΔH_τ). A degree of conversion (α) is determined as the ratio of the amount of heat released within the time $\langle 0, \tau_\alpha \rangle$ to the total heat (ΔH_∞) evolved during the process $\langle 0, \tau_\infty \rangle$.

$$\frac{d\alpha}{d\tau} = k(T) \times f(\alpha) \qquad (3.16)$$

where:

$k(T)$ is the rate constant given by the Arrhenius equation: $k(T) = A \exp(-E_a/RT)$, with A and E_a as a preexponential factor and activation energy, respectively

$f(\alpha)$ is a conversion function, that is, a mathematical representation of a kinetic model reflecting the mechanism of the monitored reaction

In most isoconversional methods the kinetic parameters are obtained from a set of experiments in which time or temperature corresponding to a fixed degree of conversion (α = constant, determined from the DSC curve) are dependent either on the temperature of experiment (in isothermal mode) or on the rate of heating (in nonisothermal mode).

3.2.1 ISOTHERMAL METHODS

Separation of variables in Equation 3.16 and integration gives the formula

$$\int_0^\alpha \frac{d\alpha}{f(\alpha)} = k(T) \int_0^{\tau_\alpha} d\tau \qquad (3.17)$$

where:

τ_α is the time when the system (undergoing the reaction at temperature T) reaches the level of conversion α

The integration of the left side of Equation 3.17 in the definite domain $[0,\alpha]$ can be denoted as $F(\alpha)-F(0)$. Thus, time τ_α can be expressed as $\tau_\alpha = [F(\alpha)-F(0)]/k(T)$ and, when combined with the Arrhenius equation, leads to the dependence of τ_α on $1/T$:

$$\tau_\alpha = \frac{F(\alpha)-F(0)}{A \exp\left[\dfrac{-E_a}{RT}\right]} = \frac{\text{const}}{A \exp\left[\dfrac{-E_a}{RT}\right]} \tag{3.18}$$

A series of isothermal measurements carried out at different temperatures gives several DSC curves. For each DSC curve the time $\tau_{\alpha=\text{const}}$ (corresponding to the same degree of conversion $\alpha=\text{const}$) can be easily determined by comparison of partial (ΔH_τ) with total heat evolved (ΔH_∞), as shown in Figure 3.1. Therefore, E_a can be obtained from the transformed Equation 3.17 taking its logarithmic form, $\ln \tau_\alpha = f(1/T)$:

$$\frac{\Delta \ln \tau_\alpha}{\Delta(1/T)} = \frac{E_a}{R} + \text{const} \tag{3.19}$$

The induction time τ_{ind} (defined as extrapolated time of the start of exothermal oxidation effect, as shown in Figure 3.1) and time of the maximal rate of the process τ_{max} (i.e., time of the maximal heat flow; see Figures 3.1 and 3.2) are the points of constant conversion at any isothermal temperature, and these two points can be easily determined from the isotherms without the calculation of ΔH_τ and ΔH_∞. Therefore, τ_{ind} and τ_{max} can be used for calculation of the kinetic parameters in a simple version of the isothermal method (Kowalski 1989, 1992). For example, from several DSC curves of oxidation recorded at $T_1, ..., T_i$ (see panel A of Figure 3.2), the induction times τ_{ind} (or times of the maximal heat flow τ_{max}) are determined and $\ln \tau_i$ (where τ_i is τ_{ind} or τ_{max}) are plotted as a function of the reciprocal of the absolute temperature of oxidation, as presented in panel B of Figure 3.2. The activation energy is calculated from the slope of $\ln \tau_i$ as a function of $(1/T)$, according to the differential form of Equation 3.18:

$$\frac{d \ln \tau_i}{dT^{-1}} = \frac{E_a}{R} + \text{const} \tag{3.20}$$

This method does not require information in the form of $f(\alpha)$ or $[F(\alpha)-F(0)]$.

The isothermal DSC method is intuitively very close to the standard accelerated tests like Rancimat or OSI, which are also performed under isothermal conditions. Since the first isothermal DSC study (Cross 1970), the induction time τ_{ind} has been regarded as a parameter qualitatively comparable to induction times determined by other methods such as AOM (Hassel 1976; Raemy et al. 1987). When induction time was too short to be determined, the time of the maximal heat flow τ_{max} was successfully applied (Raemy et al. 1987) or the isothermal conditions were reached by a very fast heating of the system (with the heating rate $\beta = 20$ K/min, up to 170°C) (Pereira and Das 1990). A good correlation between isothermal τ_{ind}

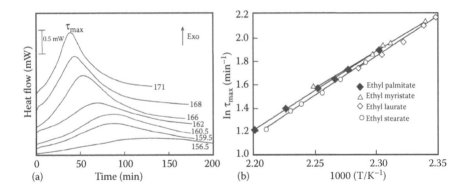

FIGURE 3.2 (a) DSC curves of isothermal oxidation of ethyl myristate at temperatures 156.5°C–171°C as indicated on the plots. (b) Straight line plots of ln τ_{max} versus reciprocal of the absolute temperatures of oxidation for isothermal oxidation of ethyl esters of lauric, myristic, palmitic, and stearic acid. Note the similar slopes for all functions, indicating similar activation parameters for oxidation of saturated fatty acids. (With kind permission from Springer: *J. Therm. Anal. Calorim.*, Oxidation of saturated fatty acids esters. DSC investigations, 54, 1998b, 211–217, Litwinienko, G. and Kasprzycka-Guttman, T., Copyright 1998 Springer.)

and OSI values for 12 edible oils was reported by Tan et al. (2002). Parameters τ_{ind} and τ_{max} depend on the oxidative stability, but τ_{ind} has the advantage of being practically independent of the mass of the oxidized sample. If τ_{max} is to be used for calculation, an approximately constant mass of samples should be used in one set of measurements (Kowalski 1989, 1992). Studies on the correlation of parameters τ_{ind} and τ_{max} with peroxide number (PN) gave linear dependence for PN ≤ 30 mmol O_2^{2-}/kg (Kowalski et al. 1997) and the authors concluded that the parameters τ_{ind} and τ_{max} are more reliable than PN because decomposition of peroxides could give an underestimated value of PN for advanced stages of autoxidation. This observation is also applicable to reprocessed oils, that is, oils heated under N_2 in order to decrease PN.

   Modeling of the kinetics of isothermal oxidation was the subject of a few studies in which degree of conversion was calculated from Equation 3.14. Differentiation of α with respect to time gives the reaction rate, which can be combined with the rate law Equation 3.16 to give a simple dependence of the rate of oxidation on α. For low degrees of conversion ($\alpha \leq 0.16$) the best agreement of experimental data was observed for a model rate equation $d\alpha/d\tau = (k_a\alpha + k_c)(1-\alpha)^2$, where k_a and k_c are the rate constants of autocatalytic and catalytic processes, respectively, and k_c is about 4–10 times lower than k_a (Kowalski 1992). Experimental data were also used for a direct calculation of E_a from the equation assuming an autocatalytic model: $\log(d\alpha/d\tau) = \log k + n\log[(1-\alpha)\alpha^{m/n}]$ (Kasprzycka-Guttman and Odzeniak 1994; Kasprzycka-Guttman et al. 1994).

   During the last decade an intensive development of research on the application of vegetable oils and animal fats as biofuels has been observed. Fuel standards include measurement of induction period (induction time) performed at 110°C. Knothe and

Dunn (2003) used OSI for assessing oxidation of monoalkyl esters of fatty acids and discussed how the structure and concentration of compounds can influence OSI induction times. Oxidative stability of biofuels is determined by the same methodology as oxidative stability of edible oils and fats. Thus, thermal analysis methods applied in food chemistry can also be applied for rapid assessment of quality and oxidative stability of biodiesel and biolubricants.

Figure 3.3 presents pressurized DSC (PDSC) employed in isothermal mode at 110°C for evaluation of the oxidative behavior of biodiesel, compared with the results of the Rancimat method carried out at the same temperature (Leonardo et al. 2012). For both methods a good correlation of data was found. PDSC has several advantages with respect to the Rancimat method: it requires a smaller sample, and it is characterized by reduced analysis time and increased sensitivity. Moreover, detection of oxidative processes by PDSC is independent of the volatility of the secondary products of oxidation. The PDSC method is relatively fast, the experimental conditions are easy to repeat, results are acquired with good precision, and the kinetic parameters can be extrapolated to lower temperatures. However, the application of τ_{ind} for determination of kinetic parameters of oxidation is not recommended for PDSC experiments with easily oxidizable samples and noninhibited oils because oxidation may occur during thermal equilibration, before a constant temperature is reached. Thus τ_{ind} might be misleading and, usually, is too short to be determined. For such easily oxidizable systems more reliable results can be obtained with a nonisothermal (dynamic) DSC.

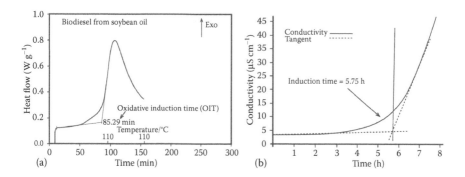

FIGURE 3.3 Determination of induction time from the isothermal PDSC curve (a) and from the Rancimat plot (b). In the PDSC experiment the sample was placed in the PDSC cell and pressurized with air (551.6 kPa); then the sample was heated from 30°C to 110°C with a heating rate of 10 K/min and maintained at this temperature. In both experiments, PDSC and Rancimat, the temperature was 110°C. (With kind permission from Springer: *J. Therm. Anal. Calorim.*, An alternative method by pressurized DSC to evaluate biodiesel antioxidants efficiency, 108, 2012, 751–759, Leonardo, R.S., Murta Valle, M.L., and Dweck, J., Copyright 2012 Springer.)

3.2.2 NONISOTHERMAL METHODS

Heating a small sample (a few milligrams) with a programmed linear heating rate (β) under an oxygen (or air) atmosphere greatly accelerates the oxidation process, and this nonisothermal mode is more frequently used than the isothermal mode. However, the increasing temperature can initiate other physical and chemical changes. Thus, every measurement of oxidative stability should be preceded by simple experiments in an inert (nitrogen, argon) atmosphere in order to check for interfering thermal effects such as, for example, melting, evaporation, or polymerization. The heat capacity of common edible oils does not vary substantially with temperature increasing from 70°C to 140°C, meaning that no baseline disturbance effect occurs (Kowalski 1988; Kasprzycka-Guttman and Odzeniak 1991). A wider temperature range (80°C–350°C, N_2) applied to triglycerides also indicates no thermal side effects (Raemy et al. 1987; Kowalski 1988). Simple methyl esters are more volatile and evaporation might occur at temperatures above 150°C. At a standard heating rate (less than 25 K/min) and under moderate oxygen pressure the edible oils and fats do not manifest a tendency to self-ignition. This process requires nonstationary conditions, with heating rate in the range 40–90 K/min and oxygen pressure of 800–2800 kPa (Raemy et al. 1983; Raemy and Loeliger 1985; Kowalski 1990). With a moderate heating rate ($\beta = 10$ K/min) a self-ignition process for linseed, safflower, corn, and peanut oils occurs above 400°C (see Figure 3.4), well above the temperature range of thermally initiated autoxidation (Baylon et al. 2008).

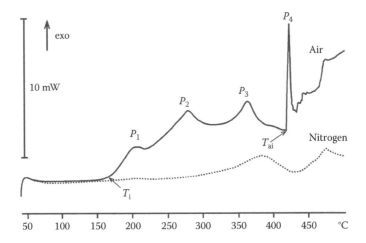

FIGURE 3.4 DSC curves recorded for a sample of linseed oil heated under air (*solid line*) and nitrogen (*dotted line*). T_i and T_{ai} represent temperatures of onset point (start of oxidation) and autoignition, respectively. P_1, P_2, and P_3 are exothermic peaks corresponding to the three exothermic events due to the autoxidation of the oil; P_4 is the exothermic peak due to the combustion of the oil. (With kind permission from Springer: *J. Forensic Sci.*, Evaluation of the self-heating tendency of vegetable oils by differential scanning calorimetry, 53, 2008, 1334–1343, Baylon, A., Stauffer, E., and Delémont, O., Copyright 2012 Springer.)

3.2.2.1 Isoconversional Methods at Linear Heating Rate

For a nonisothermal mode, the temperature is increased with linear heating rate β:

$$T_i = T_0 + \beta\tau \tag{3.21}$$

This expression, applied to Equation 3.16, gives a dependence of α on the increasing temperature:

$$\frac{d\alpha}{dT} = \frac{1}{\beta}\frac{d\alpha}{d\tau} = \frac{A}{\beta}\exp\left(-\frac{E_a}{RT}\right)f(\alpha) \tag{3.22}$$

Depending on the mathematical transformations, Equation 3.22 is a basis for two families of isoconversional methods (differential and integral) for the calculation of the kinetic parameters.

3.2.2.2 Differential Methods

In logarithmic form, Equation 3.22 can be applied for calculation of the activation parameters A and E_a:

$$\ln\frac{\frac{d\alpha}{dT}}{f(\alpha)} = \ln\frac{A}{\beta} + \left(-\frac{E_a}{R}\right)\left(\frac{1}{T}\right) \tag{3.23}$$

Carroll and Manche proposed a method (Carroll and Manche 1970) whereby $d\alpha/dT$ is determined for $\alpha =$ const in a series of experiments carried out at various heating rates β. The activation energy E_a can be obtained from the slope of $\ln(d\alpha/dT)_{\alpha = \text{const}}$ versus $(T^{-1})_{\alpha = \text{const}}$. In the Freeman and Carroll method the quantities $(d\alpha/dT)$ and $(1-\alpha)$ are determined from the experimental thermoanalytical curve and plotted versus T^{-1} (Freeman and Carroll 1958, 1969). For a given $\alpha =$ const, the value $f(\alpha)$ is constant and for the kinetic model of nth order reaction, $f(\alpha) = (1-\alpha)^n$, the equation

$$\Delta\ln\frac{d\alpha}{dT} = n\ln(1-\alpha) - \frac{E_a}{R}\Delta\left(\frac{1}{T}\right) + \text{const} \tag{3.24}$$

can be used directly. However, the presence of α and $d\alpha/dT$ in the same equation is controversial because of the possibility of autocorrelation (Agrawal 1992). Additionally, due to errors generated during differentiation, the differential methods often lead to scattered results (Flynn and Wall 1966).

3.2.2.3 Integral Methods

Separation of variables in Equation 3.22 leads to the equation

$$\frac{d\alpha}{f(\alpha)} = \frac{A}{\beta} \exp\left(-\frac{E_a}{RT}\right) dT \tag{3.25}$$

Integration gives the function $g(\alpha)$:

$$g(\alpha) = \int_{T_0}^{T} \frac{d\alpha}{f(\alpha)} = \int_{T_0}^{T} \left[\frac{A}{\beta} \exp\left(-\frac{E_a}{RT}\right)\right] dT \tag{3.26}$$

where T_0 is an initial temperature of the process. Calculation of the integral form of $g(\alpha)$ with substitution $x = E_a/RT$ yields the expression

$$\frac{A}{\beta} \int_{T_0}^{T} \exp\left(\frac{-E_a}{RT}\right) dT = \frac{A}{\beta} \frac{E_a}{R} \int_{x_0}^{x} \frac{\exp(-x)}{x^2} dx$$

$$= \frac{A}{\beta} \frac{E_a}{R} \left[\frac{\exp(-x)}{x} - \int_{x}^{\infty} \frac{\exp(-x)}{x^2} dx\right]_{x_0}^{x} = p(x) \tag{3.27}$$

The temperature integral cannot be expressed in a closed form, and several expressions and semiempirical approximations of $p(x)$ have been introduced. For $x > 20$ the approximation $\log p(x) = -2.315 - 0.4567x$ (Doyle 1965) was introduced into Equation 3.27 by Ozawa (1965, 1970), and, independently, by Flynn and Wall (1966). According to the Ozawa–Flynn–Wall method for a given degree of conversion ($\alpha =$ const), a plot of $\log \beta$ versus $1/T_\alpha$ should be a straight line.

$$\log \beta = -\frac{0.4567E_a}{RT_\alpha} - 2.315 + \log\left(\frac{AE_a}{R}\right) \tag{3.28}$$

The activation energy can be calculated from the slope and the preexponential factor A can be calculated from the intercept.

The Kissinger–Akahira–Sunose method employs the dependence:

$$\ln\left[\frac{g(\alpha)}{T^2}\right] = \ln\frac{AR}{\beta E_a} CF - \frac{E_a}{RT} \tag{3.29}$$

with the correction factor $CF = 1$ for $50 \geq x \geq 20$. Therefore, a plot of $g(\alpha)/T^2$ versus T^{-1} gives a straight line with a slope $-E_a/R$:

$$\ln\frac{\beta}{T^2} = -\frac{E_a}{RT} + \text{const} \tag{3.30}$$

Originally, this method was proposed for the maximum rate of one-step reaction (temperature T_{max}) because differential thermal analysis curves always reach the same constant α at the peak maximum (Kissinger 1957). However, this isoconversional method was extended to other temperatures of constant conversion, $T_{\alpha=const}$ (Akahira and Sunose 1971).

Calorimetric studies of lipid oxidation require careful interpretation, and such interpretation of calorimetric data has been done by our research team in a series of publications summarized in a book chapter (Litwinienko 2005). The problem of interpretation of DSC curves of nonisothermal oxidation was solved by modeling of the nonisothermal process and comparison of experimental and theoretical data (Litwinienko and Kasprzycka-Guttman 1998a). The best fit of experimental and predicted degrees of conversion was obtained for a process of the general scheme

$$a \xrightarrow{\ b\ } b \longrightarrow c \tag{3.31}$$

where the first step, $a{\to}b$, is autocatalyzed by b. Figure 3.5 presents the DSC curve of nonisothermal oxidation and, according to Equation 3.31, the first exothermal peak on the DSC curve is assigned to the process of formation of hydroperoxides. The second peak results from the decomposition of the peroxides to further products.

The above interpretation was confirmed experimentally by several other studies including oxidation of partially oxidized oils (Litwinienko 2001), and the kinetic parameters determined for a nonisothermal autoxidation were in good agreement with accessible literature data (Litwinienko et al. 1999, 2000; Litwinienko and Kasprzycka-Guttman 2000). The most important consequence of this interpretation is that the activation parameters determined for the observed process will be a mean value of individual reactions occurring at the same moment. For a complex process, the overall E_a calculated for different degrees of conversion can vary (Opfermann and Kaisersberger 1992); therefore, the isoconversional methods should be used with great caution for determination of the kinetic parameters of complex processes, since

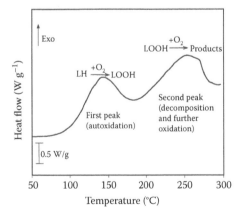

FIGURE 3.5 Schematic representation of the processes causing the first and second peaks on the DSC curve of nonisothermal oxidation of lipids.

the activation parameters are averaged over the whole conversion range. The mathematical incorrectness of the isoconversional methods can be overcome by assuming that activation energy does not vary substantially within the limited, narrow range of conversion interval (Šimon 2004).

During early studies of nonisothermal oxidation of fats and oils by DSC, the temperatures of the maximal heat flow (T_{max}) determined from oxidation curves were used for the calculation of kinetic parameters. Temperature of peak maximum is easily determined and it can be regarded as isoconversional for simple one-step reactions (i.e., at this point the reacting system always reaches the same degree of conversion) (Kissinger 1957; Ozawa 1965, 1970). However, a typical DSC oxidation curve contains more than one thermal peak, indicating that the observed process is more complex than a one-step reaction, and significant disagreement was observed between the values of E_a calculated from temperatures of the maximal heat flow on the DSC curve and those obtained from temperatures of extrapolated start of oxidation (Litwinienko et al. 1995). The interpretation of DSC curves with the assumption that thermoxidation consists of two main consecutive processes—formation of hydroperoxides (first visible thermal effect) and their decomposition (giving further thermal effects at higher temperatures) (Litwinienko and Kasprzycka-Guttman 1998a)—implies that temperatures of the start of oxidation (the temperature of the end of the induction period or the onset temperature of the oxidation peak, usually denoted as T_e and T_{ON}, respectively) should be utilized for qualitative (induction period) and quantitative (kinetic parameters) assessment of oxidative stability of organic materials. A good illustration of such a procedure was given by Šimon et al. (2000), who studied nonisothermal oxidation of rapeseed and sunflower oils by DSC and compared the results with the data obtained by Oxidograph under isothermal conditions for various temperatures. In both methods the induction periods (time or onset temperature) were applied to obtain the parameters of the Arrhenius-like equation. The authors concluded that the calorimetric method gives parameters unaffected by systematic errors, and some discrepancies between the results obtained by DSC and Oxidograph can be accounted for by oxygen diffusion within the samples. More detailed discussion on the transport of oxygen within the oxidized sample in DSC and Oxidograph was presented by the same group (Šimon and Kolman 2001).

Oils derived from unmodified and genetically modified vegetables have been subjected to PDSC studies (Adhvaryu et al. 2000), and the results (onset temperatures of oxidation [T_{ON}], temperatures of start of oxidation [T_S] [see Figure 3.6] and kinetic parameters calculated from these temperatures) were correlated with ^{1}H and ^{13}C nuclear magnetic resonance (NMR) analysis by means of multicomponent correlation models.

The statistical analysis led to the equation ($R^2 = 0.96$): $T_S/°C = 54.8 - 13.5A + 1.37B - 9.93C + 21.7D$, where A is the percentage of olefin carbons, B the percentage of saturated methylene groups, C the percentage of bisallylic methylene groups, and D the percentage of ω-2 carbons of saturated, mono- and n-6 poly unsaturated fatty acids (generally linoleic acid). Similar correlations for T_{ON}, E_a, and k as functions $f(A, B, C)$ were also described in the same work. In conclusion, the statistical models developed on the start and onset temperature, and kinetic parameters like E_a and k, can be used as predictive tools for quick assessment of vegetable oil oxidation.

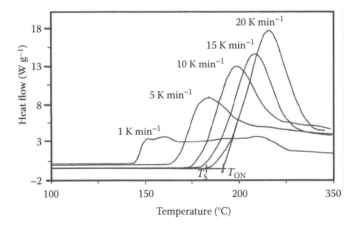

FIGURE 3.6 PDSC exothermic plots for thermoxidation of high-oleic sunflower oil at different heating rates with extrapolated temperatures of start (T_S) and onset temperature (T_{ON}). (Reprinted from *Thermochim. Acta*, 364, Adhvaryu, A., Erhan, S.Z., Liu, Z.S., and Perez, J.M., Oxidation kinetic studies of oils derived from unmodified and genetically modified vegetables using pressurized differential scanning calorimetry and nuclear magnetic resonance spectroscopy, 87–97, Copyright (2000), with permission from Elsevier.)

The structural-oxidative stability of fatty acids and their esters was the subject of many studies reviewed in a book chapter (Litwinienko 2005), but some recent works have added new insight into the structural trends and correlations of PDSC and Rancimat/OSI methods (Moser 2009; Pillar et al. 2009). The role of alkyl chain unsaturation in the formation of residue during thermooxidative decomposition of lubricants has been also studied (Pillar et al. 2009) The results obtained by a combination of thermogravimetric analysis, PDSC, Fourier transform infrared spectroscopy (FTIR), and attenuated total reflectance FTIR (ATR) indicate that polyunsaturated methyl esters form significantly more residue than saturated methyl esters during oxidative decomposition at high temperature (up to 500°C).

A complex multistep decomposition was reported for thermal oxidation of extra-virgin olive oils, with the start of oxidation differing the most among the samples (Vecchio et al. 2009). Two exothermal peaks of oxidation were deconvoluted and the kinetic parameters were calculated from the onset temperature determined for each peak separately. Kinetic parameters calculated for thermal oxidation of several samples of olive oil were consistent with the kinetic parameters calculated from induction times determined by isothermal Rancimat method at 110°C –140°C (Ostrowska-Ligeza et al. 2010). Thermal analysis methods (DSC, PDSC, and thermogravimetry) were also used for assessment of oxidative stability of modified oils (hydroxylated corn oil). Combination with FTIR, NMR, and rheometry is an excellent example of research oriented toward design of new fatty materials for nonedible applications, such as personal care, cosmetics, and some therapeutic applications (Harry-O'kuru et al. 2011). An interesting example of the application of nonisothermal DSC to study the oxidative stability of a microencapsulated fatty material was described by Pérez-Alonso et al. (2008). The activation energies for the oxidation process of microencapsulated red chili oleoresin depend on water activity content of the microcapsules, composition of biopolymer(s) in the encapsulating matrix, and geometric factors.

The thermal curve of oxidation of soy lecithin is different from the curve of oxidation of simple lipids (esters of fatty acids; Figure 3.7) and exhibits a well-defined, sharp first peak of oxidation (Ulkowski et al. 2005). The activation parameters of lecithin oxidation calculated by the Ozawa–Flynn–Wall method ($E_a = 98 \pm 6$ kJ/mol and $A = 9.1 \times 10^{10}$ s^{-1}) were used for calculation of the overall oxidation rate constants $k(T)$ and compared with those obtained for linolenic acid oxidation. The comparison indicates the inversion of their oxidative stabilities; that is, below 167°C lecithin is more resistant toward oxidation than linolenic acid (LNA) and above that temperature (termed the isokinetic temperature) the oxidative stability of lecithin is worse than the oxidative stability of LNA ($k_{\text{lecithin}} > k_{\text{LNA}}$). This kind of inversion of

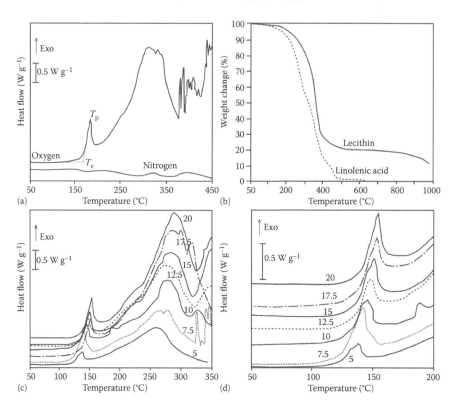

FIGURE 3.7 (a) Comparison of thermal effect of lecithin heated with $\beta = 5$ K/min under an oxygen atmosphere (*upper line*) and under N$_2$ (*bottom line*). (b) Thermogravimetric curve of weight loss of lecithin and linolenic acid heated with $\beta = 10$ K/min under nitrogen. (c) Curves of nonisothermal oxidation of lecithin. Numbers denote heating rates in K/min. DSC curves have been shifted vertically (without rescaling of the plots) in order to obtain a clearer view. (d) Figure 3.7c limited to first peak of oxidation. (Reprinted with permission from Ulkowski, M., Musialik, M., and Litwinienko, G., *J. Agric. Food Chem.*, 53, 9073–9077. Copyright (2005) American Chemical Society.)

oxidative stabilities may be a reason for false results obtained by accelerated tests at temperatures above 100°C compared with the results obtained at temperatures below 100°C.

3.3 APPLICATION OF DSC FOR STUDIES OF ANTIOXIDANT ACTIVITY

Some compounds are able to suppress lipid autoxidation even if they are present at very low concentration compared with the lipid matrix. These compounds are called antioxidants and, depending on the general mechanism of inhibition, they can be divided into two main groups: preventive and chain-breaking antioxidants. The first group of compounds inhibit the initiation process (Equation 3.1), whereas the second group inhibit the propagation step by direct reaction with the peroxyl radical (LOO• + AntioxH→LOOH + Antiox•), competing with Equation 3.3.

Since the early applications of thermal analysis in studies on the oxidative stability of fats and polymers, the same methods have been used for assessment of antioxidant activity. The results published before 2000 are reviewed in a book chapter by one of us (Litwinienko 2005). In general, the induction times determined by isothermal DSC are sufficiently correlated with data obtained by the OSI and Rancimat methods. Recently, the oxidative stability of biofuels containing various amounts of natural and synthetic antioxidants was monitored by isothermal DSC (and PDSC) and compared with the results obtained by the Rancimat method (Leonardo et al. 2012). The authors pointed out that the Rancimat method can lead to false results for some volatile antioxidants, whereas PDSC gives more reliable data. Isothermal PDSC was also employed for the determination of antioxidant activity of organic sulfides, showing good correlation with the results of the rotary bomb oxidation test (Qiu et al. 2006).

Although induction times obtained by the OSI/Rancimat methods are correlated with τ_{ind} from DSC/PDSC measurements, some discrepancies were noticed during early studies on isothermal oxidation of rapeseed oil containing 2,6-di-*tert*-butyl-4-methylphenol (BHT), 2-*tert*-butyl-4-hydroxyanisole (BHA), and propyl gallate (Kowalski 1993). In the presence of antioxidant the induction time was prolonged but the activation energies (calculated from the peaks of the maximum heat flow) were almost the same for inhibited and noninhibited oxidation. This can be explained by a misleading procedure of calculation of activation parameters. The only part of the DSC curve that can be used for assessment of antioxidant activity is the onset point (i.e., the temperature, when oxidation starts), because antioxidants are active during the induction period. Therefore, the recommended method for isothermal oxidation uses the dependency of ln τ_{ind} on T^{-1} (i.e., Equation 3.20, with τ_{ind} instead of τ_i) for calculation of the Arrhenius parameters of inhibited oxidation.

Significant development of nonisothermal DSC during the last 15 years has resulted in the dominance of this method over isothermal DSC, and today the majority of the studies on antioxidant activity are carried out in a nonisothermal mode. Research on antioxidant activity by thermal analysis is focused on two main purposes. The first is to test the oxidative stability of fats, lipids, and polymers in the presence of well-established natural and synthetic antioxidants (or their mixtures) in order to optimize their inhibitory effect at elevated temperatures. A typical example is the oxidative stability of edible oils at high temperatures, and there are attempts to increase their resistance toward oxidation by testing extracts of natural compounds or antioxidants active during frying at 180°C (Reda 2011). Other DSC studies are focused on new compounds tested as potential inhibitors of autoxidation occurring at lower temperatures (relevant to storage temperatures of oils and fats). In this case simple lipids (esters of polyunsaturated fatty acids) are used as a lipid matrix that undergoes oxidation. Both types of studies are well represented in the recent literature on thermal analysis.

The recent boom in research on biofuels made from vegetable oils or animal fats is mirrored by the increasing number of papers on the physical and chemical properties of biodiesel and biolubricants, presented in the excellent review that also describes the methodological aspects of DSC (Dunn 2008). Thermal analysis allows the oxidative stability of biodiesel, as well as the influence of antioxidants on the phase equilibrium, to be determined. The general conclusions from several studies on the stabilization of biodiesel against oxidation are that synthetic antioxidants exhibit better activity than natural antioxidants when applied to stabilize methyl soyate (Dunn 2005). Moreover, distilled biodiesel significantly loses its oxidative stability due to the loss of antioxidant content during distillation (Polavka et al. 2005).

Nonisothermal DSC can be applied not only for studies on thermal oxidation of bulk-phase lipids but also for determination of the oxidative stability of biphasic water/lipid systems such as liposomes. An example of such research was given in a series of papers on the antimicrobial and antioxidant activity of extracts from *Origanum dictamnus* encapsulated in liposomes assembled from phosphatidylcholine (Gortzi et al. 2007, 2008). Isothermal and nonisothermal DSC was also used to study the accelerated oxidation of vanillin (4-hydroxy-3-methoxybenzaldehyde). This process causes a partial pyrolysis of vanillin, giving a mixture of vanillin and vanillic acid, which shows stronger antioxidant properties than the starting compound (Mourtzinos et al. 2009).

Induction times (or induction temperatures) are valuable parameters indicating the oxidative stability of inhibited lipid matrix (i.e., lipid containing antioxidants). However, the Arrhenius parameters of inhibited autoxidation have a broader meaning, and they can be applied to predict the behavior of inhibited lipid at temperatures lower than the temperatures of accelerated tests. Since the temperatures of start of oxidation (denoted as T_e or T_{ON}) and temperatures of the first peak (T_{p1} or T_{max1}) correlate with the inhibitory effect, these isoconversional points are used in nonisothermal DSC methods to obtain the overall kinetic parameters of the nonisothermal oxidation inhibited by chain-breaking antioxidants. For this purpose a careful choice of lipid matrix is recommended, because noninhibited oxidation of edible oils usually starts at rather high temperatures (ca. 150°C), and more volatile antioxidants (even BHT and BHA) indicate no activity under such experimental conditions (Kowalski 1993). Thus, linolenic acid (18:3), which begins nonisothermal oxidation at temperatures below 100°C, can be applied as a lipid matrix to give more reliable results (Litwinienko et al. 1997; Litwinienko and Kasprzycka-Guttman 1998a). Since the first peak of the DSC curve is assigned to formation of peroxides (see above) while second and further peaks are caused by decomposition of the peroxides (see Figure 3.5), the addition of antioxidant should alter the first peak and should be manifested as an increase in onset temperature and temperature of the first maximum of the heat flow (T_e and T_{p1}), while the further thermal effects of oxidation (i.e., the further peaks) are not correlated with the concentration of the inhibitor (Litwinienko and Kasprzycka-Guttman 1998a). Oxidation of linolenic acid is described by relatively low activation energy E_a, within the range 62–70 kJ/mol, with the preexponential factor ca. $10^{10}\,s^{-1}$ (Litwinienko et al. 2000; Ulkowski et al. 2005; Musialik and Litwinienko 2007), in reasonable agreement with the activation energy of isothermal oxidation of this polyunsaturated fatty acid. An increased value of the activation energy of nonisothermal oxidation of linolenic acid in the presence of antioxidants was demonstrated for natural compounds like olivetol and dehydrozingerone (Musialik and Litwinienko 2007), proving that this kind of lipid matrix is suitable even for studies on relatively volatile antioxidants. DSC measurement of linoleic acid oxidation was also used for structure–property studies on the antiradical activity of a series of alkyl esters of protocatechuic acid (Reis et al. 2010) and synapic acid esters (Gaspar et al. 2010). The results obtained by thermal analysis were consistent with other analytical techniques (quenching of 2,2-diphenyl-1-picrylhydrazyl radical, the fluorescence recovery after photobleaching ferric reducing antioxidant power [FRAP] method, and electrochemical measurements). A literature survey also indicates several attempts to study the effect of antioxidant extracts and blends on the oxidation of oils. For example, the activation energy of oxidation of chia oil (72 kJ/mol) increases after the addition of rosemary and green tea extracts, tocopherols, and ascorbyl palmitate, reaching the highest value of 88 kJ/mol for 2500 ppm of ascorbyl palmitate (Ixtaina et al. 2012).

The simplicity and relatively short time of nonisothermal DSC analysis make this method very useful if a large number of quick measurements need to be performed for a series of samples, for example to study the activity of an antioxidant as a function of its concentration. The rather obvious observation that oxidative stability depends on the concentration of antioxidant is usually supplemented by counterintuitive results showing that at high concentrations of antioxidants the inhibitory effect disappears or turns into a prooxidative effect (Litwinienko et al. 1997; Musialik and Litwinienko 2007). The parameters E_a and A allow rate constants to be calculated at temperatures lower than temperatures of accelerated tests in order to check whether the effectiveness of an antioxidant depends on the temperature. Figure 3.8 presents the values of log k for oxidation of linolenic acid stabilized by BHT in the concentration range 1–11 mmol BHT/mol lipid calculated for three temperatures: 25°C, 90°C, and 180°C. It is clear that at high temperatures (90°C and 180°C) the oxidative stability is less sensitive to concentration of this antioxidant than at room temperature, where the rate of oxidation is strongly dependent on the concentration of BHT.

3.4 PITFALLS OF AUTOXIDATION STUDIES BY THERMAL ANALYSIS

Regardless of the large number of works describing the analysis of oxidative stability of fats, oils, and polymers by DSC, thermoanalytical methods are still regarded as supplementary compared with conventional accelerated tests. Thus, a common

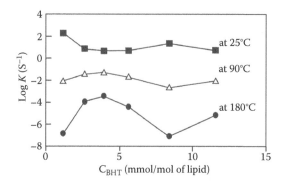

FIGURE 3.8 Plots of the logarithms of the overall rate constant (k) of thermoxidation of linolenic acid containing various concentrations of 2,6-di-tert-butyl-6-methylphenol (BHT). Values of k were calculated for temperatures 25°C, 90°C, and 180°C. (With kind permission from Springer: *J. Therm. Anal. Calorim.*, DSC study of linolenic acid autoxidation inhibited by BHT, dehydrozingerone and olivetol, 88, 2007, 781–785, Musialik, M. and Litwinienko, G., Copyright 2007 Springer.)

experimental scheme includes a comparison of the results obtained by DSC with data from Rancimat and OSI methods. Usually the conclusion is that DSC gives results that are fully consistent with results obtained by conventional accelerated tests. However, caution needs to be exercised during thermoanalytical studies of the oxidative stability of edible oils and fats. Apart from routine calibrations of the calorimeter, DSC needs careful standardization of the procedure in terms of mass of a sample, oxygen flow, and choice of an appropriate lipid matrix (if the activity of an antioxidant is to be tested). One of the most common errors during studies on oxidative stability is lack of knowledge of the thermal and oxidation history of a sample. Oxidative stability of edible oils can be measured for samples as received; however, thermal history of a sample becomes important for biphasic systems such as vesicles, powders, or formulated products and for studies on the antioxidant activity of new compounds. Another important issue is the possibility that the thermal effect of nonoxidative degradation interferes with the oxidation process. This situation was described for biodiesel from soybean oil studied by nonisothermal PDSC (however, without purge gas flow): noninhibited oil showed a clear thermal oxidation effect, but addition of antioxidants redirected the decomposition path into nonoxidative degradation (Dunn 2012).

Length of induction period (expressed as induction time in isothermal mode and onset temperature in nonisothermal mode) is a reliable quantitative parameter that can be applied for a direct comparison of the oxidative stability of a series of samples or can be converted into the Arrhenius kinetic parameters of oxidation. Two main problems appear if DSC methods are directly compared with conventional accelerated tests like Rancimat or OSI. DSC results applied for prediction of induction times give much shorter times than those determined by OSI (Šimon and Kolman 2001; Šimon et al. 2000), which was ascribed to a better oxygen saturation of the sample during the DSC experiment. The length of the induction period for a concentration of oxygen C lower than the saturated concentration (C^0) is given by the following expression (Šimon et al. 2000):

$$\tau_{ind} = \tau_{ind}^0 \frac{c_0}{c} = \tau_{ind}^0 \left(1 + \frac{kVz}{DS}\right) \qquad (3.32)$$

where:
> τ_{ind}^0 is the induction period for a sample of oil saturated with oxygen
> S is the area of the oxygen/oil interface
> V is the volume of the oil sample
> k is the rate constant of the oxidation process during the induction period
> D is the diffusion coefficient
> z is the thickness of the diffusion film in the oil

The ratio S/V is two orders of magnitude higher for DSC ($S/V = 10^4$ m^{-1}) than for Oxidograph ($S/V = 150$ m^{-1}); therefore, for a large oil sample the induction times measured in the Oxidograph and their change with temperature ($d\tau_{ind}/dT$) are limited by the sample size and by diffusion of oxygen. The authors concluded that differences in diffusion coefficients of oxygen in oils and fats (and in polymers) lead to errors in determination of the order of oxidative stability by conventional accelerated tests. In contrast, the DSC techniques do not have such serious limitations.

The temperature of accelerated tests differs from the typical temperature of storage and use of lipids (apart from the frying process). The difference in temperature may cause significant changes in the reaction mechanism; thus, data determined at higher temperatures cannot be easily extrapolated to predict oxidation behavior at lower temperatures. Some authors have tried to overcome this problem by introducing the ratio of induction times for stabilized and nonstabilized oil samples: $PF = \tau_{ind}(\text{inhibited})/\tau_{ind}(\text{non-inhibited})$. This idea is based on the assumption that the same structural units are responsible for oxidation processes in both measured samples (Polavka et al. 2005). Unfortunately, the induction times obtained at temperatures above 100°C are not good predictors of oxidative stability at lower temperatures. The kinetic parameters E_a and A determined from DSC measurements can be better descriptors of the process than the induction times. Indeed, simple kinetic considerations about the temperature dependence of rate constants $k(T)$ indicate the possibility of inversion of two processes below and above the isokinetic temperature (Litwinienko 2005; Ulkowski et al. 2005). If the rate constants are calculated as functions of temperature for two processes described by various sets of activation energy (E_{ai}) and preexponential factors (A_i): $k_i (T) = A_i \exp[-E_{ai}/RT]$, the isokinetic temperature (T_{iso}) can be calculated when $k_1 (T_{iso}) = k_2 (T_{iso})$. For temperatures above (T_{iso}) the process of lower activation energy is slower, whereas for temperatures below T_{iso} the same process is faster than the process with higher E_a. An example of such a process is the oxidation of neat linolenic acid ($E_{a1} = 74{,}600$ J/mol, $A_1 = 1.97 \times 10^8$ s^{-1}) and linolenic acid containing 7.5 mmol of antioxidant ($E_{a2} = 104{,}900$ J/mol and $A_2 = 9.35 \times 10^{12}$ s^{-1}; all experimental data are from Litwinienko 2005). T_{iso} can be calculated as: $T = (E_{a1} - E_{a2})/[R \log (A_1/A_2)] = 338$ K (65°C). Thus, at $T > 80$°C, a process of higher activation energy is faster, whereas at $T < 50$°C the process of higher E_a is slower. Assuming that the process will be governed by the same mechanism at room temperature and at 110°C, the prediction/extrapolation of the rate constants obtained from DSC measurements gives more reliable data than the accelerated tests based on measurements of induction times.

The inversion of oxidative stability for a series of several edible and pharmaceutical oils was reported 25 years ago in isothermal PDSC studies (Kowalski 1989): the order of relative oxidative stability at temperatures below 110°C was different from the sequence determined at higher temperatures. Surprisingly, the kinetic parameters calculated from the times τ_{on} and τ_{max} gave, in the temperature range 70°C–140°C, rate constants that were in full agreement with such inversion of oxidative stability below and above the isokinetic temperature (100°C). Similar observations were reported for isothermal and nonisothermal oxidation of corn oil stabilized by sulfide additives, which exhibited either prooxidative or antioxidative effects, depending on the temperature of oxidation, above and below T_{iso} (Bantchev et al. 2011). The above examples give clear evidence that DSC is a valuable tool for the assessment of oxidative stability of stabilized and nonstabilized lipid systems. Moreover, this is a caution for researchers, who often carelessly extrapolate the results of accelerated tests (Rancimat, OSI) from high temperatures (above 100°C) to much lower temperatures.

Another critical point connected with interpretation of DSC results is the physical meaning of the activation parameters calculated for complex reactions. As raised by Šimon (2004), the parameter called activation energy has a different meaning from that commonly accepted in chemical kinetics (where it is the energy barrier of a simple reaction, being the rate-determining step for the whole observed process). Autoxidation is a free radical chain process; thus, the activation energy should be supplemented with the adjective "global" or "apparent," as being an effect of several steps occurring simultaneously but dependent on each other (initiation, propagation, etc.). Consequently, the parameters E_a and A give information on the kinetics of the whole complex process and can be used for calculation and prediction of the rate constants of oxidation, but considerations about mechanistic aspects of the observed thermal processes are not justified.

Recent studies on the aging of polymers indicated more complex, non-Arrhenius behavior during oxidative degradation of polymers and their composites. Measurements of the long-term aging process conducted over a wide range of temperatures (37°C–108°C) showed downward curvature of the activation energy of oxidation of polypropylene and some commercial chlorosulfonated polyethylene composites (Gillen et al. 2005; Celina et al. 2005). This peculiarity was explained by the predominance of different types of processes at low and high temperatures, with diffusion-limited oxidation responsible for such non-Arrhenius behavior at high temperatures. Although this phenomenon was observed for polymers in the solid phase and might be less relevant to thermal oxidation of small samples of oils and fats in the liquid phase, it needs to be investigated in the future.

REFERENCES

Adhvaryu, A., Erhan, S. Z., Liu, Z. S., and Perez, J. M. Oxidation kinetic studies of oils derived from unmodified and genetically modified vegetables using pressurized differential scanning calorimetry and nuclear magnetic resonance spectroscopy. *Thermochimica Acta* 364 (2000): 87–97.

Agrawal, R. K. Analysis of non-isothermal reaction-kinetics: Part 1. Simple reactions. *Thermochimica Acta* 203 (1992): 93–110.

Akahira, T. and Sunose, T. Joint convention of four electrical institutes. Research Report (Chiba Institute of Technology). *Science and Technology* 16 (1971): 22–31.

Bantchev, G. B., Biresaw, G., Mohamed, A., and Moser, J. Temperature dependence of the oxidative stability of corn oil and polyalphaolefin in the presence of sulfides. *Thermochimica Acta* 513 (2011): 94–99.

Baylon, A., Stauffer, E., and Delémont, O. Evaluation of the self-heating tendency of vegetable oils by differential scanning calorimetry. *Journal of Forensic Science* 53 (2008): 1334–1343.

Carroll, B. and Manche, E. P. Kinetic parameters from thermogravimetric data. Comments. *Analytical Chemistry* 42 (1970): 1296–1297.

Celina, M., Gillen, K. T., and Assink, R. A. Accelerated aging and lifetime prediction: Review of non-Arrhenius behaviour due to two competing processes. *Polymer Degradation and Stability* 90 (2005): 395–404.

Chan, H. W. S. *Autoxidation of Unsaturated Lipids.* Waltham, MA: Academic Press (1987).

Cross, C. K. Oil stability—A DSC alternative for active oxygen method. *Journal of the American Oil Chemists' Society* 47 (1970): 229–230.

Doyle, C. D. Series approximations to equation of thermogravimetric data. *Nature* 207 (1965): 290–291.

Dunn, R. O. Effect of antioxidants on the oxidative stability of methyl soyate (biodiesel). *Fuel Processing Technology* 86 (2005): 1071–1085.

Dunn, R. O. Antioxidants for improving storage stability of biodiesel. *Biofuels, Bioproducts and Biorefining* 2 (2008): 304–318.

Dunn, R. O. Thermal oxidation of biodiesel by pressurized differential scanning calorimetry: Effects of heating ramp rate. *Energy and Fuels* 26 (2012): 6015–6024.

European Committee for Standards (CEN). EN 14112—Fat and oil derivatives. Fatty Acid Methyl Esters (FAME). Determination of oxidation stability (accelerated oxidation test), Belgium (2003).

Firestone, D. *Official Methods and Recommended Practices of the American Oil Chemists' Society.* Champaign: AOCS Press (2003).

Flynn, J. H. and Wall, L. A. A quick direct method for determination of activation energy from thermogravimetric data. *Journal of Polymer Science Part B: Polymer Letters* 4 (1966): 323–328.

Frankel, E. N. *Lipid Oxidation.* Cambridge: Oily Press, Woodhead Publishing (2005).

Freeman, E. S. and Carroll, B. The application of thermoanalytical techniques to reaction kinetics: The thermogravimetric evaluation of the kinetics of the decomposition of calcium oxalate monohydrate. *Journal of Physical Chemistry* 62 (1958): 394–397.

Freeman, E. S. and Carroll, B. Interpretation of the kinetics of thermogravimetric analysis. *Journal of Physical Chemistry* 73 (1969): 751–752.

Gaspar, A., Martins, M., Silva, P., Garrido, E. M., Garrido, J., Firuzi, O., Miri, R., Saso, L., and Borges, F. Dietary phenolic acids and derivatives. Evaluation of the antioxidant activity of sinapic acid and its alkyl esters. *Journal of Agricultural and Food Chemistry* 58 (2010): 11273–11280.

Gillen, K. T., Bernstein, R., and Celina, M. Non-Arrhenius behavior for oxidative degradation of chlorosulfonated polyethylene materials. *Polymer Degradation and Stability* 87 (2005): 335–346.

Gortzi, O., Lalas, S., Chinou, I., and Tsaknis, J. Evaluation of the antimicrobial and antioxidant activities of *Origanum dictamnus* extracts before and after encapsulation in liposomes. *Molecules* 12 (2007): 932–945.

Gortzi, O., Lalas, S., Chinou, I., and Tsaknis, J. Reevaluation of bioactivity and antioxidant activity of *Myrtus communis* extract before and after encapsulation in liposomes. *European Food Research and Technology* 226 (2008): 583–590.

Halliwell, B. and Gutteridge, J. M. C. *Free Radicals in Biology and Medicine.* New York: Oxford University Press (2007).

Harry-O'kuru, R. E., Mohamed, A., Xu, J., and Sharma, B. K. Synthesis and characterization of corn oil polyhydroxy fatty acids designed as additive agent for many applications. *Journal of the American Oil Chemists' Society* 88 (2011): 1211–1221.

Hassel, R. L. Thermal analysis—Alternative method of measuring oil stability. *Journal of the American Oil Chemists' Society* 53 (1976): 179–181.

Ixtaina, V. I., Nolasco, S. M., and Tomás, M. C. Oxidative stability of chia (*Salvia hispanica* L.) seed oil: Effect of antioxidants and storage conditions. *Journal of the American Oil Chemists' Society* 89 (2012): 1077–1090.

Kamal-Eldin, A. and Pokorný, J. *Analysis of Lipid Oxidation.* Champaign, IL: AOCS Press (2005).

Kasprzycka-Guttman, T. and Odzeniak, D. Specific heats of some oils and fats. *Thermochimica Acta* 191 (1991): 41–45.

Kasprzycka-Guttman, T. and Odzeniak, D. Antioxidant properties of lignin and its fractions. *Thermochimica Acta* 231 (1994): 161–168.

Kasprzycka-Guttman, T., Odzeniak, D., and Supera, M. Thermokinetic properties of inhibited vegetable oils. *Thermochimica Acta* 237 (1994): 207–211.

Kissinger, H. E. Reaction kinetics in differential thermal analysis. *Analytical Chemistry* 29 (1957): 1702–1706.

Knothe, G. and Dunn, R. O. Dependence of oil stability index of fatty compounds on their structure and concentration and presence of metals. *Journal of the American Oil Chemists' Society* 80 (2003): 1021–1026.

Kowalski, B. Determination of specific heats of some edible oils and fats by differential scanning calorimetry. *Journal of Thermal Analysis* 34 (1988): 1321–1326.

Kowalski, B. Determination of oxidative stability of edible vegetable oils by pressure differential scanning calorimetry. *Thermochimica Acta* 156 (1989): 347–358.

Kowalski, B. Determination of spontaneous ignition temperatures of edible oils and fats by pressure differential scanning calorimetry. *Thermochimica Acta* 173 (1990): 117–127.

Kowalski, B. Thermokinetic analysis of vegetable oil oxidation as an autocatalytic reaction. *Polish Journal of Food and Nutrition Sciences* 1/42 (1992): 51–59.

Kowalski, B. Evaluation of activities of antioxidant in rapeseed oil matrix by pressure differential scanning calorimetry. *Thermochimica Acta* 213 (1993): 135–146.

Kowalski, B., Ratusz, K., Miciula, A., and Krygier, K. 1997. Monitoring of rapeseed oil autoxidation with a pressure differential scanning calorimeter. *Thermochimica Acta* 307 (1997): 117–121.

Leonardo, R. S., Murta Valle, M. L., and Dweck, J. An alternative method by pressurized DSC to evaluate biodiesel antioxidants efficiency. *Journal of Thermal Analysis and Calorimetry* 108 (2012): 751–759.

Litwinienko, G. Autoxidation of unsaturated fatty acids and their esters. *Journal of Thermal Analysis and Calorimetry* 65 (2001): 639–646.

Litwinienko, G. Analysis of lipid oxidation by differential scanning calorimetry. In A. Kamal-Eldin and A. Pokorný (eds), *Analysis of Lipid Oxidation*, pp. 152–193. Champaign: AOCS Press (2005).

Litwinienko, G., Daniluk, A., and Kasprzycka-Guttman, T. A differential scanning calorimetry study on the oxidation of saturated C_{12}-C_{18} fatty acids and their esters. *Journal of the American Oil Chemists' Society* 76 (1999): 655–657.

Litwinienko, G., Daniluk, A., and Kasprzycka-Guttman, T. Study on autoxidation kinetics of fats by differential scanning calorimetry. 1. Saturated C_{12}-C_{18} fatty acids and their esters. *Industrial and Engineering Chemistry Research* 39 (2000): 7–12.

Litwinienko, G. and Kasprzycka-Guttman, T. A DSC study on thermoxidation kinetics of mustard oil. *Thermochimica Acta* 319 (1998a): 185–191.

Litwinienko, G. and Kasprzycka-Guttman, T. Oxidation of saturated fatty acids esters. DSC investigations. *Journal of Thermal Analysis and Calorimetry* 54 (1998b): 211–217.

Litwinienko, G. and Kasprzycka-Guttman, T. Study on autoxidation kinetics of fats by differential scanning calorimetry. 2. Unsaturated fatty acids and their esters. *Industrial and Engineering Chemistry Research* 39 (2000): 13–17.

Litwinienko, G., Kasprzycka-Guttman, T., and Jarosz-Jarszewska, M. Dynamic and isothermal DSC investigation of the kinetics of thermoxidative decomposition of some edible oils. *Journal of Thermal Analysis* 45 (1995): 741–750.

Litwinienko, G., Kasprzycka-Guttman, T., and Studzinski, M. Effects of selected phenol derivatives on the autoxidation of linolenic acid investigated by DSC non-isothermal methods. *Thermochimica Acta* 307 (1997): 97–106.

Moser, B. R. Comparative oxidative stability of fatty acid alkyl esters by accelerated methods. *Journal of the American Oil Chemists' Society* 86 (2009): 699–706.

Mourtzinos, I., Konteles, S., Kalogeropoulos, N., and Karathanos, V. T. Thermal oxidation of vanillin affects its antioxidant and antimicrobial properties. *Food Chemistry* 114 (2009): 791–797.

Musialik, M. and Litwinienko, G. DSC study of linolenic acid autoxidation inhibited by BHT, dehydrozingerone and olivetol. *Journal of Thermal Analysis and Calorimetry* 88 (2007): 781–785.

Opfermann, J. and Kaisersberger, E. An advantageous variant of the Ozawa-Flynn-Wall analysis. *Thermochimica Acta* 203 (1992): 167–175.

Ostrowska-Ligeza, E., Bekas, W., Kowalska, D., Lobacz, M., Wroniak, M., and Kowalski, B. Kinetics of commercial olive oil oxidation: Dynamic differential scanning calorimetry and Rancimat studies. *European Journal of Lipid Science and Technology* 112 (2010): 268–274.

Ozawa, T. A new method of analyzing thermogravimetric data. *Bulletin of the Chemical Society of Japan* 38 (1965): 1881–1885.

Ozawa, T. Kinetic analysis of derivative curves in thermal analysis. *Journal of Thermal Analysis* 2 (1970): 301–324.

Pereira, T. A. and Das, N. P. The effects of flavonoids on the thermal autoxidation of palm oil and other vegetable oils determined by differential scanning calorimetry. *Thermochimica Acta* 165 (1990): 129–137.

Pérez-Alonso, C., Cruz-Olivares, J., Barrera-Pichardo, J. F., Rodríguez-Huezo, M. E., Báez-González, J. G., and Vernon-Carter, E. J. DSC thermo-oxidative stability of red chili oleoresin microencapsulated in blended biopolymers matrices. *Journal of Food Engineering* 85 (2008): 613–624.

Pillar, R., Ginic-Markovic, M., Clarke, S., and Matisons, J. Effect of alkyl chain unsaturation on methyl ester thermo-oxidative decomposition and residue formation. *Journal of the American Oil Chemists' Society* 86 (2009): 363–373.

Polavka, J., Paligová, J., Cvengroš, J., and Šimon, P. Oxidation stability of methyl esters studied by differential thermal analysis and Rancimat. *Journal of the American Oil Chemists' Society* 82 (2005): 519–524.

Porter, N. A. Mechanisms for the autoxidation of polyunsaturated lipids. *Accounts of Chemical Research* 19 (1986): 262–268.

Porter, N. A., Caldwell, S. E., and Mills, K. A. Mechanisms of free radical oxidation of unsaturated lipids. *Lipids* 30 (1995): 277–290.

Qiu, C., Han, S., Cheng, X., and Ren, T. Determining the antioxidant activities of organic sulfides by rotary bomb oxidation test and pressurized differential scanning calorimetry. *Thermochim. Acta* 447 (2006): 36–40.

Raemy, A., Froelicher, I., and Loeliger, J. Oxidation of lipids studied by isothermal heat flux calorimetry. *Thermochimica Acta* 114 (1987): 159–164.

Raemy, A., Hurrell, R. F., and Loeliger, J. Thermal-behavior of milk powders studied by differential thermal analysis and heat-flow calorimetry. *Thermochimica Acta* 65 (1983): 81–92.

Raemy, A. and Loeliger, J. Self-ignition of powders studied by high-pressure differential thermal analysis. *Thermochimica Acta* 95 (1985): 343–346.

Reda, S. Y. Evaluation of antioxidants stability by thermal analysis and its protective effect in heated edible vegetable oil. *Ciência e Tecnologia de Alimentos* 31 (2011): 475–480.

Reis, B., Martins, M., Barreto, B., Milhazes, N., Garrido, E. M., Silva, P., Garrido, J., and Borges, F. Structure-property-activity relationship of phenolic acids and derivatives. Protocatechuic acid alkyl esters. *Journal of Agricultural and Food Chemistry* 58 (2010): 6986–6993.

Šimon, P. Isoconversional methods. Fundamentals, meaning and application. *Journal of Thermal Analysis and Calorimetry* 76 (2004): 123–132.

Šimon, P. and Kolman, L. DSC study of oxidation induction periods. *Journal of Thermal Analysis and Calorimetry* 64 (2001): 813–820.

Šimon, P., Kolman, L., Niklová, I., and Schmidt, S. Analysis of the induction period of oxidation of edible oils by differential scanning calorimetry. *Journal of the American Oil Chemists' Society* 77 (2000): 639–642.

Tan, C. P., Che Man, Y. B., Selemat, J., and Yusoff, M. S. A. Comparative studies of oxidative stability of edible oils by differential scanning calorimetry and oxidative stability index method. *Food Chemistry* 76 (2002): 385–389.

Ulkowski, M., Musialik, M., and Litwinienko, G. 2005. Use of differential scanning calorimetry to study lipid oxidation. 1. Oxidative stability of lecithin and linolenic acid. *Journal of Agricultural and Food Chemistry* 53 (2005): 9073–9077.

Vecchio, S., Cerretani, L., Bendini, A., and Chiavaro, E. Thermal decomposition study of monovarietal extra virgin olive oil by simultaneous thermogravimetry/differential scanning calorimetry: Relation with chemical composition. *Journal of Agricultural and Food Chemistry* 57 (2009): 4793–4800.

4 DSC Application to Vegetable Oils
The Case of Olive Oils

Alessandra Bendini, Lorenzo Cerretani, Emma
Chiavaro, and Maria Teresa Rodriguez-Estrada

CONTENTS

4.1 INTRODUCTION TO OLIVE OIL

Virgin olive oils (VOO) are defined by the European Community as those "oils obtained from the fruit of the olive tree solely by mechanical or other physical means under conditions that do not lead to alteration in the oil" (EEC Reg. 2568/91). The VOO extracted from fresh and healthy olive fruits (*Olea europaea* L.), properly processed and adequately stored (see Figure 4.1), is characterized by a unique and measurable combination of aroma and taste. This flavor is characteristic and markedly

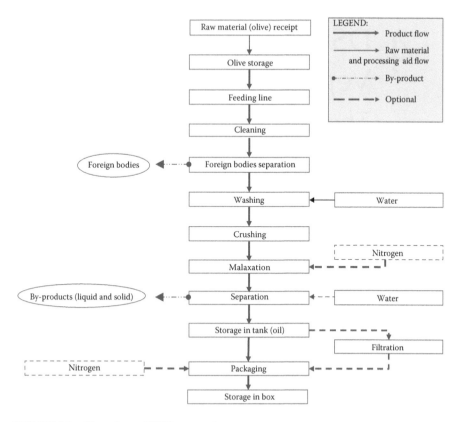

FIGURE 4.1 Flow chart of VOO processing.

different from those of other edible fats and oils that undergo refining, which leads to the loss of most minor components. The flow sheet of olive processing is shown in Figure 4.1.

Once it is stated that any oil extracted by mechanical means from olive fruit is a virgin oil, the quality can be distinguished as extra-virgin, virgin, ordinary (nowadays outside the European Union [EU]), and lampante, in decreasing order of quality, according to different limits established for selected physicochemical (percentage of free acidity, peroxide number, specific extinction at ultraviolet wavelengths, fatty acid alkyl ester content) and sensory (median value of fruity, intensity of the mean perceived defect) characteristics.

4.1.1 COMPOSITION OF OLIVE OIL

Olive oil (OO) is primarily composed of triacylglycerols (TAG) (~98%) and secondarily of free fatty acids (FFA), mono- and diacylglycerols (DAG), and a plethora of minor compounds such as hydrocarbons, sterols, aliphatic alcohols, tocopherols, pigments, and phenolic and volatile compounds. In Table 4.1, the percentage ranges of the fatty acid (FA) composition of olive oil and the most common vegetable oils are shown (Codex Alimentarius, 2013).

TABLE 4.1

Fatty Acid (FA) Composition (Expressed as Percentage of Total Fatty Acids) of Olive Oil and Other Vegetable Oils

Fatty Acid	Olive Oil	Peanut Oil	Coconut Oil	Maize Oil	Palm Oil	Palm Kernel Oil	Canola Oil	Soybean Oil	Sunflower Seed Oil	Sunflower Seed Oil[a]
C6:0	ND	ND	ND-0.7	ND	ND	ND-0.8	ND	ND	ND	ND
C8:0	ND	ND	4.6-10.0	ND	ND	2.4-6.2	ND	ND	ND	ND
C10:0	ND	ND	5.0-8.0	ND	ND	2.6-5.0	ND	ND	ND	ND
C12:0	ND	ND-0.1	45.1-53.2	ND-0.3	ND-0.5	45.0-55.0	ND	ND-0.1	ND-0.1	ND
C14:0	0.0-0.05	ND-0.1	16.8-21.0	ND-0.3	0.5-2.0	14.0-18.0	ND-0.2	ND-0.2	ND-0.2	ND-0.1
C16:0	7.5-20.0	8.0-14.0	7.5-10.2	8.6-16.5	39.3-47.5	6.5-10.0	2.5-7.0	8.0-13.5	5.0-7.6	2.6-5.0
C16:1	0.3-3.5	ND-0.2	ND	ND-0.5	ND-0.6	ND-0.2	ND-0.6	ND-0.2	ND-0.3	ND-0.1
C17:0	0.0-0.3	ND-0.1	ND	ND-0.1	ND-0.2	ND	ND-0.3	ND-0.1	ND-0.2	ND-0.1
C17:1	0.0-0.3	ND-0.1	ND	ND-0.1	ND	ND	ND-0.3	ND-0.1	ND-0.1	ND-0.1
C18:0	0.5-5.0	1.0-4.5	2.0-4.0	ND-3.3	3.5-6.0	1.0-3.0	0.8-3.0	2.0-5.4	2.7-6.5	2.9-6.2
C18:1	55.0-83.0	35.0-69.0	5.0-10.0	20.0-42.2	36.0-44.0	12.0-19.0	51.0-70.0	17.0-30.0	14.0-39.4	75.0-90.7
C18:2	3.5-21.0	12.0-43.0	1.0-2.5	34.0-65.6	9.0-12.0	1.0-3.5	15.0-30.0	48.0-59.0	48.3-74.0	2.1-17.0
C18:3	≤1.0	ND-0.3	ND-0.2	ND-2.0	ND-0.5	ND-0.2	5.0-14.0	4.5-11.0	ND-0.3	ND-0.3
C20:0	0.0-0.6	1.0-2.0	ND-0.2	0.3-1.0	ND-1.0	ND-0.2	0.2-1.2	0.1-0.6	0.1-0.5	0.2-0.5
C20:1	0.0-0.4	0.7-1.7	ND-0.2	0.2-0.6	ND-0.4	ND-0.2	0.1-4.3	ND-0.5	ND-0.3	0.1-0.5
C20:2	ND	ND	ND	ND-0.1	ND	ND	ND-0.1	ND-0.1	ND	ND

(continued)

TABLE 4.1 CONTINUED

Fatty Acid (FA) Composition (Expressed as Percentage of Total Fatty Acids) of Olive Oil and Other Vegetable Oils

Fatty Acid	Olive Oil	Peanut Oil	Coconut Oil	Maize Oil	Palm Oil	Palm Kernel Oil	Canola Oil	Soybean Oil	Sunflower Seed Oil	Sunflower Seed Oil[a]
C22:0	0.0–0.2	1.5–4.5	ND	ND–0.5	ND–0.2	ND–0.2	ND–0.6	ND–0.7	0.3–1.5	0.5–1.6
C22:1	ND	ND–0.3	ND	ND–0.3	ND	ND	ND–2.0	ND–0.3	ND–0.3	ND–0.3
C22:2	ND	ND	ND	ND	ND	ND	ND–0.1	ND	ND–0.3	ND
C24:0	0.0–0.2	0.5–2.5	ND	ND–0.5	ND	ND	ND–0.1	ND–0.5	ND–0.5	ND–0.5
C24:1	ND	ND–0.3	ND	ND	ND	ND	ND–0.3	ND	ND	ND

Source: Codex Alimentarius. Codex standard for named vegetable oils Codex Stan 210–1999. Adopted 1999. Revisions 2001, 2003, 2009. Amendments 2005, 2011, 2013 (2013).

[a] High-oleic sunflower oil.

C6:0, caproic or hexanoic acid; C8:0, caprylic or octanoic acid; C10:0, capric or decanoic acid; C12:0, lauric or dodecanoic acid; C14:0, myristic or tetradecanoic acid; C16:0, palmitic or hexadecanoic acid; C16:1, palmitoleic or hexadecenoic acid; C17:0, margaric or heptadecanoic acid; C17:1, heptadecenoic acid; C18:0, stearic or octadecanoic acid; C18:1, oleic or octadecenoic acid; C18:2, linoleic or octadecadienoic acid; C18:3, linolenic or octadecatrienoic acid; C20:1, arachic or eicosanoic acid; C20:1, gadoleic or eicosenoic acid; C20:2, eicosadienoic acid; C22:0, behenic or docosanoic acid; C22:1, erucic or docosenoic acid; C22:2, docosadienoic acid; C24:0, lignoceric or tetracosanoic acid; C24:1, nervonic or tetracosenoic acid. ND, not detected.

FA composition may significantly differ depending on the production area, the latitude, the climate, the variety, and the maturity stage of the fruit. Diversity in environment and cultivar characteristics, due, for example, to the expansion of olive tree cultivation in countries of the Southern hemisphere (Australia, Argentina, New Zealand, and South Africa), has been evidenced as the cause of "anomalous composition" with wide ranges for the four major FAs (palmitic [7.8%–18.8%], oleic [58.5%–83.2%], linoleic [2.8%–21.1%], and linolenic [0.42%–1.91%]) as well as slight deviations from the 1,3-random-2-random distribution of FA in the glycerol moiety of TAG (Boskou et al., 2006). In the latest EU regulation (EU Reg. 1348/2013), in order to facilitate trade and guarantee olive oil authenticity, more restrictive parameters have been introduced: specifically, a lower limit has been fixed for myristic acid ($\leq0.03\%$ instead of $\leq0.05\%$), while the other limits for FA have not been changed (linolenic $\leq1.00\%$, arachidic $\leq0.60\%$, eicosenoic $\leq0.40\%$, behenic $\leq0.20\%$, lignoceric $\leq0.20\%$). Other FAs can be present in the following percentage ranges: palmitic 7.50%–20.00%, palmitoleic 0.30%–3.50%, heptadecanoic $\leq0.30\%$, heptadecenoic $\leq0.30\%$, stearic 0.50–5.00%, oleic 55.00–83.00%, and linoleic 3.50–21.00%.

Table 4.2 shows the percentage ranges for the TAG composition of olive oil and other vegetable oils (Codex Alimentarius, 2013).

The most representative TAG in olive oil (see the footnotes to Table 4.2 for explanation of acronyms for TAG) are OOO (21%–48%), POO (13%–25%), OOL (21%–23%), PLO (11%–14%), and SOO (3%–8%), together with smaller amounts of POP, POS, OLLn, OLL, PLL, and LLL (Boskou et al., 2006). ECN 42 TAG content is the sum of the amounts of LLL, tripalmitolein (PoPoPo), dilinolenoyl-stearoyl-glycerol (SLnLn), dipalmitoleoyl-linoleoyl-glycerol (PoPoL), palmitoyl-palmitoleoyl-linolenoyl-glycerol (PPoLn), OLLn, PLLn, and palmitoleoyl-oleoyl-linolenoyl-glycerol (PoOLn) (positional isomers included) and is an official authenticity parameter (IOC/T20/Doc. 20, 2010; EU Reg. 1348/2013) useful for the detection of illegal addition of hazelnut or almond oil to olive oil.

Newly produced VOO contains a low amount of DAG (1%–3%), due to either incomplete TAG biosynthesis or hydrolytic reactions (Frega et al., 1993; Spyros et al., 2004). During storage, many changes may occur in DAG composition due to isomerization of 1,2-diacylglycerols (1,2-DAGs), the predominant form in fresh extra virgin olive oil (EVOO), to 1,3-diacylglycerol (1,3-DAGs), and the subclasses with 34 and 36 carbon atoms prevail (Frega et al., 1993; Caponio et al., 2013b). The content of DAG and the ratio between isomers might be considered as useful criteria to monitor quality and to evaluate the freshness state of a VOO (Spyros et al., 2004). As previously reported, the content of free acidity (FFA), expressed as grams of oleic acid per 100 g of oil, is a basic parameter to attribute a specific commercial quality grade to VOO. FFA is the oldest parameter used for evaluating olive oil quality, since it represents the extent of lipolysis and it is tightly related to the quality of olives. Oils obtained from healthy fruits, processed just after harvesting, show very low values of FFA, but, if olives are damaged by *Bactrocera oleae* attacks or molds or are submitted to prolonged preservation before processing, hydrolytic enzymes become active and the free acidity of the oil slightly increases. Specifically, samples having up to 0.8% FFA can be defined as EVOO, whereas those containing up to 2% are classified as VOO. In addition, according to EU or extra EU Regulations (IOC,

TABLE 4.2

Triacylglycerol (TAG) Composition (Expressed as Percentage) of Olive Oil and Other Vegetable Oils from Authentic Samples

Triacylglycerols	Olive Oil	Rapeseed Oil	Palm Oil	Soybean Oil	Sunflower Seed Oil (High-Oleic Acid)
LLnLn	–	<1.0	ND	1.3	–
LLLn	–	1.0	ND	8.0	–
OLnLn	–	2.5	ND	ND	
LLL	2.2–5.0	1.0	0.4	17.4	1–5.6
OLLn	0.8–1.3	7.1	ND	4.8	ND
PLLn	–	1.1	ND	3.5	–
MPL	ND	ND	<1.0	ND	ND
OLL	1.0–15.5	7.8	<1.0–2.2	16.4	15.3
PLL	3.1–9.5	ND	2.1–8.3	12.5	6.6
OOLn	ND	12.3	ND	ND	ND
POLn	ND	1.8	ND–1.9	2.1	ND
OOL	21.4–22.8	22.1	1.8–5.3	8.1	16.3
PLO	11.4–13.6	4.2–5.4	10.3	8.7–11.4	4.7
PPL	ND	<1.0	10.3	2.2	ND
MPO	ND	ND	<1.0	ND	ND
OOO	20.8–48.2	16.8–28.7	4.0	0.8–2.5	30.9
POO	13.5–25.5	4.3–5.7	23.9	2.32–5.5	6.1
POP	1.0–4.6	ND–0.8	11.2–31.3	0.2–1.9	ND
PPP	ND	ND	5.6	ND	ND
OOE	ND	1.2–1.4	ND	ND	ND
SOO	3.4–8.2	2.5	1.7–2.7	0.2–1.3	4.4
POS	0.4–2.5	ND	ND–1.9	0.3–1.1	ND
PPS	ND	ND	1.1	ND	ND

Source: Codex Alimentarius. Codex standard for named vegetable oils Codex Stan 210–1999. Adopted 1999. Revisions 2001, 2003, 2009. Amendments 2005, 2011, 2013 (2013).

Abbreviations: LLnLn, dilinolenoyl-linoleoyl-glycerol; LLLn, dilinoleoyl-linolenoyl-glycerol; OLnLn, dilinolenoyl-oleoyl-glycerol; LLL, trilinolein; OLLn, oleoyl-linoleoyl-linolenoyl-glycerol; PLLn, palmitoyl-linoleoyl-linolenoyl-glycerol; MPL, myristoyl-palmitoyl-linoleoyl-glycerol; OLL, dilinoleoyl-oleoyl-glycerol; PLL, dilinoleoyl-palmitoyl-glycerol; OOLn, dioleoyl-linolenoyl-glycerol; POLn, palmitoyl-oleoyl-linolenoyl-glycerol; OOL, dioleoyl-linoleoyl-glycerol; PLO, palmitoyl-oleoyl-linoleoyl-glycerol; PPL, dipalmitoyl-linoleoyl-glycerol; MPO, myristoyl-palmitoyl-oleoyl-glycerol; OOO, triolein; POO, dioleoyl-palmitoyl-glycerol; POP, dipalmitoyl-oleoyl-glycerol; PPP, tripalmitin; OOE, dioleoyl-erucoyl-glycerol; SOO, dioleoyl-stearoyl-glycerol; POS, palmitoyl-stearoyl-oleoyl-glycerol; PPS, dipalmitoyl-stearoyl-glycerol. ND, not detected.

Codex), there is a distinction to define lower-quality commercial classes: ordinary VOO having a free acidity from 2% to 3.3% (the EU does not include this category) and virgin lampante olive oil with free acidity higher than 3.3%.

The health value of VOO has been attributed not only to the presence of oleic acid at a high percentage (generally between 70% and 80%), but also to the occurrence of minor components such as phytosterols, carotenoids, tocopherols, and hydrophilic phenolic molecules (Bendini et al., 2007; Lercker et al., 2011). These natural antioxidants, both lipophilic (carotenoids and tocopherols) and hydrophilic (phenol and polyphenols), even in small quantities, are fundamental for protecting the lipid matrix from oxidation. According to Montedoro et al. (1992), the total amount of phenolic compounds, evaluated by spectrophotometric testing using the Folin-Ciocalteu reagent and calculated as mg of gallic acid per kg of oil, can be considered as poor when it is lower than 200 mg kg^{-1}, medium when it varies between 200 and 500 mg kg^{-1}, and elevated when it is higher than 500 mg kg^{-1}. Obviously, these ranges are different if the total amount of phenols is expressed using a different standard to build the calibration curve (e.g., caffeic acid or tyrosol). VOO is characterized by a higher oxidative stability with respect to other vegetable oils, mainly due to its FA composition, with a particularly high ratio of monounsaturated FA (MUFA) to polyunsaturated FA (PUFA), and to the presence of the aforementioned minor compounds that play a key role in preventing oxidation. The evaluation of the degree of olive oil oxidation is based on the determination of both primary and secondary oxidation products: the formation of hydroperoxides from unsaturated FA through a free radical mechanism is known as the primary stage of oxidation. In particular, the level of hydroperoxides is determined as peroxide value (PV) and expressed as milliequivalents of active oxygen per kilogram of oil (meq O_2/kg); a limit value of 20 meq O_2/kg has been established for EVOO and VOO commercial categories. The evaluation of the degree of olive oil oxidation can also be performed by means of the determination of the p-anisidine value (PAV) or specific extinction coefficients (K) measured in the UV region at the maximum absorption wavelengths of conjugated dienes and trienes (232 and 270 nm, respectively). Conjugated dienes and trienes are formed from hydroperoxides of unsaturated FA and their fragmentation products during autoxidation. The absorption around 270 nm could also be caused by substances formed during bleaching in the refining process. In official regulations, specific limits are fixed for each olive oil category. During autoxidation, hydroperoxides decompose and give rise to hydroxy, keto, epoxy, and epodioxy derivatives of FA. The keto-derivatives of FA are the most abundant oxidized fatty acids (OFA) in olive oil, and samples characterized by a total OFA content higher than 4% must be considered "expired" (Rovellini and Cortesi, 2004). Other oxidation products from TAG could be present, such as oligopolymers and oxidized TAG, which are of major interest due to both their potentially harmful effects on consumers' health and their prooxidant activity.

4.1.2 THE CONCEPT OF AUTHENTICITY FOR OLIVE OIL

The analytical control of oil obtained from olives is based on the evaluation of its purity and quality. Purity or genuineness deals with characteristics depending on the botanical origin of raw matter, as well as on the use of correct (legally admitted) extraction and production technology and on the actual correspondence with the characteristics declared in the label (legal designation). For instance, according to the legislation and international trade standards, a VOO is pure (or genuine) if it has been extracted from olive fruits using the permitted technological means (by mechanical means only) without the presence of other types of vegetable oils.

More than 95% of the world's olive oil production is concentrated in the Mediterranean countries, with 75% being produced by Mediterranean member states of the European Union (Kavallari et al., 2010). However, the adulteration of an EVOO by total or partial replacement with a lower-grade and cheaper oil may be very attractive and lucrative for the oil industry or dishonest suppliers of raw material (Gallina Toschi et al., 2013). It is estimated that, in the European Union, four million euros per year are lost because of illegal admixtures of EVOO with cheaper oil. The most common detected adulterants are refined olive oil, olive-pomace oil, seed oils (sunflower, soy, corn, and rapeseed), and nut oils (hazelnut and peanut oil), also mixed with chlorophylls and β-carotene (Frankel, 2010). Among the ways in which EVOO can be misdescribed, it is possible to consider: misdeclaration of one or all of the phases of the technological process (e.g., mixture between EVOO and refined olive oil or refined pomace olive oil, or mixture with a VOO subjected to soft deodorization), partial or total substitution with an oil having a different botanical origin (e.g., mixture of EVOO and other vegetable oils from oilseeds or fruits), and declaration of a false geographical origin.

Geographical origin and cultivar play an important role in ensuring fair competition, and can be considered an important issue for oil producers, regulatory authorities, and consumers. In this way, the European Community has introduced specific certifications for olive oil categories, such as protected denomination of origin (PDO) and protected geographical indication (PGI), which allow certain products to be labeled with the names of their geographical area of production (EU Regulation 29/2012). The increase in demand for high-quality EVOO has led to the appearance in the market of certain oils having a particular flavor due to specific olive variety composition, such as *coupage* or monovarietal EVOO. The introduction of certifications of origin and quality and the higher added value for these products have made necessary the implementation of traceability procedures. Problems of authenticity and misdescriptions of EVOO have been dealt with using traditional and innovative analytical tools, such as chromatographic or derived approaches, spectrometric, spectroscopic, and DNA-based methods. Thermogravimetric and calorimetric procedures could offer new perspectives in this field.

4.2   OLIVE OIL DSC CURVES: RELATION TO CHEMICAL COMPOSITION

4.2.1   DSC PROTOCOL

Figure 4.2 reports typical differential scanning calorimetry (DSC) cooling and heating transitions of an EVOO sample obtained at a scanning rate of 2°C/min (Chiavaro, 2013). Olive oil showed a distinctive cooling profile with two well-defined exothermic events, the major event (A of Figure 4.2) peaking at approximately −40°C and the minor event (B of Figure 4.2) with a maximum at a higher temperature (about −13°C).

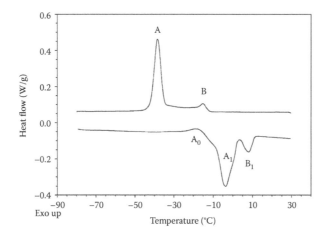

FIGURE 4.2 Representative DSC cooling and heating transitions of an EVOO sample obtained at a scanning rate of 2°C/min. Main exothermic and endothermic events are indicated with letters: A, B for cooling and A_0, A_1, and B_1 for heating. (Reprinted from Chiavaro, E: Crystal polymorph structure determined for extra-virgin olive oil. *European Journal of Lipid Science and Technology*, 2013, 115, 267–269. With kind permission from Wiley.)

The heating profile was complex due to the presence of melting–recrystallization phenomena of the original fat crystals, well known as polymorphism, and present in several vegetable oils (Tan and Che Man, 2000). Olive oils showed, at first, a minor exothermic peak (A_0 of Figure 4.2) and, successively, two endothermic events occurring over the −18°C/12°C temperature range (A_1 and B_1 of Figure 4.2). The first exothermic event was recently attributed by Barba et al. (2013) to an exothermic molecular rearrangement of metastable β' crystals into a more thermodynamically stable β form on heating, while the complex endothermic events occurring at higher temperatures were related to the melting of $\beta'a$ and $\beta'b$ crystal structures (Barba et al., 2013). Additional endothermic events were sometimes observed in some samples, displaying themselves as shoulder peaks embedded in the major event at lower or higher temperature (at ~−14°C and ~−3°C).

Starting from the pioneering paper of Kaiserberger (1989), and then from the paper of Tan and Che Man (2000), in which both crystallization and melting profiles obtained by DSC were described for an olive oil sample, different analytical protocols have been applied to characterize the olive oil transitions. Regarding crystallization, the protocol used by several authors started from 60°C (Barba et al., 2013) or 50°C (Jiménez Márquez and Beltrán Maza, 2003; Angiuli et al., 2006; Ferrari et al., 2007; Chatziantoniou et al., 2014), in order to erase crystallization memory and provide a homogeneous sample. Some authors also applied a pretreatment of the oils by stirring in a water bath at 50°C for some minutes, as this sample handling seems to give highly reproducible results (Chatziantoniou et al., 2014). Chiavaro et al. (2007, 2008c) decided from the beginning to adopt a different protocol, using a starting temperature of 30°C to avoid depletion of oil antioxidant molecules such as polyphenols (Mancebo-Campos et al., 2008). Obtaining the crystallization curves involved cooling to −30°C (Ferrari et al., 2007), −60°C (Barba et al., 2013), or −80°C (Chiavaro et al., 2007, 2008c). Tan and Che Man (2000) cooled the sample to −100°C, as other vegetable oils were analyzed with the same protocol. The same temperature was chosen by Jiménez Márquez and Beltrán Maza (2003). More recently, Chatziantoniou et al. (2014) cooled the samples to −40°C, according to Angiuli et al. (2006). The chosen temperature was the lowest at which the authors observed the effective end of the crystallization peaks and where the heat flow became null.

The heating protocol is generally carried at a temperature higher than −85°C for olive and other vegetable oils. Tan and Che Man (2000) reported that, in the temperature region from −100°C to −85°C, the melting process appeared to be out of control, due to the limitation of DSC in conducting heat through the oils at such low temperatures. Generally, a first isothermal step at the lowest temperature reached during the cooling cycle is carried out according to the protocol applied. The length of the isotherm (from 3 to 10 min) should be such as to ensure a complete solidification of the sample (Ferrari et al., 2007). The samples are then heated to obtain a melting transition and this heating step ends in the temperature range from 30°C to 60°C, according to the different authors cited above.

Different scanning rates have also been applied in the literature for both cooling and heating protocol development. Commonly, a scanning rate of 5°C/min was applied (Tan and Che Man, 2000; Jiménez Márquez and Beltrán Maza, 2003). Some authors selected a lower scanning rate (2°C/min) to minimize the instrumental lag in output response and to obtain a transition close to the chemical equilibrium (Chiavaro et al., 2007, 2008c; Barba et al., 2013). This scanning rate permitted a detailed evaluation of cooling and heating curves for olive oil in relation to chemical components (discussed in Sections 4.2.2 and 4.2.3). Others reported that the use of a higher scanning rate (10°C/min) could be a good compromise between analysis time and obtaining accurate details from the curves (Ferrari et al., 2007; Chatziantoniou et al., 2014). Figures 4.3 and 4.4 show the curves obtained for an olive oil at different scanning rates (Che Man and Tan, 2002; Tan and Che Man, 2002).

High cooling rates caused a broadening of the whole transition, an increase of breadth of the main exothermic event that also shifted toward the lower-temperature region of the thermograms. This was largely due to the low cooperativeness of the chemical species taking part in the crystallization process and the instrumental limitations in conducting heat through the sample. On the other hand, at the lowest scanning rate (1°C/min), it appears that only one exothermic event was present (Figure 4.3, curve A), indicating a complete cocrystallization of the sample in a narrow temperature range. High scanning rates (Figure 4.4) resulted in high melting points and a smoothing of the curve shape, possibly due to insufficient heat transmission from the DSC to the sample (Tan and Che Man, 2002). Thus, detailed information was lost as melting of different chemical species took place simultaneously, resulting in melting transitions that were broad, overlapping, or both (Tan and Che Man, 2002).

4.2.2 RELATION TO MAJOR COMPONENTS

DSC crystallization curves are well known to be mainly influenced by the composition of TAG (and the FA that constitute them) and not by the initial crystalline state,

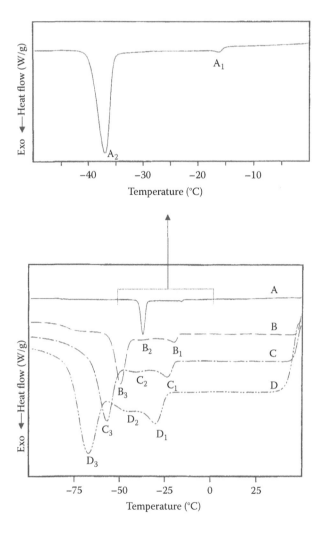

FIGURE 4.3 DSC cooling curves of olive oil at different scanning rates. A, 1°C/min; B, 5°C/min; C, 10°C/min; D, 20°C/min. (Reprinted from Che Man, Y.B. and Tan, C.P: Comparative differential scanning calorimetric analysis of vegetable oils: II. Effects of cooling rate variation. *Phytochemical Analysis*, 2002, 13, 142–151. With kind permission from Wiley.) Main events are also indicated with letters.

while melting curves are not easily interpretable due to the polymorphism of oils and fats, which is strongly dependent on the thermal history of the sample.

In the papers of Tan and Che Man (2000, 2002) and Che Man and Tan (2002), cooling and heating profiles of olive oil were related to TAG composition. In particular, the authors related the main exothermic event ($A_{1,2}$ of Figure 4.3) of cooling to the cocrystallization of OOO, POO, and OOL and the main endotherm of heating profiles ($A_{1,2,3}$ of Figure 4.4) to their melting.

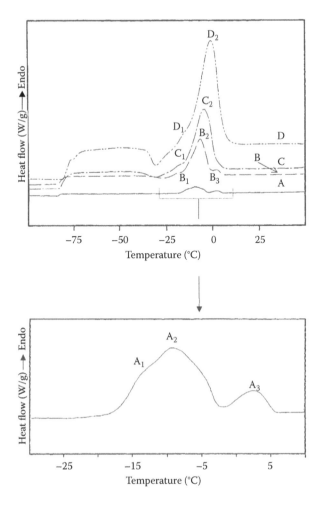

FIGURE 4.4 DSC heating curves of olive oil at different scanning rates. A, 1°C/min; B, 5°C/min; C, 10°C/min; D, 20°C/min. (Reprinted from Tan, C.P. and Che Man, Y.B: Comparative differential scanning calorimetric analysis of vegetable oils: I. Effects of heating rate variation. *Phytochemical Analysis*, 2002, 13, 129–141. With kind permission from Wiley.) Main events are also indicated with letters.

Starting with the papers of Jiménez Márquez et al. (2003, 2007), the relation of DSC curves and obtained thermal properties with the main FA and TAG was more deeply investigated. The authors reported that cooling profiles were mainly influenced by the ratio between oleic and linoleic acids, as well as the related TAG profiles. Oils with higher content of oleic acid and OOO presented a high crystallization temperature and a narrow range of transition, as well as a high melting temperature for peak A_1 of Figure 4.2. In addition, oils with a high content of disaturated (DSTAG) and monosaturated triacylglycerols (MSTAG), in particular POL + POO and PPL + PPO, were found to exhibit a high offset temperature of melting, as well as a high maximum peak temperature (T_{max}) for peak B_1 of Figure 4.2.

In the first papers of Chiavaro et al. (2007, 2008c), in which three monovarietal Sicilian EVOO samples were analyzed, these previous results were partially confirmed. In Figure 4.5, representative DSC cooling curves obtained on two EVOO samples from two different Italian regions and having different contents of both oleic and linoleic acids (about 71% and 9%, respectively, upper curve; about 78% and 5%, respectively, lower curve), are shown.

The first papers of Chiavaro et al. (2007, 2008c) introduced the novel application of deconvolution analysis to thermograms to separate and characterize overlapping contributions to the transitions as a whole, relating obtained thermal properties to chemical components. In Figure 4.5b and c, the three peaks obtained by deconvoluting cooling curves in Chiavaro et al. (2007) and in other, more recent papers are shown. The predominant peak (peak 1) was resolved with an asymmetric double Gaussian, and peaks 2 and 3 with asymmetric double sigmoid functions. Peak 1 was characterized by a quite symmetrical line shape and a narrow profile as the transition developed over a limited temperature range, indicating a highly cooperative transition and homogeneity of crystallizing molecules. Peaks 2 and 3 had more complex, asymmetrical line shape, suggesting a more heterogeneous crystallization process. The area of peaks 1 and 2 was found to be highly correlated with percentage TAG areas obtained by high-pressure liquid chromatography (HPLC): peak 1 with triunsaturated TAG (TUTAG) ($R = .96$, $p \le .01$) and peak 2 with MSTAG ($R = .87$, $p \le .01$), respectively (Chiavaro et al., 2007). On the other hand, the correlation between percentage area of peak 3 and DSTAG was not completely established.

Deconvoluted peaks obtained by heating transitions (Chiavaro et al., 2008c) are shown in Figure 4.6. All peaks were asymmetric double Gaussian functions. Peak 1 was an exothermic transition, whereas the other peaks were all endothermic events. Three different events (peaks 2, 3, and 4) were involved in the first major endothermic event A ($-18°C/2°C$) observed in the thermograms, while the smaller peak shoulder B ($2°C/12°C$) was resolved with two endothermic transitions. Chiavaro et al. (2008c) found that the sum of area percentages of deconvoluted peaks 2, 3, and 4 correlated well with TUTAG ($R = .78$, $p \le .01$). Similarly, total area percentages of peaks 5 and 6 correlated with MSTAG ($R = .90$, $p \le .01$).

Chiavaro et al. (2010) studied in-depth the relation among thermal parameters of 13 monovarietal EVOO samples and chemical composition by applying Pearson correlation analysis to cooling thermograms and their deconvoluted peaks. Enthalpy of crystallization was not influenced by the main chemical components of EVOO. On the contrary, all the other thermal properties obtained from the cooling thermograms were found to be statistically correlated with TAG and FA. In particular, OOO was highly correlated with both T_{on} (negatively) and T_{off} (positively) of crystallization, leading to a negative correlation with range of transition. In addition, onset and offset temperatures of transition (and range of crystallization, as a consequence) were largely influenced by other minor TAG (i.e., dilinoleoyl-palmityol-glycerol [LLP] + dioleoyl-linolenoyl-glycerol [OLnO], palmitoyl-oleoyl-linoleoyl-glycerol [OLP] + dioleoyl-palmitoleoyl-glycerol [OOPo], and palmitoyl-palmitoleoyl-oleoyl-glycerol [POPo]). Crystallization shifted toward higher temperature, occurring over a wider range of temperature, when saturated fatty acids (SFA) and palmitic and stearic acids increased (high positive correlation). In contrast, a shift of transition

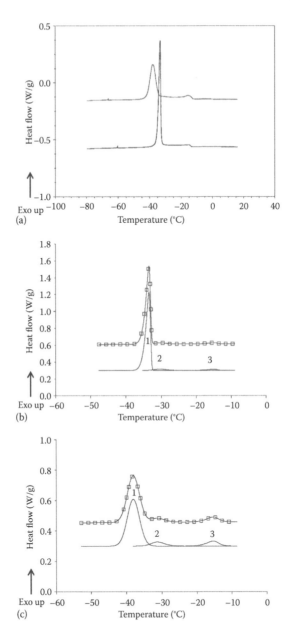

FIGURE 4.5 Representative DSC cooling thermograms of two Italian EVOO samples of different chemical composition: (a) upper curve, sample from Sicily, lower curve, sample from Marche. Deconvoluted thermograms for samples from Marche (b) and from Sicily (c): experimental data (□), fitted curve (solid line), and the constituent peaks (peaks 1–3) are shown. (Reprinted from Chiavaro, E., Cerretani, L., Di Matteo, A., Barnaba, C., Bendini, A. and Iacumin, P: Application of a multidisciplinary approach for the evaluation of traceability of extra-virgin olive oil. *European Journal of Lipid Science and Technology*, 2011b, 113, 1509–1519. With kind permission from Wiley.)

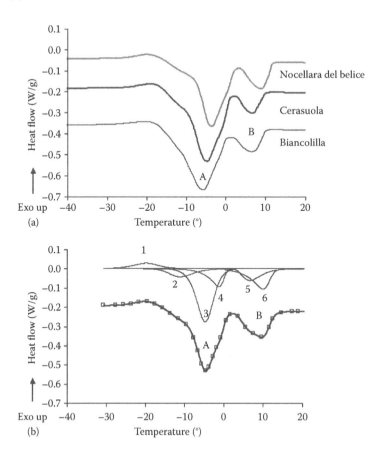

FIGURE 4.6 (a) Representative DSC heating thermograms of three monovarietal EVOO; (b) deconvolution of the heating thermogram of a Biancolilla sample: experimental data (□), fitted curve (solid bold line), and the constituent peaks are shown. (Reprinted from Chiavaro, E., Vittadini, E., Rodriguez-Estrada, M.T., Cerretani, L., Bonoli, M. and Bendini, A., *Journal of Agricultural and Food Chemistry*, 56, 496–501. Copyright [2008c], with kind permission from American Chemical Society.)

toward lower temperature and a narrower range of crystallization showed a high negative correlation with oleic acid and MUFA contents. Crystallization also shifted toward lower temperature and occurred in a narrower range with increasing linoleic acid and PUFA contents (high negative correlation).

Table 4.3 summarizes the coefficients obtained for the deconvoluted peaks. FA showed a high correlation with thermal properties after deconvolution. In particular, oleic and linoleic acids exhibited high correlation values with all thermal properties of the deconvoluted peaks, having a similar trend to those displayed by OOO and both LLP+OLnO and OLP+OOPo, respectively. Palmitic and stearic acids highly correlated with thermal properties of peak 3, shifting the onset of this transition toward higher temperature. Consequently, SFA, MUFA, and PUFA percentages were also highly correlated with all thermal properties of the deconvoluted peaks.

TABLE 4.3

Pearson Correlation Coefficients between Main Chemical Indices and Thermal Properties from Deconvoluted Peaks of Cooling Thermograms

	Deconvoluted Peak 1					Deconvoluted Peak 2					Deconvoluted Peak 3				
	Area (%)	T_p (°C)	T_{on} (°C)	T_{off} (°C)	Range[a] (°C)	Area (%)	T_p (°C)	T_{on} (°C)	T_{off} (°C)	Range[a] (°C)	Area (%)	T_p (°C)	T_{on} (°C)	T_{off} (°C)	Range[a] (°C)
TAG (% of Total TAG)															
LLL + LLPo	–	*	**	*	–	–	–	–	–	–	–	–	–	–	–
OLL + OLPo	–	*	*	*	–	–	–	–	–	–	–	–	–	–	–
LLP + OLnO	**	**	**	**	–	**	**	**	**	**	–	**	**	**	–
OLP + OOPo	**	**	**	**	–	**	**	**	**	**	–	**	**	*	–
POPo	**	–	–	*	–	*	*	–	**	*	*	*	*	–	–
OOO	**	**	*	*	–	**	*	*	**	**	*	**	**	*	–
FA (%)															
Palmitic acid	**	–	–	–	–	**	*	*	**	**	*	*	**	*	–
Stearic acid	–	–	*	**	–	*	*	–	–	*	–	*	**	–	–
Oleic acid	**	**	**	**	–	**	**	**	**	**	**	**	**	*	–

Linoleic acid	**	*	**	**	–	**	**	**	*	**	**	*	**	–
SFA	*	**	**	**	–	**	**	**	**	**	**	**	**	–
MUFA	**	**	**	**	–	**	**	**	**	**	**	**	**	–
PUFA	**	**	**	**	–	**	**	**	**	*	**	**	*	–
DAG (% of total DAG)														
1,2-PO	–	–	–	–	–	**	–	–	–	–	–	–	*	**
1,2-PL	–	–	–	–	**	**	–	*	**	–	–	–	**	*
1,3-PO	–	–	–	–	–	*	–	–	–	–	–	–	**	*
1,2-OO	*	–	–	*	**	**	*	**	**	–	–	–	**	*
1,2-OL	–	–	–	–	*	–	–	–	–	–	–	–	*	*
1,2-DAG	*	–	–	**	*	–	**	**	–	–	–	–	**	–
OSI	**	*	**	**	*	–	**	**	*	**	–	–	–	–

Source: Chiavaro, E., Rodriguez-Estrada, M.T., Bendini, A. and Cerretani, L., *European Journal of Lipid Science and Technology*, 112, 580–592, 2010.

Notes: OLPo, palmitoleoyl-oleoyl-linoleoyl-glycerol; 1,2-PO, 1-palmitoyl-2-oleyl-sn-glycerol; 1,2-PL, 1-palmitoyl-2-linoleoyl-sn-glycerol; 1,3-PO, 1-palmitoyl-32-oleyl-sn-glycerol; 1,2-OO, 1,2-diolein; 1,2-OL, 1-oleoyl-2-linoleoyl-sn-glycerol.

[a] Temperature difference between T_{on} and T_{off}.

* Significance at the 0.05 level ($p < .05$).

** Significance at the 0.01 level ($p < .01$).

One year later, Ilyasoglu and Ozcelik (2011) chose to analyze heating thermograms of four Turkish EVOO from two different seasons and to relate thermal properties to TAG and FA by means of statistical correlations. Table 4.4 summarizes the obtained results. In particular, the authors found correlations among all the thermal parameters (enthalpy, onset and offset temperatures of transition, and its range), the main FA (palmitic, oleic, and linoleic acids), and TAG (OOO and POO).

Laddomada et al. (2013) applied a simple calorimetric assay on 18 monovarietal Italian EVOO samples from the Apulia region (Italy) by using modulated scanning adiabatic calorimetry (MASC) in a temperature-scanning mode and with a tailor-made time–temperature protocol. Compared with DSC, the presented method offered advantages in terms of rapid measurements, easy to read and interpret outcome data, and higher instrumental sensitivity (thanks to the use of glass capillary flame-sealed tubes that enabled testing an oil sample mass 40 times larger than with DSC). The preliminary results presented in the paper revealed good correlations among oleic and linoleic acids, their ratio (O/L), and the ratio between unsaturated and saturated (U/S) fatty acids, with the area and width obtained from cooling and heating peaks. The correlation values were negative for oleic acid, O/L, and U/S

TABLE 4.4

Pearson Correlation Coefficients between Main Chemical Indices and Thermal Properties from Heating

	T_{on} (°C)	T_{off} (°C)	Range[a] (°C)	Enthalpy (J/g)
TAG (% of Total TAG)				
POP	−0.279	0.386	**0.602**	−0.725
POO	−0.300	**0.696**	**0.899**	−0.917
OOO	**0.657**	−0.267	−0.841	**0.784**
DSTAG	−0.279	0.386	**0.602**	−0.725
MSTAG	−0.428	**0.564**	**0.898**	−0.905
TUTAG	0.392	−0.504	−0.811	**0.856**
FA (%)				
Palmitic acid	−0.204	**0.781**	**0.888**	−0.909
Stearic acid	−0.883	−0.234	**0.601**	−0.456
Oleic acid	**0.740**	−0.188	−0.847	**0.762**
Linoleic acid	−0.864	−0.137	**0.670**	−0.541
Linolenic acid	**0.699**	0.125	−0.529	0.379
SFA	−0.393	**0.645**	**0.939**	−0.924
MUFA	**0.757**	−0.161	−0.839	**0.750**
PUFA	−0.858	−0.135	**0.666**	−0.540

Source: Ilyasoglu, H. and Ozcelik, B., *Journal of the American Oil Chemists' Society*, 88, 907–913, 2011.

[a] Temperature difference between T_{on} and T_{off}.

Significant correlations at $p < .05$ in bold.

and both crystallization peak and minor heating endotherm, while they presented a positive value for the major melting peak. The opposite was found for linoleic acid.

A different and novel approach, based on the application of partial least square (PLS) methodology to unfolded differential scanning calorimetric curves, was reported in the papers of Cerretani et al. (2011) and Maggio et al. (2014), in which thermal data were related to FA and TAG composition, respectively. In the former paper, high correlations between gas chromatographic data and predicted DSC-PLS values were found for SFA, palmitic and oleic acids, and a large set of olive oil samples. Satisfactory results were also obtained for MUFA, PUFA, stearic and linoleic acids, and O/L ratio, as shown in Figure 4.7.

More recently, these encouraging results incited the same authors to apply a similar procedure for the evaluation of the relation of TAG composition and DSC cooling profiles, analyzing 69 samples of EVOO. Findings obtained by the application of the PLS algorithm to TAG concentration (%), obtained by means of HPLC and digitized DSC curves, showed that cooling transitions were markedly influenced by OOO, OLO, and OOP + SLO, which are the most representative TAG of EVOO, as shown in Figure 4.8.

Finally, when TAG were grouped according to degree of unsaturation, high correlation coefficients (>0.80) and low relative standard deviations (<11%) were found for TUTAG in both calibration and validation sets.

4.2.3 RELATION TO MINOR COMPONENTS

The relation of thermal properties obtained from cooling and heating thermograms to minor components of olive oil (such as DAG, oxidation compounds and related stability indices, FFA, and acidity values), has been debated in the literature in past and recent years.

In the papers of Chiavaro et al. (2007, 2008c), discussed above in relation to the influence of TAG and FA on thermograms, the three EVOO samples presented different values of acidity, degree of oxidation, and amounts and profiles of DAG, according to the different harvesting periods of drupe picking. Regarding cooling thermograms (Chiavaro et al., 2007), the authors found that the presence of TAG lysis and lipid oxidation products shifted the crystallization toward higher temperatures and the phase transition developed over a larger temperature range. This behavior was theorized to be related to hindrance of TAG crystallization due to FFA and lipid oxidation products having a polar and "irregular" conformational structure (Chiavaro et al., 2007). The same authors also hypothesized that DAG, in particular 1,2-PL, 1,2-OL, and 1,3-OO, which increase greatly during ripening, may have played an important role in influencing the crystallization process for olive oil, probably being incorporated into the crystal lattice of TAG (Chiavaro et al., 2007). An influence of these minor components on heating thermograms obtained on the same oil samples has also been reported (Chiavaro et al., 2008c). In particular, the authors found that the thermal heating transition started at lower temperatures and developed over a narrower range in oil samples richer in DAG, lipid oxidation products, and FFA, probably due to the insertion of these molecules derived from TAG lysis into the crystal lattice (Chiavaro et al., 2008c). The mixed crystals so formed were more

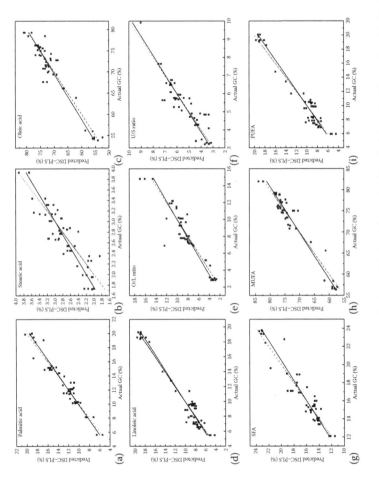

FIGURE 4.7 GC actual versus DSC–PLS predicted values in the validation (■) sets for (a) palmitic acid, (b) stearic acid, (c) oleic acid, (d) linoleic acid, (e) O/L ratio, (f) U/S ratio, (g) SFA, (h), MUFA, (i) PUFA. Ideal fitting (intercept = 0, slope = 1, solid line) and actual fitting curves (dashed line) for predicted values of (a) palmitic acid, (b) stearic acid, (c) oleic acid, (d) linoleic acid, (e) O/L ratio, (f) U/S ratio, (g) SFA, (h) MUFA, (i) PUFA are shown. (Reprinted from *Food Chemistry*, 127, Cerretani, L., Maggio, R.M., Barnaba, C., Gallina-Toschi, T. and Chiavaro, E., Application of partial least square regression to differential scanning calorimetry data for fatty acid quantitation in olive oil, 1899–1904, Copyright [2011], with kind permission from Elsevier.)

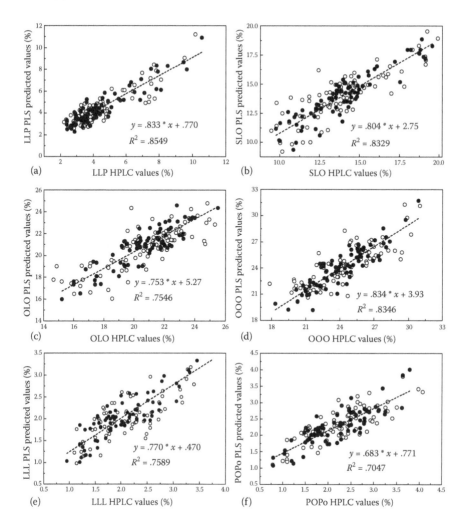

FIGURE 4.8 Percentage values calculated by HPLC–DAD (diode array)-MS (mass spectrometer) versus PLS calculated in calibration (black circle) and validation (white circle) (dilinoleoyl-palmitoleyl-glycerol) sets for individual TAG (LLP+OLnO, a; OOP+SLO, b; OLO, c; OOO, d; LLL+LLPo, e; and POPo, f). (Reprinted from *Journal of Thermal Analysis and Calorimetry*, Study of the influence of triacylglycerol composition on DSC cooling curves of extra-virgin olive oil by chemometric data processing, 115, 2014, 2037–2044, Maggio, R.M., Barnaba, C., Cerretani, L., Paciulli, M. and Chiavaro, E., with kind permission from Springer.)

easily disrupted on heating than pure TAG. Deconvoluted peaks from both transitions were also shown to be influenced by these molecules (Chiavaro et al., 2007, 2008c). In particular, this was found for the thermal properties of the deconvoluted peaks displaying themselves at the highest temperature regions of both transitions: peak 3 from the cooling deconvolution (Figure 4.5) (Chiavaro et al., 2007) and peaks 5 and 6 from the heating deconvolution (Figure 4.6) (Chiavaro et al., 2008c).

In Chiavaro et al. (2010), Pearson correlation coefficients were also calculated among cooling thermal parameters obtained by analyzing both the whole crystallization and the deconvoluted peaks and those minor components already determined in the previous papers (Chiavaro et al., 2007, 2008c) (Table 4.3). DAG (particularly 1,2-OO, which is the most representative) and FFA values were found to influence crystallization patterns and process kinetics of peaks 2 and 3 (Figure 4.5b, c), even though these minor compounds did not correlate with thermal properties of the whole cooling transition. Regarding the presence of oxidation compounds, the oxidative stability measured by the oxidative stability index (OSI) (which depends on the major [TAG] and minor [phenols] compounds and is negatively influenced by the presence of oxidation products) affected the thermal properties of the whole crystallization (onset and offset temperatures of transition and its range), suggesting an active role of lipid oxidation products in hindering TAG crystallization. This effect was also obtained on the deconvoluted peaks (particularly peak 1), as shown in Table 4.3.

More recently, FFA and quality parameters (K_{232} and K_{270}) were found not to be statistically correlated with thermal properties obtained from heating thermograms on Turkish EVOO by Ilyasoglu and Ozcelik (2011). These findings were partially confirmed in the cooling thermograms of the paper by Laddomada et al. (2013). In this latter study instead, the authors found high Pearson statistical correlation coefficients between area and width of the main peak of crystallization and oxidative indices (OSI and PV), as well as between these two chemical parameters and area, width, and T_{max} of the melting peaks obtained by means of MASC. In particular, a higher degree of oxidation was found to raise the temperature range of both whole crystallization and melting of the minor peak (B_1 of Figure 4.2) from heating thermograms.

The influence of phenolic compounds has been less debated in the literature. Jiménez Márquez et al. (2007) analyzed the effect of phenolic compounds on cooling and heating thermograms for two common Spanish variety samples. The oils were subjected to a water extraction to diminish the polyphenol content, finally resulting in four samples with polyphenol amounts ranging from 36 to 520 mg kg^{-1}. In Figure 4.9, the cooling and heating thermograms obtained at different scanning rates for the Picual variety of the original sample (a and b) and the sample depleted of phenolic compounds (c and d), are shown. The influence of polyphenols on thermograms was stressed by the authors for the oil variety richer in these molecules (Picual variety), relating increasing quantities of polyphenols to a slight shift of the crystallization point and its onset toward high temperature as well as a delay of the melting transition. In more recent papers (Chiavaro et al., 2011c; Cerretani et al., 2012), however, it was observed that different contents of phenolic compounds in EVOO samples did not affect the crystallization profiles and the related thermal properties, in contradiction to the previous study. Laddomada et al. (2013) confirmed the absence of a direct influence of phenols on both crystallization and melting transitions for EVOO. No statistical correlations were found by these authors between either total phenols or o-diphenol contents and thermal properties obtained by means of MASC.

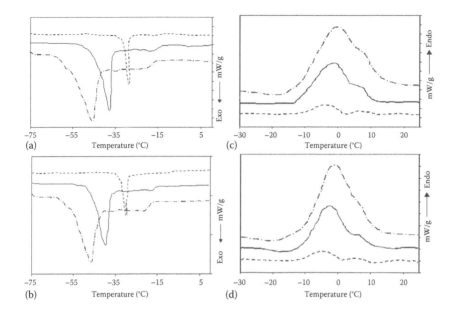

FIGURE 4.9 Representative DSC cooling (a and b) and heating (c and d) thermograms for a VOO of a Picual variety at different scanning rates (1°C/min dotted line; 5°C/min solid line; 10°C/min dashed line): A and B, original sample; C and D, sample extracted with water. (Adapted from Jiménez Márquez, A., Beltrán Maza, G., Aguilera Herrera, M.P. and Uceda Ojeda, M., *Grasas y Aceites*, 58, 122–129, 2007. (With kind permission from the journal *Grasas y Aceites*.)

4.2.4 DSC OXIDATIVE STABILITY

DSC could be used to determine the oxidative stability of an oil or fat by applying an isothermal treatment and passing oxygen or air through the sample. The DSC induction time (T_o) is calculated at the intersection of the extrapolated baseline and the tangent line of the exotherm (Tan et al., 2002). Few papers deal with the calculation of DSC induction time for olive oils. Tan et al. (2002) obtained T_o values in the range 110°C–140°C for different vegetable oils (including an olive oil sample), showing good correlation of DSC with OSI time. Jiménez Márquez and Beltrán Maza (2003) applied a different approach. These authors heated different monovarietal EVOO samples at a scanning rate of 5°C/min from 50°C to 300°C under air, calculating the onset temperature of transition and comparing values with the induction time obtained by Rancimat. EVOO samples with a higher content of linoleic acid showed lower induction times and a good correlation was obtained with the Rancimat method.

More recently, Ostrowska-Ligeza et al. (2010) analyzed four samples of commercial olive oils under dynamic conditions in the cell of a normal-pressure DSC and in a Metrohm Rancimat apparatus, extrapolating the onset DSC temperatures to assess the stability of the oils. Using the Arrhenius-type correlation between the inverse of the induction times and the absolute temperature of the measurements, the authors calculated the kinetic parameters for oil oxidation under Rancimat conditions, showing that the data obtained by both methods were qualitatively consistent for olive oil but not as reliable as for other vegetable oils.

4.3 OLIVE OIL DSC CURVES: AUTHENTICITY, BOTANICAL AND GEOGRAPHICAL ORIGIN, ADULTERATION

4.3.1 OLIVE OIL DSC CURVES: DISCRIMINATING AMONG GEOGRAPHICAL AND BOTANICAL ORIGINS

Angiuli et al. (2006) verified the utility and effectiveness of calorimetry for screening the traceability of EVOO produced in Tuscany (six samples from different Tuscan provinces and different cultivars). In this article, the authors report preliminary results from DSC, high-sensitivity isothermal calorimetry, and MASC. The obtained results evidenced that DSC can discriminate rapidly between commercial olive oil and guaranteed-origin EVOO. The following year, Jiménez Márquez et al. (2007) found that profiles and peak temperature of melting transitions were useful to discriminate between Spanish EVOO samples, according to the variety. Three years later, Angiuli et al. (2009) showed a strong similarity between the melting curves of six EVOO (blends of the cultivars Leccino, Moraiolo, Frantoio, and Pendolino from Tuscany) and supported the possibility of defining a mean curve (MC), calculated as the mean of the melting curves of samples. The MC and the melting curves of three commercial EVOO (one blend of other cultivars [from Apulia], one monocultivar [Taggiasca] from Liguria and a commercial Protected Geographical Indication [PGI] blend from Tuscany), were also calculated. Clear differences in the MC and the curves between the Apulian and Ligurian oils were found, whereas the PGI Tuscan EVOO agrees very well with the MC of the six selected EVOOs from Tuscany. Moreover, a third peak (P1) was observed at $-20°C$ in the melting curve of Apulian and Ligurian EVOO, rather than the major and minor classic endothermic events. In addition, the authors stated that a melting curve whose R value (ratio between the heights of the two peaks in the MC; A_1 and B_1 of Figure 4.2) was too high could indicate a different origin of the sample. The results obtained by applying this procedure to six samples of Tuscan EVOO (produced in three different provinces) and to several commercial EVOO (whose origin is not declared on the label) are very promising: the six Tuscan EVOOs, four from private production and two bought at the market with PDO declaration, fulfilled the above criterion for genuineness, while the commercial EVOO frequently showed a peak at $-20°C$ and the R value was too high. Kotti et al. (2009) characterized seven monovarietal EVOO samples obtained from the two main Tunisian cultivars (Chetoui and Chemlali) and grown in different regions of Tunisia for thermal properties and chemical composition. In this preliminary work, the authors evaluated the potential of DSC to differentiate samples on the basis of cultivar–environment interaction, relating their cooling and heating profiles to TAG and FA composition. It appeared to be possible to differentiate between samples from different cultivars by the analysis of crystallization profiles: in particular, onset and offset temperatures of transition were distinctive on the basis of cultivar–environment interaction as related to differences in macrocomponents (TAG and FA). In addition, all thermal properties obtained from heating thermograms differed for samples from the same cultivar but of different geographical provenance (the north, central, and coastal regions of Tunisia). Chiavaro et al. (2010), analyzing thermal properties of monovarietal EVOO oils from two Italian regions in terms of correlation with major and minor chemical components, also observed that cooling thermal properties were unable to discriminate among samples from two regions and different varieties; however, it was possible to differentiate oils by applying a deconvolution analysis to the transition. On the contrary, samples from the same variety (i.e., Leccino) and different geographical provenance could be distinguished on the basis of onset temperature and range of transition, as shown in Figure 4.10.

In a more recent publication (Chiavaro et al., 2011b), a multidisciplinary approach for the evaluation of EVOO traceability (geographical provenance and botanical differentiation) was developed on the basis of the comparison among thermal properties obtained from cooling curves and their deconvoluted peaks, chemical composition (TAG and FA), and stable isotopic ratio ($^{13}C/^{12}C$ in combination with $^{18}O/^{16}O$). These different analytical strategies were applied to 53 samples from different Italian regions and two Mediterranean areas (Italy and Croatia). Data were analyzed by means of linear discriminant analysis (LDA) and the obtained models revealed a good resolution among categories (different Italian regions and cultivars) based on

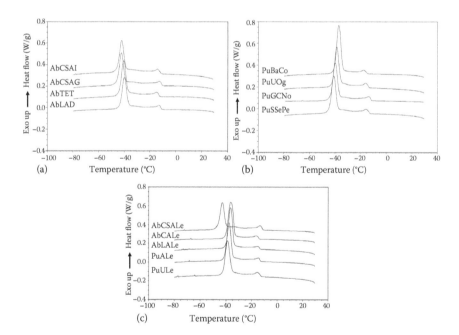

FIGURE 4.10 Representative DSC cooling thermograms of EVOO samples obtained from Dritta, Tortiglione, Gentile, and Intosso in the Abruzzo region (a); Coratina, Ogliarola, Nociara, and Peranzana in the Apulia region (b); and Leccino (c) in both Italian regions. (Reprinted from Chiavaro, E., Rodriguez-Estrada, M.T., Bendini, A. and Cerretani, L: Correlation between thermal properties and chemical composition of Italian virgin olive oils. *European Journal of Lipid Science and Technology*, 2010, 112, 580–592. With kind permission from Wiley.)

selected TAG, $\delta^{18}O$ values, and thermal properties of the deconvoluted peaks set at the highest temperature of cooling transition (peaks 2 and 3 of Figure 4.5b, c), employing three discriminant functions. Italian and Croatian samples were also distinguished according to two discriminant functions by means of TAG, SFA, MUFA, and onset temperature of deconvoluted peak 1 (Figure 4.5b, c). In Figure 4.11, the score plots show the good discrimination achieved by applying the different analytical methods.

More recently, Chatziantoniou et al. (2014) studied thermal properties of 27 EVOO samples from different Greek varieties (Koroneiki, Lianolia, Adramitiani, and Thasitiki) and geographical provenance, relating them to FA composition by means of LDA. FA composition, alone or in combination with thermal parameters from DSC, did not prove to be effective in discriminating between samples as related to cultivar or geographical origin. On the contrary, a good resolution was obtained by means of some thermal parameters obtained by both cooling and heating protocols (without the insertion of thermal properties obtained by the deconvolution of heating transition), with an effective classification of all samples to the assigned cultivar groups, geographical origins, or both.

In the same year, Laddomada et al. (2013) showed that the overall thermogram (crystallization and melting curves) obtained by MASC can provide characteristic fingerprints of different monovarietal EVOO, discriminating between samples of the same cultivar and growing area, but milled differently (in a large-scale continuous system or in a small batch-scale one).

4.3.2 OLIVE OIL DSC CURVES: DISCRIMINATING AMONG COMMERCIAL CATEGORIES AND CHEAPER OILS

Starting with the preliminary paper of Jiménez Márquez (2003), in which an EVOO sample, a refined olive oil, a sample obtained with a second centrifugation, and their mixtures were discriminated by means of heating thermograms and related thermal properties, DSC proposed itself as a valuable tool to establish compositional differences in nonadulterated and adulterated olive oils, being also of great interest for the characterization of samples by botanical origin and geographical provenance and for disclosing adulteration.

In the aforementioned study, Angiuli et al. (2006) verified the utility and effectiveness of different calorimetric apparatus for screening the quality of olive oils, in particular concerning the presence of defects (according to the judgment of a qualified panel of tasters), resulting either from natural degradation due to time–temperature storage history or from adulteration and fraud (addition of low-cost edible oils). The solid/liquid phase transitions monitored during cooling and heating cycles in a wide temperature range with both DSC and MASC provided useful information to discriminate rapidly between olive oil and illegal mixtures with other edible oils. Isothermal calorimetry (TAM) turned out to be time-consuming, but it was able to provide parameters, such as IT (induction time measured by the intercept of the tangent at the flex point of the freezing curve with the baseline) and H (the freezing enthalpy), that are particularly sensitive to EVOO defects.

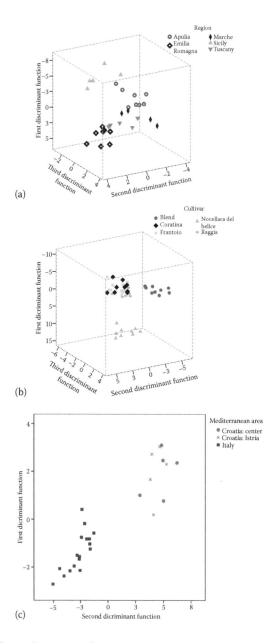

FIGURE 4.11 Score plots on an oblique plane of the 3D space defined by the three discriminant functions constructed to classify EVOO samples according to different Italian region (a) ($n=30$) and cultivar (b) ($n=42$), and on the plane defined by the two discriminant functions constructed to classify EVOO samples according to different Mediterranean area (c) ($n=27$). (Reprinted from Chiavaro, E., Cerretani, L., Di Matteo, A., Barnaba, C., Bendini, A. and Iacumin, P: Application of a multidisciplinary approach for the evaluation of traceability of extra-virgin olive oil. *European Journal of Lipid Science and Technology*, 2011b, 113, 1509–1519. With kind permission from Wiley.)

Ferrari et al. (2007) strongly promoted calorimetry for olive oil authentication. They affirmed that the DSC melting and freezing curves of olive oil and other edible oils could be correlated with quality, origin, and storage history of the oil. Moreover, the addition of low-cost oils to the EVOO, the application of thermal and mechanical treatments, or both (refinement, deodorization, filtration, etc.), can be assessed by a first-sight analysis of the thermograms.

Chiavaro et al. (2008a) verified the potential application of DSC to discriminate among different commercial categories of olive oils (an EVOO, a refined olive oil, an olive oil, a refined pomace oil, and a pomace oil), evaluating the relationship between thermal properties (obtained on cooling and heating and their deconvolution) and chemical composition in terms of major compounds (TAG and FA composition) and minor components (DAG, FA, PV, and other parameters for monitoring the oxidation status). The evaluation of the cooling thermal properties appeared promising for discriminating between olive and olive-pomace oils. In particular, T_{on} and T_{off} of the crystallization showed significant differences among different commercial categories, while enthalpy could differentiate between olive and olive-pomace oils. In addition, the minor endothermic event (B_1 of Figure 4.2) presented different lineshapes among oil samples. Application of deconvolution analysis to DSC thermograms provided additional information for olive oil classification.

In the same year, Chiavaro et al. (2008b) evaluated the potential use of DSC to detect adulteration of EVOO with refined hazelnut oil (HaO), by establishing possible relationships between the thermal properties of cooling and heating thermograms and the TAG and FA composition of the oils and their admixtures. Thermal properties of cooling thermograms (i.e., enthalpy and T_{on} of transition) were affected by HaO addition at a concentration as low as 5%, but it was not possible to determine the exact percentage of adulterant added. Concerning the heating profiles, a broadening of the major endothermic peak (A_1 of Figure 4.1) was shown, while the minor endothermic peak shifted toward lower temperatures at HaO amounts $\geq 5\%$ and disappeared at HaO addition of $\geq 40\%$. The authors concluded that the addition of HaO to EVOO at economically remunerative levels (20%–40%) was clearly detectable by DSC analysis. The heights of the major and minor peaks from cooling and heating thermograms, respectively, were also useful parameters for establishing fraudulent HaO addition. Changes in thermal profiles are shown in Figure 4.12.

With a similar analytical approach, the same authors (Chiavaro et al., 2009b) investigated the application of DSC for the recognition of high-oleic sunflower oil as an adulterant of olive oil. Addition of high-oleic sunflower oil to olive oil appeared to be clearly recognized (in both cooling and heating profiles) only at the highest percentage of adulterant added (40%), probably due to the similarity of the two oils in terms of TAG and FA composition (e.g., OOO and oleic acid content). Deconvolution of cooling thermograms appeared to be more useful than the analysis of original thermograms to detect fraudulent high-oleic sunflower oil addition, but its recognition at added levels lower than 20% was not possible even by applying this analysis. In particular, thermal properties of the two lower-temperature exothermic events (peaks 1 and 2 of Figure 4.5) were found to be significantly changed. In the paper of Angiuli et al. (2009), the solid–liquid phase transitions of olive oil and seed oils, in which the resulting melting peak behavior correlated with the type, quality, and composition

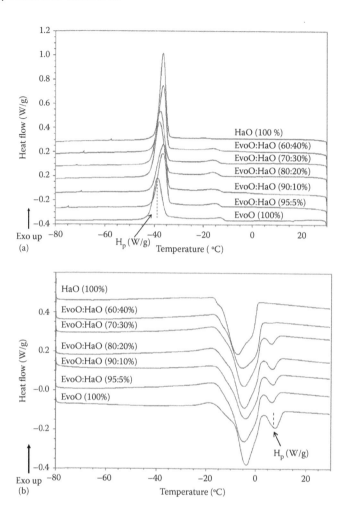

FIGURE 4.12 Representative DSC cooling (a) and heating (b) thermograms of EVOO and HaO and their admixtures. (Adapted from *Food Chemistry*, 110, Chiavaro, E., Vittadini, E., Rodriguez-Estrada, M.T., Cerretani, L. and Bendini, A., Differential scanning calorimeter application to the detection of refined hazelnut oil in extra-virgin olive oil, 248–256, Copyright [2008b], with kind permission from Elsevier.)

of the oil, were presented. In this paper, the authors studied the melting curves of selected EVOO and their mixtures with some seed oils and discussed the potential of calorimetry for the authentication of high-quality EVOO; moreover, they suggested the setting-up of a calorimetric "identity card" of the oil to be used in commerce for rapid conformity tests and the detection of light and oxygen exposure effects to evaluate oil storage conditions and aging processes. A melting curve whose *R* value (ratio between the height of the two peaks in melting curve) is too high could indicate that seed oils, refined olive oil, or both have been incorporated, and the simultaneous presence of a peak at about −20°C in the melting curve could indicate the addition of refined olive oil or peanut oil, or both. Concerning the control of the addition of seed oil or refined olive oil to a known EVOO sample, in spite of its sensitivity limit (about 2%), the calorimetric method is not suitable for an absolute authentication test when more than one oil has been added. In the same year, the detection of adulteration of olive oil with soybean, sunflower, and canola oils was investigated by Jafari et al. (2009) using different analytical tools (such as gas chromatography [GC], nuclear magnetic resonance [NMR], and DSC) to evidence differences in the FA and TAG compositions. Satisfactory results were achieved from quantitation of DSC parameters, showing differences in offset crystallization temperature and onset melting temperature in samples with higher amounts of unsaturated FA.

More recently, Chiavaro et al. (2011a) characterized cooling and heating thermogram profiles of a set of 18 samples of different products from olive processing (EVOOs, repaso oils obtained from the second centrifugation step, lampante olive oils, and refined olive oils) and the thermal properties were related to their FA compositions. Deconvolution was also applied to the crystallization transition, and a good discrimination of EVOO from the other products of minor nutritional and economic values was shown. A LDA model, which used three oil typologies as categories (extra-virgin, lampante-repaso oils, and refined) and 23 predictors that corresponded to the cooling and heating thermal properties, was built and an excellent resolution among categories was obtained. In the same year, a very interesting approach was proposed by Torrecilla et al. (2011) to detect and quantify EVOO adulteration by calculating only one type of chaotic parameter, known in the statistical field as lag-k autocorrelation coefficients (LCCs), which can be calculated easily, extracting the essential information from huge databases such as DSC scans. Complex DSC profiles can be transformed into simple LCC profiles, which maintain the information necessary to detect slight adulterations of EVOO with inferior edible oils (refined olive oil and refined olive pomace, corn, or sunflower oils). This quantification was carried out using successful linear correlation of LCCs and different concentrations of adulterants in a set of 462 adulterated samples. The LCCs were calculated from DSC scans of EVOOs adulterated with agents at concentrations <14% w/w.

4.4 OLIVE OIL DSC CURVES: AUTOXIDATION, PHOTOOXIDATION, THERMOOXIDATION

4.4.1 DSC AND OLIVE OIL AUTOXIDATION

Chiavaro et al. (2013) evaluated the relationship between thermal and chemical parameters in naturally autoxidized virgin olive oils. Twenty one samples of VOO (obtained from drupe batches of different cultivars that are commonly grown in two different central Italian regions) were stored under ideal conditions after production (production period: 1991–2005) to allow the spontaneous development of autoxidation and were then subjected to DSC and chemical analysis. PCA was run to correlate the chemical data and thermal properties; 23 variables were selected for the PCA analysis. Thermal properties of transitions were differently influenced by degradation compounds; the onset of both cooling and heating profiles was particularly influenced by DAG and oxidized lipids (oxidized TAG and OFA). On the other hand, triacylglycerol oligopolymers did not seem to markedly influence oil transitions from PCA, probably as a consequence of their low quantity, the presence of a high level of oxidation compounds that could diminish their influence, or both. The application of the deconvolution analysis of transitions also supported these findings. In fact, several thermal properties were found to contribute to PC1, which clearly segregated groups according to their different degrees of hydrolysis or oxidation, or both, in particular those obtained by the deconvolution of the crystallization profile that occurred in the highest temperature region of the cooling thermogram. The deconvolution of the complex heating profiles seemed to be less useful, with the exception of peak 2, which indicated the real onset temperature of melting of VOO crystals that influenced PC1. In the projection of the cases on the factor plane defined by the two PCs, PC1 clearly segregated samples into two groups according to the year of storage.

To better understand the oxidative stability of oils, accelerated oxidation tests can be performed at high temperatures, with or without oxygen flow. Chiavaro et al. (2011c) evaluated the DSC cooling thermal properties of an Italian EVOO (Tuscan cultivar blend) in the presence and absence of its phenolic fraction, at different times of accelerated storage treatment (up to four weeks at 60°C with direct air contact in open bottles); the DSC properties were correlated with lipid oxidation compounds (measured with K_{232} and K_{270} indices) and total phenol content. Phenols did not appear to directly influence crystallization of EVOO, as neither cooling profiles nor thermal properties significantly differed between the two oils at the beginning of storage. However, oil samples deprived of the phenolic fraction exhibited more significant changes at the longest storage time as compared with the untreated oil. Cooling transitions were all deconvoluted into three peaks. Changes in cooling thermal properties were more evident for the two transition peaks at the highest temperature in both oil samples. A marked influence of lipid oxidation products on the crystallization pattern of these two peaks was hypothesized by the authors.

In a later study, the same authors (Cerretani et al., 2012) assessed the combined effect of FA composition and phenol contents on DSC parameters of EVOO stability subjected to accelerated oxidation during storage. Three Italian EVOO, obtained from different cultivar blends of diverse geographical origin and having different FA compositions and total phenol contents, were stored at 60°C for up to 21 weeks. Their oxidative status, evaluated by PV and total phenolic content, was related to DSC cooling profiles and thermal properties. Figure 4.13 shows representative DSC cooling curves of EVOO samples (a–c) at different storage times (0, 4, 12, and 21 weeks). At time zero, all samples showed the classical crystallization profile of EVOO, with two exothermic events: a major peak (peak 1, Figure 4.13a) at lower temperature and a minor peak (peak 2, Figure 4.13a) at higher temperature of the DSC curves, as reported in previous studies (Chiavaro et al., 2007, 2008a,b, 2010). Increasing amounts of phenolic compounds do not seem to have influenced the crystallization profiles or related thermal properties in unheated EVOO under these experimental conditions. Changes in crystallization profiles were consistent from 12-week storage onward for the two oil samples (b and c) with a higher content of linoleic acid and medium/low amounts of phenols, respectively, whereas they became detectable only at the end of the storage in the other oil (sample a, less unsaturated [low PUFA, high MUFA] and highest phenol content). Decrease of crystallization enthalpy and shift

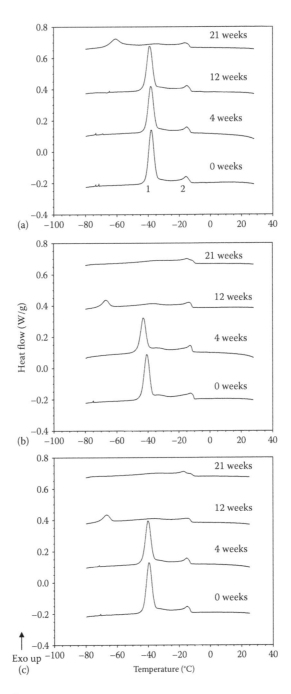

FIGURE 4.13 Representative DSC cooling curves of EVOO samples A (a), B (b), and C (c) at different storage times (weeks). (Reprinted from Cerretani, L., Bendini, A., Rinaldi, M., Paciulli, M., Vecchio, S. and Chiavaro, E., *Journal of Oleo Science*, 61, 303–309, 2012. With kind permission from Japan Oil Chemists' Society.)

of transition toward lower temperature were also evident at four weeks of storage for samples b and c, whereas similar changes in the transition profile were only noticeable at 12 weeks for sample a. Considering these results, DSC is apparently able to discriminate between EVOO with diverse oxidative status and FA composition.

In a subsequent work, Caponio et al. (2013a) evaluated the influence of refining on chemical and thermal parameters (cooling and heating curves and related thermal properties) of VOO having different initial oxidative levels (PV = 15 and 30 meq O_2/kg). This is the first work that discusses DSC curves and thermal properties obtained at each step of olive oil refining, considering the degradation compounds generated during refining (TAG oxidation, polymerization, and hydrolysis) that can affect the overall quality of the refined oil. A VOO was refined before and after being subjected to accelerated oxidation in the presence of air and light. A regular refining process was run in a pilot laboratory plant, except for the degumming step, as olive oil has negligible amounts of phospholipids. Thermal properties were statistically correlated with oxidation, polymerization, and lipolysis products by means of PCA. During oil bleaching, a simultaneous significant increase of K_{270} and polymerized TAG, as well as a general decrease of oxidation, was observed. From the bleaching step onward, crystallization and the heating onset temperature significantly shifted toward lower temperatures, thus enlarging the transition range in both oils. PCA was performed to correlate the thermal and chemical data to tentatively discriminate among samples according to the refining step, the initial level of oxidation, or both (Figure 4.14). The first principal component clearly clustered samples according to different refining steps, where thermal parameters were correlated with polymerized TAG and conjugated dienes and trienes (K_{232} and K_{270}, respectively). These molecular species seem to be able to influence DSC curves and thermal properties, like other lipid oxidation compounds that have been recognized to induce changes in transitions and related parameters in vegetable oils and fats (Cerretani et al., 2012; Chiavaro et al., 2011d; Giuffrida et al., 2007). DSC curves and related thermal properties, together with the PCA application, appeared to be suitable in differentiating samples from diverse refining steps according to two groups, crude and neutralized on one side and bleached and deodorized on the other, regardless of the initial oxidation level. The preliminary findings of this study promote the application of DSC for discriminating between other vegetable oils during refining, to better control their quality as well as the efficacy and efficiency of the refining process applied.

4.4.2   DSC and Olive Oil Photooxidation

The effects of light exposure on thermal parameters (freezing and melting) of EVOO were studied by Angiuli et al. (2007). A typical Tuscan EVOO produced from a blend of three cultivars (about 25% Muraiolo, 30% Leccino, and 45% Frantoio) was divided into two portions, one of which was exposed to diffuse light while the other was kept in the dark. The melting thermograms of EVOO kept in darkness showed an exothermic peak at about –21°C, a central high peak, and an additional peak at $T > 0$°C. During the first 11 weeks of the oil's life, the central peak slightly shifted toward higher temperatures, while in the successive six weeks appreciable changes

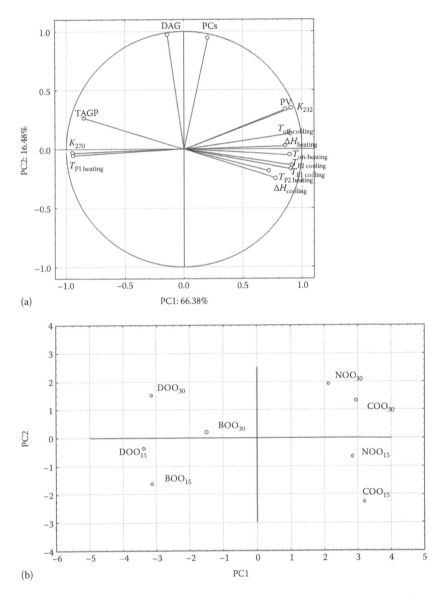

(a)

(b)

FIGURE 4.14 PCA results obtained for the two principal components: (a) Projection of the variables on the factor plane (1 × 2). (b) Projection of the cases on the factor plane (1 × 2). Abbreviations: PV, peroxide value; K_{232} and K_{270}, specific absorption at 232 and 270 nm; TAGP, TAG oligopolymers; DAG, diacylglycerols; PCs, polar compounds; ΔH, enthalpy change for transition; T_{on}, onset temperature of transition; T_{off}, offset temperature of transition; T_{p}, peak temperature at the maximum; COO, crude olive oil; NOO, neutralized olive oil; BOO, bleached olive oil; DOO, deodorized olive oil. (Reprinted from Caponio, F., Chiavaro, E., Paradiso, V.M. et al: Chemical and thermal evaluation of olive oil refining at different oxidative levels. *European Journal of Lipid Science and Technology*, 2013b, 115, 1146–1154. With kind permission from Wiley.)

were observed in the thermogram, which then remained approximately constant until the 31st week. The waiting time (WT) measured from the beginning of the isotherm to the exothermic (negative) peak seemed to increase slightly with the storage time under darkness; however, no regular trend was noted. The crystallization curve showed changes in both the waiting time and the peak minimum (PM), being more and more relevant as the time of light exposure increased. Such effects were much larger than those occurring in the sample stored in the dark. Changes induced even by a few days of light exposure could be detected by DSC, which is thus particularly sensitive to the compositional changes caused by photooxidation. However, it should be noted that the concomitant action of light and oxygen in the atmosphere in contact with the oil sample was renewed at each calorimetric test, so a possible amplification of the light effects under these experimental conditions should be considered. Both melting and freezing thermograms of EVOO stored in the dark at room temperature maintained their more relevant features up to at least seven months after oil pressing. However, some changes were observed during the first six to eight weeks of storage in the dark, which could likely be attributed to the "maturation" process occurring in this first period of the EVOO's life. The WT slightly increased up to a limiting value during storage in darkness, whereas the PM and the enthalpy variation (ΔH) remained almost constant. On the other hand, all the solidification parameters (WT, PM, and ΔH) exhibited marked changes with increasing storage time, showing a nearly linear behavior. Moreover, the height of the last peak of the melting thermogram decreased with increasing light exposure, while the exothermic (negative) peak at $-21°C$ progressively disappeared. A new endothermic (positive) peak, whose intensity increased with the exposure time, was observed in the temperature range $-30°C < T < -15°C$.

4.4.3   DSC AND OLIVE OIL THERMOOXIDATION

Vittadini et al. (2003) performed the first study that compared thermooxidation with autoxidation in EVOO, by evaluating autoxidation at 50°C and its thermal oxidation at 93°C and 180°C in 10-ml airtight vials. Figure 4.15 shows the DSC thermograms of EVOO autoxidized at 50°C (a) and thermooxidized at 93°C (b) and 180°C (c). A shift of the crystallization peak toward lower temperatures was observed in all samples as a consequence of oxidation and varied according to the oxidation conditions. Figure 4.16 shows the effects of storage time at 50°C, 93°C, and 180°C on the peak enthalpy (a) and the initial temperature of crystallization (b) of EVOO. The enthalpy of the crystallization of EVOO decreased with increasing length of high-temperature treatment (Figure 4.16a), indicating that a smaller amount of sample was crystallized under these experimental conditions. A similar trend was also found for the initial crystallization temperature of EVOO (Figure 4.16b), since it decreased as the olive oil was oxidized, more perceptibly and quickly in samples held at 180°C. The decrease in crystallization enthalpy (Figure 4.16a) and the shift of the crystallization peak to lower temperatures (Figure 4.16b) could be due to the depletion of TAG, the increase of the FFA content, the increase of viscosity during oxidation, or a combination of these. DSC peak enthalpy and peak crystallization temperatures were compared with headspace oxygen depletion and headspace

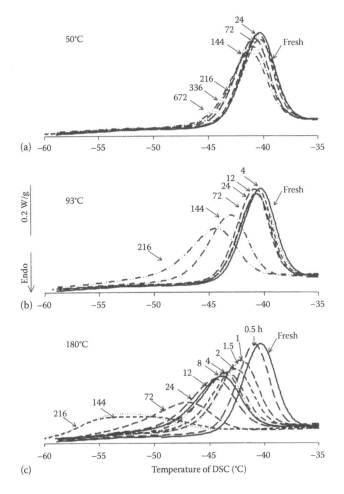

FIGURE 4.15 DSC thermograms of EVOO during storage at 50°C (a), 93°C (b), and 180°C (c) in the dark. The number of hours of storage at 50°C, 93°C, and 180°C is shown on each thermogram. (Reprinted from Vittadini, E., Lee, J.H., Frega, N.G., Min, D.B. and Vodovotz, Y., *Journal of the American Oil Chemists' Society*, 80, 533–537, 2003. With kind permission from AOCS Press.)

volatiles. DSC peak enthalpy correlation with headspace oxygen depletion increased with higher oxidation temperatures ($R^2 = .84$, .91, and .95 at 50°C, 93°C, and 180°C, respectively). A lower correlation between DSC initial peak temperature and head-space oxygen depletion was found ($R^2 = .53$, .87, and .95 at 50°C, 93°C, and 180°C, respectively), even though it followed a similar temperature-dependent trend as the one observed for DSC peak enthalpy. When thermooxidized at 180°C, correlations of DSC peak enthalpy and initial peak temperature with headspace volatiles were very high ($R^2 = .95$ and .97, respectively), which confirms that DSC could be a suitable alternative method to determine the oxidative stability of olive oil at frying temperature.

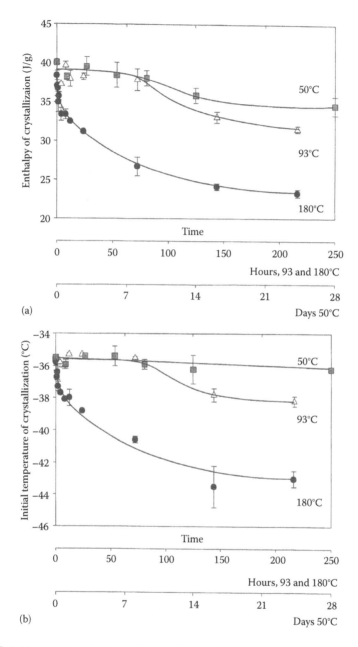

FIGURE 4.16 Effects of storage days at 50°C, and hours at 93°C and 180°C, on the peak enthalpy of EVOO (a) and on the initial temperature of crystallization of the oil (b). (Adapted from Vittadini, E., Lee, J.H., Frega, N.G., Min, D.B. and Vodovotz, Y., *Journal of the American Oil Chemists' Society*, 80, 533–537, 2003. With kind permission from AOCS Press.)

The effect of oxidative treatments on EVOO has also been evaluated by modulated differential scanning calorimetry (M-DSC) and correlated with flavor and off-flavor compounds deriving from oil oxidation (Kanavouras and Selke, 2004). In this study, Spanish EVOO samples were subjected to three oxidation treatments: (a) purged with air at two flow rates (15 and 200 ml/min), (b) heated in a conventional oven at two area/oil mass ratios (in Petri dishes [20 g] and in 2-cm i.d. test tubes [2 g]), and (c) heated in a microwave (MW) oven at two area/oil mass ratios (same as b). All oil samples were analyzed by M-DSC during cooling from 25°C to 260°C at 7°C/min, and heating back to 40°C at 10°C/min. The main crystallization peak moved toward lower temperatures, while the first transition remained rather stable in position or slightly decreased, depending on the oxidation treatment. Oxidation evolution in the heat-treated olive oil further enhanced these differences. As the TAG molecular weight decreased due to their decomposition during oxidation, their crystallization temperature dropped, while the shape of the curve, which is related to the size of the molecules, was also influenced. Similar decreases were recorded for the crystallization enthalpy and the onset point of the curve during oxidation. During oxidation, a loss of the melting curves' sharpness was observed, which may be due to the alteration of the polymorphism of the mixed-fatty acid TAG, giving rise to less stable crystals. EVOO treated inside test tubes exhibited similar results, regardless of the heating method, and so did the oil treated in Petri dishes. EVOO subjected to 15 ml/min airflow was not greatly altered, while modifications in the oil treated at 200 ml/min were close to those recorded for oil samples heated in Petri dishes. In addition, a greater reduction of the total crystallization enthalpy (as a percentage of the total melting peak enthalpy) was recorded for heated compared with air-treated samples. Furthermore, the crystallization exothermic energy per gram of oil as a percentage of the melting energy was reduced over time in all treated samples, but at different rates. A good correlation ($R^2 > .92$) was obtained between the temperature and enthalpy values corresponding to the main crystallization peak (*cis*-triolein) and the off-flavor compounds known to derive from oleic acid oxidation. A poor correlation ($R^2 < .6$) was obtained between the thermograph parameters and the characteristic off-flavors deriving from linoleic and linolenic acids. M-DSC is thus able to identify these oxidative alterations quite sufficiently to provide a rough estimate of the oxidative status of the oil.

Chiavaro et al. (2009a) also evaluated the influence of MW heating on the DSC thermal properties of different commercial categories of olive oils for human consumption (EVOO, olive-pomace oil, and olive oil). Figure 4.17 shows the DSC cooling thermograms of unheated and microwave-heated oils for EVOO (a), olive-pomace oil (b), and olive oil (c). The cooling thermogram of untreated EVOO exhibited two well-distinguishable exothermic events, with the minor exotherm peaking at about −15°C and the major at −37°C, as already reported (Chiavaro et al., 2007, 2008c). Marked changes of DSC cooling profiles were found for EVOO and olive-pomace oil subjected to microwaving, with the major exotherm shifting toward lower temperature and decreased height with increasing treatment time. Slight differences were noted between the temperatures at which major and minor events peaked in the two oils, which could be related to the different degree of unsaturation of their lipid composition, as olive-pomace

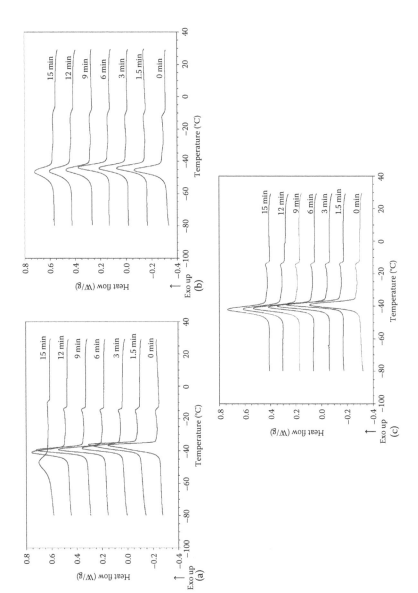

FIGURE 4.17 DSC cooling thermograms of EVOO (a), olive-pomace oil (b), and olive oil (c) at different microwave heating times. (Reprinted from *Food Chemistry*, 115, Chiavaro, E., Barnaba, C., Vittadini, E., Rodriguez-Estrada, M.T., Cerretani, L. and Bendini, A., Microwave heating of different commercial categories of olive oil: Part II. Effect on thermal properties, 1393–1400, Copyright [2009b], with kind permission from Elsevier.)

oil exhibited a higher amount of unsaturated lipids (i.e., linoleic acid, PUFA, MSTAG) and SFA. During DSC cooling analysis, all samples displayed a significant decrease in crystallization enthalpy, with a phase transition that developed over a wider temperature interval; the latter was mainly due to the more heterogeneous chemical composition of all oils that arose from triacylglycerol lysis and the formation of lipid oxidation products. Figure 4.18 shows the DSC heating thermograms of unheated and microwave-heated oils for EVOO (a), olive-pomace oil (b), and olive oil (c). MW also modified the DSC heating profiles of EVOO and olive-pomace oil, as the minor endotherm progressively disappeared, significantly shifting the offset temperature of transition toward lower temperature. Olive oil did not show changes of thermal properties and phase transition profiles such as those observed for EVOO and olive-pomace oil. This may be mainly related to the lower water content of olive oil, which might have made it more stable even though it contained smaller amounts of antioxidant (polyphenols) as compared with the other oils evaluated. These preliminary results suggest that DSC cooling thermograms can be useful, not only for monitoring modifications of chemical composition with increasing MW treatment time, but also to discriminate among olive oils according to their response to MW exposure. Moreover, such resistance to MW thermooxidation seems to be related not only to major (e.g., TAG, FA) but also to minor (e.g., water, phenols, lipid oxidation products) chemical components, as shown through the evaluation of both cooling and heating thermograms. However, due to their complexity, DSC heating profiles seemed to be less informative, as changes at lower times of treatment were clearly shown only by the minor endotherm peaking at the highest temperature region of the thermograms.

A similar approach was followed by Chiavaro et al. (2011d) to assess the potential of DSC to discriminate among microwaved Italian EVOO from different olive cultivars and origins (three monovarietal, one bivarietal, and one commercial sample), considering changes in thermal properties (on cooling and heating) and traditional oxidative stability indices (PV, PAV, and total oxidation [TOTOX] values). As reported in the previous study on MW carried out by the same authors (Chiavaro et al., 2009a), crystallization enthalpies significantly decreased and the major exothermic peak shifted toward lower temperature, leading to enlargement of the transition range in all samples due to the formation of weak and mixed crystals among TAG and lipid degradation products. The analysis of DSC thermal properties (enthalpy, T_{off}, range of transition, and T_p of the major exothermic event) on cooling seemed to clearly discriminate among different EVOO samples after microwaving. On the contrary, thermal properties on heating vary similarly among samples, thus making EVOO samples less distinguishable; this is probably due to the complexity of the heating thermograms, as different polymorphic forms of crystals melted simultaneously. The results of this study confirmed the suitability of DSC for the evaluation of thermooxidation of EVOO by microwave heating, and this could be of great interest for the oil and fat industries, as this oil is often employed as the lipid phase in several food formulations and as a filling medium for canned products.

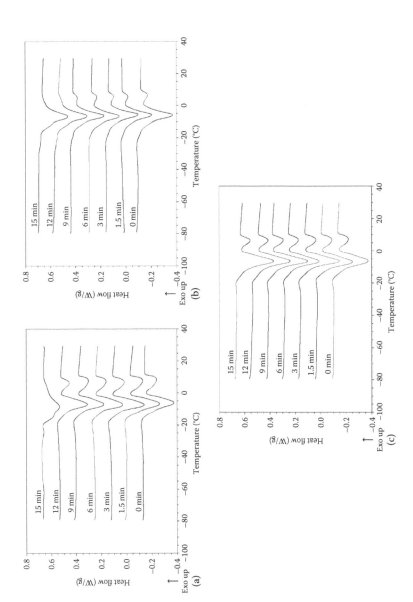

FIGURE 4.18 DSC heating thermograms of EVOO (a), olive-pomace oil (b), and olive oil (c) at different microwave heating times. (Reprinted from *Food Chemistry*, 115, Chiavaro, E., Barnaba, C., Vittadini, E., Rodriguez-Estrada, M.T., Cerretani, L. and Bendini, A., Microwave heating of different commercial categories of olive oil: Part II. Effect on thermal properties, 1393–1400, Copyright [2009b], with kind permission from Elsevier.)

An innovative DSC–PCA coupled procedure to evaluate thermooxidation in EVOO was applied by Maggio et al. (2012). An Italian bivarietal EVOO was subjected to thermal stress under convectional or microwave heating treatments at different heating times up to 1440 and 15 min, respectively; oxidative degradation was monitored by PAV determination. The entire DSC profiles obtained on cooling in the range from 30°C to −80°C and subsequent reheating to 30°C (at different times and under cooking procedures) were subjected to PCA data analysis. Thermooxidized oil samples were differentiated by PCA according to changes in the main DSC signals in both cooling and heating profiles. In particular, the shift in the major exothermic transition in cooling thermograms toward lower temperature and the decrease of height of the major endothermic transition of heating thermograms appeared to be most helpful to discriminate samples according to three different treatment times (short, medium, and long) by means of PC1 and PC2. These three groups also accounted for a different degree of thermooxidation as related to PAV values. A discrimination considering different heating processes was also possible by means of PC2 and PC3. The proposed procedure may be useful to select appropriate heating conditions to be applied to EVOO, as related to both the degree of thermooxidation and the EVOO composition.

Vecchio et al. (2009) evaluated the thermal decomposition of 12 monovarietal EVOO from different geographical origins (eight from Italy, two from Spain, and the others from Tunisia) by simultaneous thermogravimetry and DSC. In this study, the thermal analysis was combined with a kinetic procedure (Kissinger equation) to better characterize EVOO considering other parameters, such as activation energy. All EVOO exhibited a complex multistep decomposition pattern; in particular, a quite different profile among samples was noted at the first step. Thermal properties of the two peaks obtained by the deconvolution of the first step of decomposition by DSC were related to the chemical composition of the samples (TAG, FA, total phenols, and antioxidant activity). Onset temperatures of the thermal decomposition transition and peak temperature (T_p) values of both deconvoluted peaks, as well as the sum of enthalpy, showed statistically significant correlations with chemical parameters, in particular palmitic and oleic acids and the related TAG. On the contrary, thermal properties of the two peaks obtained by deconvolution of the first thermal degradation event were not statistically correlated with chemical parameters related to the antioxidant content of EVOO (2,2′-azinobis-(3-ethylbenzothiazoline-6-sulfonic acid radical cation [ABTS$^{\bullet+}$], total phenols, and o-diphenols) for all samples. The kinetic procedure was thus applied to obtain the activation energy values from the deconvoluted peaks of the first step of degradation. Activation energy values of the second deconvoluted peak were also found to be highly correlated with the chemical composition, and a stability scale among the samples was proposed on the basis of its values. In this case, EVOO from the Chemlali variety appeared to be the most stable oil among the samples, while, in general, the Italian oil samples from the Chetoui cultivar exhibited higher stability than both Spanish and Tunisian oils. The application of a kinetic model using thermal analysis seems to be an interesting approach for the creation of a stability scale of monovarietal EVOO on the basis of activation energy values for the most important decomposition step, considering differences of chemical composition due to geographical proveniences of olives.

4.5 FUTURE PERSPECTIVES

DSC use for olive oil quality determination presents undisputable advantages, such as the absence of time-consuming manipulation practices or sample treatment and toxic chemicals that could be hazardous for the analyst and the environment. Its application, however, has not been as consistent as other techniques (e.g., spectroscopy, chromatography) in this field of research, but a trend toward change has been noted in the last two or three years. Fortunately, a growing interest has been shown by several research teams, starting with the first studies of Tan and Che Man and Chiavaro and coauthors, and an increase in literature data now available in research databases. Nevertheless, some important aspects still deserve to be explored.

Considering the relation with chemical components, the influence of phenolic compounds on cooling curves deserves to be investigated in greater depth, as they are known to be hydrophilic molecules partially dispersed in the water contained in olive oils and showing a great influence on its quality and oxidative stability. Novelty and usefulness could also be offered by a wider application of the chemometric processing of digitized DSC curves in several aspects of olive oil quality, including: (a) the evaluation of different degrees of oxidation, (b) the relation with chemical components for discrimination according to different botanical origins, geographical provenances, or both, and (c) the detection of olive oil adulteration with cheaper oils from other vegetable sources or lower-quality olive oils, which is the new fraud frontier in this field. Considering the importance of the global market in olive oil, DSC could be a particularly useful technique for characterization of the curves and thermal properties of olive oils from new producer countries and regions (such as North Africa, Chile, America, and Australia), as these data are still lacking.

REFERENCES

Angiuli, M., Bussolino, G. C., Ferrari, C., et al. Calorimetry for fast authentication of edible oils. *International Journal of Thermophysics* 30 (2009): 1014–1024.

Angiuli, M., Ferrari, C., Lepori, L. et al. On testing quality and traceability of virgin olive oil by calorimetry. *Journal of Thermal Analysis and Calorimetry* 84 (2006): 105–112.

Angiuli, M., Ferrari, C., Righetti, M. C., Tombari, E., and Salvetti, G. Calorimetry of edible oils: Isothermal freezing curve for assessing extra-virgin olive oil storage history. *European Journal of Lipid Science and Technology* 109 (2007): 1010–1014.

Barba, L., Arrighetti, G., and Calligaris, S. Crystallization and melting properties of extra virgin olive oil studied by synchrotron XRD and DSC. *European Journal of Lipid Science and Technology* 115 (2013): 322–329.

Bendini, A., Cerretani, L., Carrasco-Pancorbo, A., et al. Phenolic molecules in virgin olive oils: A survey of their sensory properties, health effects, antioxidant activity and analytical methods. An overview of the last decade. *Molecules* 12 (2007): 1679–1719.

Boskou, D., Blekas, G., and Tsimidou, M. Z. Olive oil composition. In: D. Boskou (ed.), *Olive Oil, Chemistry and Technology*, 2nd edn., pp. 41–72. Boca Raton: AOCS Press (2006).

Calligaris, S., Sovrano, S., Manzocco, L., and Nicoli, M. C. Influence of crystallization on the oxidative stability of extra virgin olive oil. *Journal of Agricultural and Food Chemistry* 54 (2006): 529–535.

Caponio, F., Chiavaro, E., Paradiso, V. M., et al. Chemical and thermal evaluation of olive oil refining at different oxidative levels. *European Journal of Lipid Science and Technology* 115 (2013a): 1146–1154.

Caponio, F., Paradiso, V. M., Bilancia, M. T., Summo, C., Pasqualone, A., and Gomes, T. Diacylglycerol isomers in extra virgin olive oil: Effect of different storage conditions. *Food Chemistry* 140 (2013b): 772–776.

Cerretani, L., Bendini, A., Rinaldi, M., Paciulli, M., Vecchio, S., and Chiavaro, E. DSC evaluation of extra virgin olive oil stability under accelerated oxidative test: Effect of fatty acid composition and phenol contents. *Journal of Oleo Science* 61 (2012): 303–309.

Cerretani, L., Maggio, R. M., Barnaba, C., Gallina-Toschi, T., and Chiavaro, E. Application of partial least square regression to differential scanning calorimetry data for fatty acid quantitation in olive oil. *Food Chemistry* 127 (2011): 1899–1904.

Chatziantoniou, S. E., Triantafillou, D. J., Karayannakidis, P. D., and Diamantopoulos, E. Traceability monitoring of Greek extra virgin olive oil by differential scanning calorimetry. *Thermochimica Acta* 576 (2014): 9–17.

Che Man, Y. B. and Tan, C. P. Comparative differential scanning calorimetric analysis of vegetable oils: II. Effects of cooling rate variation. *Phytochemical Analysis* 13 (2002): 142–151.

Chiavaro, E. Crystal polymorph structure determined for extra virgin olive oil. *European Journal of Lipid Science and Technology* 115 (2013): 267–269.

Chiavaro, E., Barnaba, C., Vittadini, E., Rodriguez-Estrada, M. T., Cerretani, L., and Bendini, A. Microwave heating of different commercial categories of olive oil: Part II. Effect on thermal properties. *Food Chemistry* 115 (2009a): 1393–1400.

Chiavaro, E., Cerretani, L., Bendini, A., Rinaldi, M., and Lercker, G. DSC characterization of different products by olive oil processing. *Rivista Italiana delle Sostanze Grasse* 88 (2011a): 182–190.

Chiavaro, E., Cerretani, L., Di Matteo, A., Barnaba, C., Bendini, A., and Iacumin, P. Application of a multidisciplinary approach for the evaluation of traceability of extra virgin olive oil. *European Journal of Lipid Science and Technology* 113 (2011b): 1509–1519.

Chiavaro, E., Cerretani, L., Paradiso, V. M., et al. Thermal and chemical evaluation of naturally auto-oxidized virgin olive oils: A correlation study. *Journal of the Science of Food and Agriculture* 93 (2013): 2909–2916.

Chiavaro, E., Mahesar, S., Bendini, A., Foroni, E., Valli, E., and Cerretani, L. DSC evaluation of olive oil during an accelerated oxidation. *Italian Journal of Food Science* 23 (2011c): 164–172.

Chiavaro, E., Rodriguez-Estrada, M. T., Barnaba, C., Vittadini, E., Cerretani, L., and Bendini, A. Differential scanning calorimetry: A potential tool for discrimination of olive oil commercial categories. *Analytica Chimica Acta* 625 (2008a): 215–226.

Chiavaro, E., Rodriguez-Estrada, M. T., Bendini, A., and Cerretani, L. Correlation between thermal properties and chemical composition of Italian virgin olive oils. *European Journal of Lipid Science and Technology* 112 (2010): 580–592.

Chiavaro, E., Rodriguez-Estrada, M. T., Bendini, A., Rinaldi, M., and Cerretani, L. DSC thermal properties and oxidative stability indices of microwave heated extra virgin olive oils. *Journal of the Science of Food and Agriculture* 91 (2011d): 198–206.

Chiavaro, E., Vittadini, E., Rodriguez-Estrada, M. T., Cerretani, L., and Bendini, A. Differential scanning calorimeter application to the detection of refined hazelnut oil in extra virgin olive oil. *Food Chemistry* 110 (2008b): 248–256.

Chiavaro, E., Vittadini, E., Rodriguez-Estrada, M. T., Cerretani, L., Bonoli, M., and Bendini, A. Monovarietal extra virgin olive oils. Correlation between thermal properties and chemical composition: Heating thermograms. *Journal of Agricultural and Food Chemistry* 56 (2008c): 496–501.

Chiavaro, E., Vittadini, E., Rodriguez-Estrada, M. T., et al. Monovarietal extra virgin olive oils: Correlation between thermal properties and chemical composition. *Journal of Agricultural and Food Chemistry* 55 (2007): 10779–10786.

Chiavaro, E., Vittadini, E., Rodriguez-Estrada, M. T., Cerretani, L., Capelli, L., and Bendini, A. Differential scanning calorimetry detection of high oleic sunflower oil as an adulterant in extra-virgin olive oil. *Journal of Food Lipids* 16 (2009b): 227–244.

Codex Alimentarius. Codex standard for named vegetable oils Codex Stan 210–1999. Adopted 1999. Revisions 2009. Amendments 2013.

EEC Reg. 2568/91 European Communities (EC). 1991. Official Journal of the Commission of the European Communities. Regulation n. 2568/91, L248, 5 September 1991.

EU Reg. 1348/2013 of 16 December 2013 amending Regulation (EEC) No 2568/91 on the characteristics of olive oil and olive-residue oil and on the relevant methods of analysis, Official Journal of the European Union, L 338/31.

EU Reg. 29/2012 of 13 January 2012 on marketing standards for olive oil, Official Journal of the European Union, L 12/14.

Ferrari, C., Angiuli, M., Tombari, E., Righetti, M. C., Matteoli, E., and Salvetti, G. Promoting calorimetry for olive oil authentication. *Thermochimica Acta* 459 (2007): 58–63.

Frankel, E. Chemistry of extra virgin olive oil: Adulteration, oxidative stability, and antioxidants. *Journal of Agricultural and Food Chemistry* 58 (2010): 5991–6006.

Frega, N., Bocci, F., and Lercker, G. Free fatty acids and diacylglycerols as quality parameters for extra virgin olive oil. *Rivista Italiana delle Sostanze Grasse* 70 (1993): 153–156.

Gallina Toschi, T., Bendini, A., Lozano-Sánchez, J., Segura-Carretero, A., and Conte, L. Misdescription of edible oils: Flowcharts of analytical choices in a forensic view. *European Journal of Lipid Science and Technology* 115 (2013): 1205–1223.

Giuffrida, F., Destaillats, F., Egart, M. H., et al. Activity and thermal stability of antioxidants by differential scanning calorimetry and electron spin resonance spectroscopy. *Food Chemistry* 101 (2007): 1108–1114.

Ilyasoglu, H. and Ozcelik, B. Determination of seasonal changes in olive oil by using differential scanning calorimetry heating thermograms. *Journal of the American Oil Chemists' Society* 88 (2011): 907–913.

Jafari, M., Kadivar, M., and Keramat, J. Detection of adulteration in Iranian olive oils using instrumental (GC, NMR, DSC) methods. *Journal of the American Oil Chemists' Society* 86 (2009): 103–110.

Jiménez Márquez, A. Preliminary results on the characterization of mixtures of olive oil by differential scanning calorimetry. *Ciencia y Tecnologia Alimentaria* 4 (2003): 47–54.

Jiménez Márquez, A. and Beltrán Maza, G. Application of differential scanning calorimetry (DSC) at the characterization of the virgin olive oil. *Grasas y Aceites* 54 (2003): 403–409.

Jiménez Márquez, A., Beltrán Maza, G., Aguilera Herrera, M. P., and Uceda Ojeda, M. Differential scanning calorimetry. Influence of virgin olive oil composition on its thermal profile. *Grasas y Aceites* 58 (2007): 122–129.

Kaiserberger, E. DSC investigations of the thermal characterization of edible fats and oils. *Thermochimica Acta* 151 (1989): 83–90.

Kanavouras, A. and Selke, S. Evolution of thermograph parameters during the oxidation of extra virgin olive oil. *European Journal of Lipid Science and Technology* 106 (2004): 359–368.

Kavallari, A., Maas, S., and Schmitz, M. Examining the determinants of olive oil demand in nonproducing countries: Evidence from Germany and the UK. *Journal of Food Products Marketing* 17 (2010): 355–372.

Kotti, F., Chiavaro, E., Cerretani, L., Barnaba, C., Gargouri, M., and Bendini, A. Chemical and thermal characterization of Tunisian extra virgin olive oil from Chetoui and Chemlali cultivars and different geographical origin. *European Food Research and Technology* 228 (2009): 735–742.

Laddomada, B., Colella G., Tufariello, M., et al. Application of a simplified calorimetric assay for the evaluation of extra virgin olive oil quality. *Food Research International* 54 (2013): 2062–2068.

Lercker, G., Caramia, G., Bendini, A., and Cerretani, L. Health effects, antioxidant activity and sensory properties of VOO. In A. Farooqui and T. Farooqui (eds), *Phytochemicals and Human Health: Pharmacological and Molecular Aspects*, Chapter IX, pp. 241–255. New York: Nova Science (2011).

Maggio, R. M., Barnaba, C., Cerretani, L., Paciulli, M., and Chiavaro, E. Study of the influence of triacylglycerol composition on DSC cooling curves of extra virgin olive oil by chemometric data processing. *Journal of Thermal Analysis and Calorimetry* 115 (2014): 2037–2044.

Maggio, R. M., Cerretani, L., Barnaba, C., and Chiavaro, E. Application of differential scanning calorimetry-chemometric coupled procedure to the evaluation of thermo-oxidation on extra virgin olive oil. *Food Biophysics* 7 (2012): 114–123.

Mancebo-Campos, V., Fregapane, G., and Salvador, D. M. Kinetic study for the development of an accelerated oxidative stability test to estimate virgin olive oil potential shelf life. *European Journal of Lipid Science and Technology* 110 (2008): 969–976.

Montedoro, G. F., Servili, M., Baldioli, M., and Miniati, E. Simple and hydrolyzable phenolic compounds in virgin olive oil. 1. Their extraction, separation, and quantitative and semiquantitative evaluation by HPLC. *Journal of Agricultural and Food Chemistry* 40 (1992): 1571–1576.

Ostrowska-Ligeza, E., Bekas, W., Kowalska, D., Lobacz, M., Wroniak, M., and Kowalski, B. Kinetics of commercial olive oil oxidation: Dynamic differential scanning calorimetry and Rancimat studies. *European Journal of Lipid Science and Technology* 112 (2010): 268–274.

Rovellini, P. and Cortesi, N. Oxidative status of extra virgin olive oils: HPLC evaluation. *Italian Journal of Food Science* 16 (2004): 335–344.

Spyros, A., Philippidis, A. M., and Dais, B. Kinetics of diglyceride formation and isomerization in virgin olive oils by employing ^{31}P NMR spectroscopy. Formulation of a quantitative measure to assess olive oil storage history. *Journal of Agricultural and Food Chemistry* 52 (2004): 157–164.

Tan, C. P. and Che Man, Y. B. Differential scanning calorimetric analysis of edible oils: Comparison of thermal properties and chemical composition. *Journal of the American Oil Chemists' Society* 77 (2000): 142–155.

Tan, C. P. and Che Man, Y. B. Comparative differential scanning calorimetric analysis of vegetable oils: I. Effects of heating rate variation. *Phytochemical Analysis* 13 (2002): 129–141.

Tan, C. P., Che Man, Y. B., Selamat, J., and Yusoff, M. S. A. Comparative studies of oxidative stability of edible oils by differential scanning calorimetry and oxidative stability index methods. *Food Chemistry* 76 (2002): 385–389.

Torrecilla, J. S., García, J., and García, S. F. Quantification of adulterant agents in extra virgin olive oil by models based on its thermophysical properties. *Journal of Food Engineering* 103 (2011): 211–218.

Vecchio, S., Cerretani, L., Bendini, A., and Chiavaro, E. Thermal decomposition study of monovarietal extra virgin olive oil by simultaneous TG/DSC: Relation with chemical composition. *Journal of Agricultural and Food Chemistry* 57 (2009): 4793–4800.

Vittadini, E., Lee, J. H., Frega, N. G., Min, D. B., and Vodovotz, Y. DSC determination of thermally oxidized olive oil. *Journal of the American Oil Chemists' Society* 80 (2003): 533–537.

Section II

Application of DSC in Oil and Fat Technology: Coupling with Other Thermal and Physical Approaches

5 Application of Thermogravimetric Analysis in the Field of Oils and Fats

Stefano Vecchio Ciprioti

CONTENTS

5.1 INTRODUCTION

Thermal analysis (TA) represents a family of techniques in which a given physical property of the sample (in the condensed phase) is recorded as a function of time, temperature, or both under a controlled static or flowing inert or reactive gas atmosphere. Thermogravimetry (TG), differential scanning calorimetry (DSC), and differential thermal analysis (DTA) are three of the most representative and most commonly used TA techniques. Mass, enthalpy, and temperature (actually the difference between the sample and a suitable reference) are the physical properties measured during a TG, DSC, or DTA experiment, respectively, mainly under nonisothermal conditions at a constant heating rate.

In order to better reveal the initial and final temperatures corresponding to each process accompanied by a mass loss recorded upon heating, very often TG data are processed to obtain the first-order derivative with respect to time (DTG).

All these three techniques have been extensively applied in the past to investigate foodstuffs or processes occurring in these very complicated systems upon heating under nonisothermal conditions (only in some cases at constant temperature).

From the literature concerning the application of TA to food systems (with particular reference to oils and fats) and related processes, it is evident that DSC (and DTA to a lesser extent) is largely applied to examine their thermal behavior by observing any change in enthalpy and related heat capacity due to physical or chemical processes (dehydration, oxidation, thermal decomposition, crystallization, glass transitions, etc.). TG is mainly applied to providing quantitative data on the temperature ranges in which all given processes occurred and assessing the relative stability of related materials on the basis of the onset decomposition temperatures.

This second aim is sometimes also achieved using a kinetic approach. In the last 30 years, in almost all fields of application of TA, many authors have adopted the procedure of processing the TG data recorded for a given step of mass loss using many different algorithms and regression procedures. Their efforts were focused mainly on obtaining reasonable parameters that could assume the physical meaning of kinetic parameters if the authors strictly followed the recommendations of the Kinetic Committee of the International Confederation for Thermal Analysis and Calorimetry (ICTAC) published recently (Vyazovkin et al. 2011).

TA techniques are very powerful tools for characterizing oils and fats, which may degrade if not properly stored or processed, oxidation being the principal decomposition reaction.

Any form of lipid that is obtained from animals can be denoted as animal oil (AO), while those extracted from plants in the liquid form at room temperature can be classified as vegetable oil (VO). By contrast, a lipid present in its solid form at room temperature will be denoted as a vegetable fat. Animal oils and fats are primarily made up of saturated fats (or saturated triglycerides), whereas vegetable oils are made up of unsaturated fats (polyunsaturated or monounsaturated).

From a chemical point of view, vegetable oils and fats, usually called triglycerides, are triesters of glycerol, the ester groups commonly consisting of fatty acid molecules from 16 to 18 carbon atoms in length.

There are many different types of vegetable oils, derived from different plants (palm, sesame, olive, almond, mustard, lemon, coconut, cottonseed, etc.). A few examples of the most commonly used vegetable oils (olive oil, coconut oil, and almond oil) have been used since the dawn of civilization for a variety of purposes, with the earliest documentation of such use dating back as far as 4000 years ago. Some of these VOs are contained in many products used every day, such as perfumes, paints, and shampoos, while others, such as lubricants, are usually purely comprised of inedible vegetable oils with some additives.

Essential oils, commonly known as aromatherapy oils, are aromatic compounds that are extracted from parts of plants including, but not limited to, roots, stems, seeds, flowers, and even leaves.

More recently, the use of VOs as biofuels is becoming increasingly important, since our supply of fossil fuels is decreasing dramatically. The most important advantage is that they are usually cheaper than conventional fossil fuels but supply almost as much energy. Unlike fossil fuels, biofuels can be mass-produced and will never run out as long as we continue to grow the plants from which they are derived. They make engines more efficient and help them run more quietly. Most importantly, they help reduce the amount of pollutants and greenhouse gases commonly released by conventional fuels.

Frying is one of the most commonly used methods of food preparation, in the home as well as in industry, and the prolonged use of oil for this purpose causes changes in its physical or chemical properties or both (Moretto and Fett 1998).

Several methods have been proposed to evaluate the quality of edible oils (Felsner and Matos 1998). Most of these methods submit a sample to conditions that accelerate the normal oxidation process. The older methods used to estimate the quality of vegetable oils induced deterioration by maintaining the sample at elevated temperatures in contact with air; the sample was then periodically weighed to verify the gain in mass or examined organoleptically for rancidity. Tests frequently employed to predict the quality of vegetable oils include the periodic determination of peroxide index, the active oxygen method (AOM), the automated AOM method, and the oxygen pump method (Lee and Min 1990).

More recently, TA methods such as DSC and TG/DTG have received considerable attention. TG/DTG curves can be used to estimate the quality of edible oil by determining the kinetic parameters (Santos et al. 2001, 2002) and the induction period for oxidation.

Many medicines and foods contain fatty acids and are often submitted to thermal treatment during processing or storage. Thus, knowledge of the thermal stability of these oils or fats obtained from TG experiments is very important from an industrial point of view (Dweck and Sampaio 2004).

This chapter deals with the development of thermogravimetric analysis (TGA) to characterize oils and fats. In order to achieve this purpose, two major objectives have to be fulfilled: (i) to establish initial and final temperatures of all steps of mass loss occurring upon heating and (ii) to assess a thermal stability scale among related materials on the basis of a kinetic analysis of TG data. Taking into account all the previous considerations on results of the TG experiments and the great variety of oils and fats that can be considered (edible and inedible VOs, AOs and VOs, bio-oils transformed into fuels, etc.), Section 5.2 will deal with TG application to edible oils, bio-oils, and fats. Section 5.3 then briefly outlines the most commonly adopted kinetic procedures and how TG data are processed to obtain reliable kinetic parameters that may (or may not) have a physical meaning and a reasonable interpretation in terms of stability. Section 5.4 discusses the kinetic results related to pyrolysis, combustion, decomposition, thermal oxidation, or other processes at relatively high temperatures for all classes of oils and fats, for stability purposes or for lifetime prediction at temperatures close to room temperature.

5.2 APPLICATION OF TG FOR CHARACTERIZATION OF EDIBLE OILS, BIO-OILS, AND FATS

Edible oils play a major role in the culinary industry. Their most common use is to heat and cook food, but they are also regularly used to add flavor or texture to the food. Seed oils such as sesame oil are commonly used for these purposes.

These oils have required the development of new analytical methods to evaluate their processing and storage conditions (Felsner and Matos 1998). More recently, a review has focused on the most important studies dealing with olive oil heating, under laboratory simulation or, preferably, real cooking conditions, while attempting to understand adequate processing conditions to preserve its quality and to reduce hazard formation (Santos et al. 2013). Moreover, current work identifies those issues that require further research to reinforce the correct use of olive oil and to potentiate its health benefits (Santos et al. 2013). Due to the high temperature and prolonged time of use with repeated frying, the oils are progressively degraded by a complex series of chemical reactions including oxidation, hydrolysis, and polymerization. These reactions, however, are not equivalent for all vegetable oils, and there is particular concern regarding olive oil, since its bioactive attributes might be lost during this process, despite being highly resistant to thermal oxidation. Oil oxidation occurs through a free radical mechanism, initially characterized by the emergence of a sweetish and unpleasant odor, which becomes progressively worse until the decomposition of hydroxides and peroxides in low-molecular weight acids and aldehydes occurs, resulting in the characteristic smell of rancid fat (Gennaro et al. 1998).

Many methodologies have been proposed to evaluate the quality of commercial extra-virgin olive and peanut oils through an artificial isothermal rancidification process, following the trends of properties more classical than those measured by TA methods, such as antioxidant capacity and total polyphenol concentration (Amati et al. 2008; Tomassetti et al. 2013a). In these studies a sample of the oil or of its components is submitted to conditions that accelerate the normal oxidation process (Campanella et al. 2008a,b).

The stability of virgin olive oil to autoxidation is mainly due to naturally occurring phenolic compounds, but contrasting data have been published on the effectiveness of the same antioxidant compounds. To solve this problem, Gennaro et al. (1998) estimated by TGA the oil's resistance to oxidation by measuring the mass gain percentage due to the reaction of sample with oxygen during oxidation, and initial and final oxidation temperatures (Figure 5.1).

The line AB indicates loss of solvent, if present, when the sample is held isothermally at 70°C, while the point B corresponds to the beginning of the temperature program. The line from B to T_i represents the static condition of oil resistance to oxidation when temperature is increased. T_i is the *initial temperature*, at which the rate of oxidation increases rapidly, as shown by a mass gain. T_f is the *final temperature*, at which the maximum mass gain is recorded. Actually, thermoanalytical methods such as DSC and, to a lesser extent, TG have received considerable attention in the past with the aim of determining the optimum conditions for examination of oxidative stability and oxidation state of edible oils (Buzás and Kurucz 1979), oil shales (Barkia

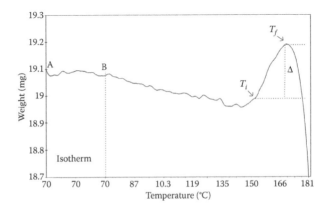

FIGURE 5.1 Typical TG curve for an olive oil sample. See text for detailed description.

et al. 2003), linseed oils (Turri et al. 2001), and food systems (Roos 2003). These methods show many advantages with respect to conventional methods because they provide higher precision and sensitivity as well as using a smaller amount of sample, and the results are obtained more quickly (Souza et al. 2004). Three different almond cultivars (Spanish Guara, Marcona, and Butte from the United States) were characterized using attenuated total reflectance Fourier transform infrared spectroscopy (ATR-FTIR) and thermal analysis techniques (DSC and TG) (García et al. 2013). TG curves showed that samples were stable up to around 220°C. Different stages of degradation were observed with increasing temperature, corresponding to the degradation of the complex matrix of the samples. The obtained results proved the suitability of the studied analytical techniques combined with linear discriminant analysis (LDA) for easy and rapid discrimination among different almond cultivars in food processing.

TG is a powerful technique, able to process the raw data derived from multiple heating rate conditions to obtain several physicochemical properties related to the behavior of the sample regarding oxidation and stability. Vecchio et al. (2008) carried out a thermal decomposition study of four different commercial extra-virgin olive oils (EVOOs) and four unsaturated or saturated esterified C18 fatty acids with glycerol (i.e., glyceryl tristearate (C18:0), trioleate (C18:1), trilinoleate (C18:2), and trilinolenate (C18:3)), whose structural details are summarized in Table 5.1, under the same experimental conditions. Thermal stability of oils should depend strictly on their chemical structure, oils with a high content of unsaturated fatty acids being less stable than the saturated ones (Dweck et al. 2004; Garcia et al. 2007). The results of the kinetic analysis of both EVOOs and unsaturated or saturated esterified C18 fatty acids reported in (Vecchio et al. 2008) will be discussed in Section 5.4.

The thermal decomposition of 12 monovarietal EVOOs from different geographical origins (eight from Italy, two from Spain, and the others from Tunisia) was evaluated by simultaneous TG and DSC analyses (Vecchio et al. 2009).

The TG and DTG curves of three samples (one from Italy (It1), one from Spain (Sp1), and one from Tunisia (Tu1)) are shown in Figure 5.2. The absence of any step of mass loss recorded up to 473 K reveals the limited water content of these samples, which would be detectable only by higher-sensitivity thermal analysis equipment, as expected in vegetable oil (Cerretani et al. 2008). The thermal oxidative decomposition processes occurred in all samples as several (mainly three or four) consecutive

---

TABLE 5.1
Triglycerides Considered in This Study and Their Contents in EVOO

Compound	Structural Characteristics No. of C Atoms and Insaturations	%$_{m/m}$ in EVOO (Range)
Glyceryl tristearate	C18:0	0.5–5.0
Glyceryl trioleate	C18:1	55.0–83.0
Glyceryl trilinoleate	C18:2	3.5–21.0
Glyceryl trilinolenate	C18:3	0.0–0.9

[a] Expressed in ranges of percent by mass.

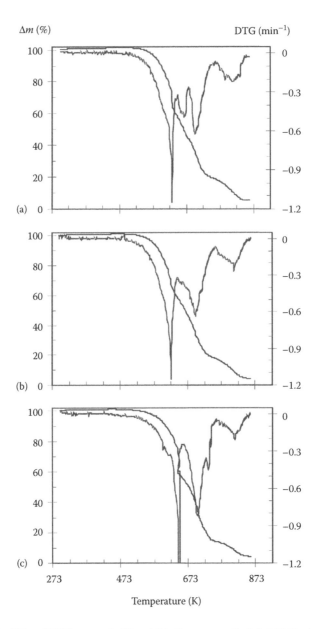

FIGURE 5.2 TG and DTG curves (solid and thin lines, respectively) at 10 K min^{-1}: (a) sample It1, (b) sample Sp1, and (c) sample Tu1.

and simultaneous steps of mass loss in the temperature range between 473 and 873 K. The first decomposition step, which occurred in the temperature range 473–640 K for all of the tested samples, showed a complex multistep decomposition pattern that exhibited quite a different profile among samples. In addition, the authors found that onset temperatures of the thermal decomposition exhibited statistically significant correlations with chemical components of the samples, in particular palmitic and oleic acids and related triacylglycerols (Vecchio et al. 2009).

Deterioration of sunflower oil under frying conditions was studied with and without antioxidants in an atmosphere of nitrogen and air, respectively, and the kinetic parameters (activation energy, frequency factor, and order of reaction) were determined using TG/DTG and DSC to assess stability with a view to better predicting the time/temperature storage conditions (Santos et al. 2004). According to the initial decomposition temperature recorded by TG, the following stability order was observed: corn (A)>corn > sunflower (A)>rice > soybean > rapeseed (A)>olive > rapeseed > sunflower ((A) denotes oils that contain artificial antioxidants), while a different order was evaluated according to kinetic analysis of the first decomposition step. The results of the kinetic analysis will be discussed in Section 5.4.

The thermal decomposition of commercial canola, sunflower, corn, olive, and soybean vegetable oils was studied by simultaneous TG/DTA equipment (Dweck et al. 2004).

Figure 5.3 shows the TG, DTG, and DTA curves of the olive oil. As can be seen from the TG curve, and more precisely from the DTG curve, the first step of decomposition begins at 200°C and ends at 350°C. The extrapolated onset temperature is 288°C and the first DTG and DTA peaks occur at 324°C and 329°C, respectively. The second mass-loss step occurs from 350°C to 410°C, also presenting an exothermal DTA peak. The third mass-loss step, which actually is composed of many sharp DTG and exothermal DTA peaks, occurs up to about 470°C and is followed by a final exothermal mass-loss step, which ends at 580°C, due to the burnout of the residual carbonaceous material of the previous steps.

As can be seen in Figure 5.4, corn oil presents a different behavior during its second thermal decomposition step, which occurs from 375°C to about 425°C, with a significantly higher DTG peak than in the same decomposition step of the other oils, probably because of its fatty acid composition. Thus, the first three steps present a more homogeneous mass-loss rate behavior than the other oils, as can be seen in the DTG curve. The last step, when the combustion of the carbonaceous residue occurs, has similar behavior to the other ones. On the basis of the decomposition

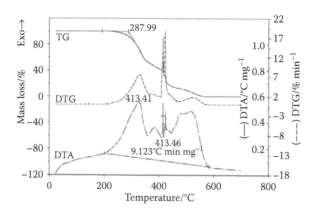

FIGURE 5.3   TG, DTG, and DTA curves of olive oil in air.

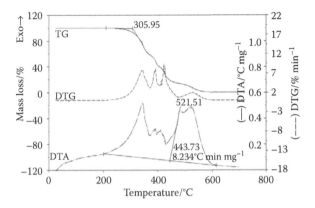

FIGURE 5.4   TG, DTG, and DTA curves of corn oil in air.

onset temperatures obtained from corresponding TG curves in air for all vegetable oils, corn oil presents the highest thermal stability, followed in decreasing stability order by sunflower, soybean, canola, and olive oils.

Recently, epoxidized cotton seed oil (ECSO) was conveniently synthesized from cotton seed oil (CSO) and then converted into carbonated cotton seed oil (CCSO) by reaction with CO_2 (Zhang et al. 2014a). The thermal and oxidation stabilities of CCSO were compared with those of CSO and ECSO by observing their TG curves (Figure 5.5).

It was found that the initial degradation temperature of CCSO is close to 350°C, which is higher than that of ECSO, suggesting that the carbonated oil possesses good thermal stability and is more suitable for high-temperature applications. CSO presented three thermal decomposition steps and a higher thermal decomposition temperature than ECSO due to the higher thermal stability of unsaturated C=C double bonds than of epoxy groups.

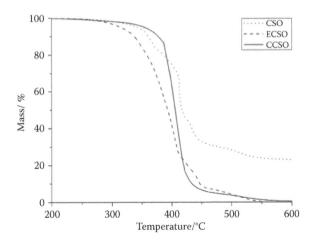

FIGURE 5.5   TG curves of cotton seed oil (CSO), epoxidized cotton seed oil (ECSO), and carbonated cotton seed oil (CCSO) at 10 K min^{-1}.

Several authors have studied pyrolysis of triglyceride materials. In particular, Maher and Bressler (2007) presented a review of these processes, comparing pyrolysis of vegetable oils with biomass pyrolysis.

Fried vegetable oils become waste cooking oils that can be transformed into liquid derivatives to produce energy or fuels by thermal decomposition (Maher and Bressler 2007). Waste cooking oils are an economical choice for biodiesel production, because of their availability and low cost (Kulkarni and Dalai 2006).

Thermal cracking of vegetable oils represents an alternative method of producing renewable biobased products for use in fuel and chemical applications. Biodiesel (fatty acid methyl ester) is a nontoxic and biodegradable alternative fuel that is obtained from renewable sources. A major hurdle in the commercialization of biodiesel from virgin oil, in comparison to petroleum-based diesel fuel, is its cost of manufacturing, primarily the raw material cost. Used cooking oil is one economical source for biodiesel production. However, the products formed during frying, such as free fatty acid and some polymerized triglycerides, can affect the transesterification reaction and the biodiesel properties. Apart from this phenomenon, biodiesel obtained from waste cooking oil gives better engine performance and lower emissions when tested on commercial diesel engines (Kulkarni and Dalai 2006).

In recent years, with increased demands on energy resources, oil shale has been looked at as an important alternative energy source for conventional oil resources. Kerogen, the main organic matter in oil shale, can be converted to shale oil through pyrolysis. Retorting is considered one of the most attractive approaches to extract oil from oil shale. Unfortunately, this technology is expensive and harmful to the environment (Wang et al. 2013). The pyrolysis kinetics of Huadian oil shale, spent oil shale obtained from near-critical water extraction experiments, and their mixtures were investigated using TGA, and it was found that thermal decomposition occurring in oil shale, spent oil shale, and their mixtures involved three degradation steps.

Soaps are an important class of surface-active compounds derived from natural oils and fats. Double decomposition reactions permit the synthesis of metallic soaps, which are long-chain carboxylates of metal ions, from alkaline soaps such as sodium, potassium, or ammonium soaps (Balkose et al. 2010). Many applications are related to the thermal properties of these compounds (driers in paints or inks, components of lubricating greases, heat stabilizers for plastics, especially PVC, catalysts and waterproofing agents, fuel additives, and cosmetic products) and the thermal behavior of metal soaps in terms of decomposition processes is of great importance. Some results of the kinetic analysis will be reported in Section 5.4.

Brazilian Cerrado is a unique biome of the midwest–center of the country and shows an enormous variety of oil-rich plant species. These plants can offer new perspectives for sustainable development of this region. They have high oleic acid content, making them useful for human consumption. For these reasons, the thermal stability of oils taken from Brazilian Cerrado plants (amburana [*Amburana cearensis* (Fr. Allem) A. C. Smith], baru [*Dypterix alata* Vog.], and pequi [*Caryocar brasiliense* Camb.] pulp) was studied in nitrogen and synthetic air using TG/DTG techniques (Garcia et al. 2007). The authors observed that the oils in the nitrogen atmosphere showed very similar thermal stability, probably due to their similar fatty acid composition. Due to the larger amount of short-chain fatty acids in pequi pulp oil, it has the lowest thermal stability in synthetic air. On the other hand, the reduction of production costs and finding a permanent oil source have remained the two main concerns for these green fuels. Production of biodiesel from acid oils (almost never categorized as edible oils) is one of the ways to reduce biodiesel production costs. Consequently, this new material has received increased interest in debates concerning the security of food, compared with other oils considered for biodiesel production (Javidialesaadi and Raeissi 2013). The use of thermoanalytical techniques makes it possible to evaluate the thermal stability of biodiesels, which is an important factor in quality control during processing, handling, and storage of biofuels.

The thermal stability of poultry fat and its ethyl (BEF) and methyl (BMF) biodiesels was determined using TG/DTG, DTA, and DSC, in different atmospheres (Ramalho et al. 2011). Poultry fat showed different behavior under synthetic air and nitrogen atmospheres, with combustion reactions and a volatilization/decomposition process, respectively. The transesterification reaction led to smaller chain lengths of the biodiesel and consequently to a lower thermal stability.

The oil content of an artificial polluted soil was estimated at different times during a biodegradation process by means of TG under nonisothermal conditions and in an oxidative atmosphere (Jurconi et al. 2007). The model pollutant of the soil was hexadecane, and a *Pseudomonas aeruginosa* culture was used for biodegradation. The results obtained seem to demonstrate that TG is a versatile method for quantitative determination of organic components in a soil sample. Therefore, this method is also recommended for evaluating the process of biodegradation of an oily soil.

The sunflower and soybean samples met the specifications of Resolution no. 7/2008 ANP, the National (Brazilian) Agency of Petroleum, while the castor biodiesel had values of density and kinematic viscosities beyond the required specifications, properties that limit its use as a fuel (Silva et al. 2011). Furthermore, from the TG curves it was observed that castor biodiesel has higher thermal stability than the others; this could be attributed to its high content of ricinoleic acid, showing the importance of this technique for this type of evaluation. These results suggest that castor biodiesel could be used as an antioxidant additive for other fuels to enhance thermal stability and reduce tendency to oxidation.

5.3 THEORETICAL BACKGROUND OF KINETIC PROCEDURES APPLIED TO THERMALLY STIMULATED PROCESSES OCCURRING IN OILS AND FATS

All steps of complex thermally activated processes (i.e., degradation, decomposition, pyrolysis, etc.) occurring in condensed-phase samples are commonly described by the explicit dependence of the reaction rate on both the absolute temperature and the extent of conversion α, according to the following general equation:

$$d\alpha/dt = k(T)f(\alpha) \tag{5.1}$$

where:

$d\alpha/dt$ is the reaction rate

$k(T)$ is the rate constant, whose temperature dependence usually assumes the form of the Arrhenius equation

According to this equation, Equation 5.1 can be rewritten in the following form:

$$d\alpha/dt = A\exp(-E/RT)f(\alpha) \tag{5.2}$$

where:

- A is the pre-exponential factor
- E is the activation energy
- $f(\alpha)$ is a function called reaction model, whose mathematical expressions, taken from the literature (Brown et al. 1980), are given in Table 5.2

The previous project by the ICTAC Kinetic Committee, focusing on a comparison of various computational methods to determine kinetic parameters, concluded (in 2000) that multiple heating rate (or temperature) programs are highly recommended to obtain reliable results (Brown et al. 2000). The latest recommendations of this committee (Vyazovkin et al. 2011) confirm the advantage of applying these methods over all those using single-heating rate experiments.

Since degradation, decomposition, or pyrolysis occurring in a material in the condensed phase during heating is usually accompanied by mass loss, the extent of conversion α is evaluated as a fraction of the total mass associated with this process. The most suitable technique to obtain these data is TG under nonisothermal (very often constant-heating rate (β)) conditions. As a consequence, the temperature dependence of the reaction rate $d\alpha/dt$ may assume the following form: $d\alpha/dt = (dT/dt)(d\alpha/dT)$. Equation 5.2 can be rewritten as follows:

$$d\alpha/f(\alpha) = (A/\beta)\exp(-E/RT)dT \tag{5.3}$$

TABLE 5.2
Most Commonly Used Kinetic Models for Kinetic Analysis of Thermally Stimulated Processes

	Reaction Model	Code	$f(\alpha)$	$g(\alpha) = \int d\alpha/f(\alpha)$
1	Power law	P4	$4\alpha^{3/4}$	$\alpha^{1/4}$
2	Power law	P3	$3\alpha^{2/3}$	$\alpha^{1/3}$
3	Power law	P2	$2\alpha^{1/2}$	$\alpha^{1/2}$
4	Power law	P2/3	$(2/3)\alpha^{-1/2}$	$\alpha^{3/2}$
5	One-dimensional diffusion	D1	$1/(2\alpha)$	α^2
6	Mampel (first-order)	F1	$1-\alpha$	$-\ln(1-\alpha)$
7	Avrami–Erofeev	A4	$4(1-\alpha)[-\ln(1-\alpha)]^{3/4}$	$[-\ln(1-\alpha)]^{1/4}$
8	Avrami–Erofeev	A3	$3(1-\alpha)[-\ln(1-\alpha)]^{2/3}$	$[-\ln(1-\alpha)]^{1/3}$
9	Avrami–Erofeev	A2	$2(1-\alpha)[-\ln(1-\alpha)]^{1/2}$	$[-\ln(1-\alpha)]^{1/2}$
10	Three-dimensional diffusion	D3	$(3/2)(1-\alpha)^{2/3}[1-(1-\alpha)^{1/3}]^{-1}$	$[1-(1-\alpha)^{1/3}]^2$
11	Contracting sphere	R3	$3(1-\alpha)^{2/3}$	$1-(1-\alpha)^{1/3}$
12	Contracting cylinder	R2	$2(1-\alpha)^{1/2}$	$1-(1-\alpha)^{1/2}$
13	Two-dimensional diffusion	D2	$[-\ln(1-\alpha)]^{-1}$	$(1-\alpha)\ln(1-\alpha)+\alpha$

Integrating both the left- and right-hand sides of Equation 5.3 yields:

$$\int_0^\alpha d\alpha/f(\alpha) = g(\alpha) = (A/\beta)\int_0^T \exp(-E/RT)dT \qquad (5.4)$$

The most important aim of kinetic analysis of thermally stimulated processes is to find a mathematical relationship among the reaction rate, the extent of conversion, and the temperature (Vyazovkin et al. 2011) by evaluating a set of three A, E, and $f(\alpha)$ (denoted the kinetic triplet) for each single-step process, according to Equations 5.2 and 5.4. It is commonly recognized that multiheating programs provide more reliable kinetic results than single-heating ones (Brown et al. 2000; Vyazovkin et al. 2011). Among the former methods, the isoconversional ones, which are based on the assumption that, at constant extent of conversion, reaction rate depends only on temperature ($f(\alpha)$ or $g(\alpha)$ is constant), seem to provide more reliable results, even if with different degrees of accuracy.

The temperature integral in Equation 5.4 has no exact analytical solution, but does have approximate solutions, which give rise to some of the most common isoconversional methods, in which the equations for each fixed extent of conversion usually assume the following general form:

$$\ln(\beta/T_\alpha^B) = \text{Const} - C(E_\alpha/R)(1/T_{\alpha,\beta}) \qquad (5.5)$$

where B and C are adjustable parameters, whose values depend on the approximation made. In particular, for the Ozawa–Flynn–Wall method (Flynn and Wall 1966; Ozawa 1965), based on the Doyle's approximation (Doyle 1962), $B = 0$ and $C = 1.052$. More accurate results can be obtained using the Kissinger–Akahira–Sunose method (Akahira and Sunose 1971), where $B = 2$ and $C = 1$, or the Starink method (Starink 2003) ($B = 1.92$ and $C = 1.008$). Each value of activation energy at each given extent of conversion is calculated from the slope of the regression line obtained by plotting the left-hand side of Equation 5.6 against the reciprocal temperature $T_{\alpha,\beta}^{-1}$. Vyazovkin developed a method that gives results with a better accuracy by numerical integration of the right-hand side of Equation 5.4 (Vyazovkin et al. 2011).

5.4   KINETIC ANALYSIS OF PROCESSES OCCURRING IN EDIBLE OILS, BIO-OILS AND FATS FOR STABILITY PURPOSES

Kinetic data regarding the oxidative process of triglycerides and extra-virgin olive oil are incomplete and are often contradictory (Dweck and Sampaio 2004; Santos et al. 2004). For this reason, Vecchio et al. (2008), after a deconvolution procedure applied only to the first two overlapping steps of EVOO and C18:1, determined the conversion dependency of activation energy for decomposition using the Ozawa–Flynn–Wall isoconversional method for the two deconvoluted steps of EVOO and C18:1, as well as for the only single step of the other three C18 triglycerides, and compared the results so obtained with those calculated using the Kissinger method, which provides a single (mean) value of activation energy for the whole process.

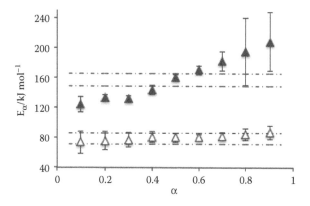

FIGURE 5.6 Conversion dependency of activation energy for the first and second steps of thermal degradation of EVOO (white and black triangles, respectively).

According to the Ozawa–Flynn–Wall (OFW) method, the authors found practically constant activation energy for the first deconvoluted step of EVOO (Figure 5.6) and C18:1 and for the single step of C18:0, in good agreement with the results obtained with the Kissinger method. A similar increasing trend was observed for the second decomposition step of EVOO and C18:1 and for the single steps of C18:2 and C18:3 triglycerides.

A thermal degradation study using TG/DTG was performed on three saturated fatty acids esterified with glycerol (i.e., glyceryl tristearate (C18), tripalmitate (C16), and trimyristate (C14)) as well as on glyceryl distearate and glyceryl monostearate at different heating rates (Tomassetti et al. 2013b). A deconvolution procedure applied to the first process, overlapping at least two steps between about 200°C and 350°C, enabled the activation energy of decomposition to be determined by both the Kissinger and the OFW isoconversional methods for the deconvoluted steps of the abovementioned fatty acids.

Figure 5.7 shows in graphical form the values of activation energy E obtained as a function of the extent of conversion α by applying the OFW method. It should be noted that, in the case of glyceryl tristearate, the values of E may be considered to be practically constant over the confidence interval in the case of both TG steps. Furthermore, the mean values (of about 59 ± 7 and 108 ± 20 kJ mol^{-1}) obtained in the case of the first and the second TG step, respectively, are found to be in reasonably good agreement with the corresponding values referring to the first two steps of the glyceryl tristearate obtained using the Kissinger method. Conversely, the E values obtained using the OFW method in the case of the thermal decomposition of both glyceryl distearate and glyceryl monostearate varied considerably with α. The authors concluded that this study proved to be useful both from a heuristic point of view and for the purpose of gaining greater insight into the thermooxidative events involving different types of glycerides, some of which are actually contained in edible oils and others directly linked to those present in food oils used daily for cooking (Tomassetti et al. 2013b).

According to the activation energy trend obtained by the Coats–Redfern method for the first thermal decomposition event that the authors ascribed to the thermal

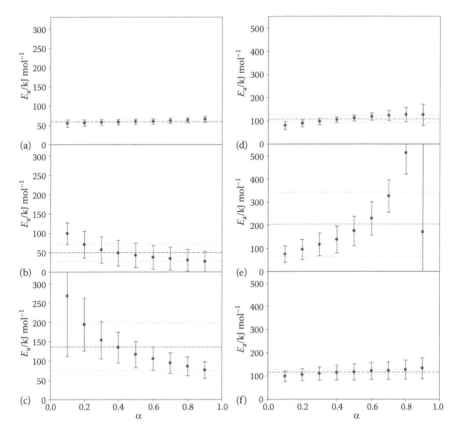

FIGURE 5.7 Conversion dependency of activation energy according to OFW method for the first (deconvoluted) peak relative to glyceryl (a) tri-, (b) di-, and (c) monostearate and for the second DTG deconvoluted peak relative to glyceryl (d) tri-, (e) di-, and (f) monostearate, respectively.

decomposition of polyunsaturated fatty acids, the following stability order is proposed: sunflower > corn > rice > soybean > rapeseed > olive (Santos et al. 2004). The authors concluded that thermoanalytical methods are interesting techniques, which permit the study of thermal stability and degradation caused by frying with a small amount of material using a kinetic approach. Results are obtained more quickly than by conventional techniques, and the presence of artificial antioxidants may also be detected.

More recently, the degradation properties and combustion performance of bio-oil compared with that of gasoline, diesel, motor oil, and fuel oils have been investigated using TGA (Ren et al. 2014). The results obtained in this study demonstrated that TG is a useful tool, able to discriminate between the different oils on the basis of the temperature ranges in which combustion occurs, as is evident from the values reported in Table 5.3. Pyrolysis kinetics and combustion index analysis were also performed to evaluate the thermal behavior and combustion performance of bio-oils compared with those of commercial petroleum oils (Ren et al. 2014). Aged bio-oil

TABLE 5.3

Comparison of Temperature Ranges for Combustion Process Occurring in Bio-Oil, Gasoline, Diesel, Motor Oil, and Fuel Oils Recorded by TG Experiments

	First Step		Second Step	
	$\Delta T_{range}/°C$	$T_{peak}/°C$	$\Delta T_{range}/°C$	$T_{peak}/°C$
No. 2 fuel oil	30–255	205		
Gasoline	25–110	65		
Raw bio-oil	20–250	148		
Motor oil	145–385	310		
No. 6 fuel oil	410–500	445	500–610	540
Diesel	30–260	208		

Source: Ren, X., Meng, J., Moore, A.M., Chang, J., Gou, J., and Park, S., *Bioresource Technology*, 152, 267–274, 2014.

was found to be more thermally unstable and generated a higher amount of carbonaceous solid. Raw bio-oil and bio-oil from torrefied wood were found to have combustion performance comparable with that of fuel oils.

A modified first-order kinetic equation with variable activation energy was considered to model the total mass loss of Ellajjun oil shale samples in the temperature range 350–550°C. Activation energy was allowed to vary as a function of oil shale conversion. The value of the activation energy increased from 98 to 120 kJ mol^{-1} (Al-Ayed et al. 2010).

Recently, TG data of oil shale obtained at the Masdar Institute (Waste to Energy laboratory) were studied to evaluate the kinetic parameters for El-Lujjun oil shale samples (Syed et al. 2011). The extent of char combustion was determined by relating TG data for pyrolysis and combustion with the ultimate analysis. Due to the distinct behavior of oil shale during pyrolysis, TG curves were divided into three separate events: moisture release; devolatilization; and evolution of fixed carbon/char. For each event, kinetic parameters, based on Arrhenius theory, were calculated. Three methods were used and compared: the integral method, the direct Arrhenius plot method, and the temperature integral approximation method.

The pyrolysis of triglyceride materials has been studied by several authors. Maher and Bressler (2007) presented a review of these processes, comparing pyrolysis of vegetable oils with biomass pyrolysis. Thermal cracking of vegetable oils represents an alternative method of producing renewable biobased products for use in fuel and chemical applications. A complete kinetic study has been carried out to determine the activation energy, reaction order, and preexponential factor, also discussing the vaporization process (Font and Rey 2013).

The pyrolysis of oil shale mixed with spent oil shale can provide fuel that can be processed into added-value chemicals. Moreover, the copyrolysis of oil shale mixed with spent oil shale contributes to solving the waste pollution problem. With these aims, it is worth studying the pyrolysis kinetics of oil shale, spent oil shale, and their mixtures. Results obtained in a recent study (Wang et al. 2013) indicated that the apparent activation energy of oil shale and spent oil shale decomposition decreased with increasing heating rates. The apparent activation energy of the decomposition of the mixtures (17.9–49.3 kJ mol^{-1}) was lower than that of oil shale (64.0 kJ mol^{-1}), and it decreased with the decreasing spent oil shale fraction in the mixture. The results proved that the spent oil shale was helpful for the pyrolysis of oil shale in the mixture.

Balkose et al. (2010) studied the thermal behavior of soaps (Ba, Ca, Cd, and Zn) of rubber seed oil for use as additives in the processing of poly(vinyl chloride) (PVC) using TG and DSC. The stability of the soaps was examined by TG up to 873 K at a constant heating rate of 10°C min^{-1}. The soaps were found to be thermally stable up to 473 K, as they recorded <5% mass loss at this temperature with values of apparent activation energy for decomposition varying from 52 to 96 kJ mol^{-1}.

Pyrolysis, combustion, and gasification behaviors of deoiled asphalt were studied using a TG unit under different atmospheric conditions: flowing N$_2$ (pyrolysis), air (combustion), or CO$_2$ (gasification) atmospheres. The kinetics was also analyzed using a multistage first-order integral model (Zhang et al. 2014b).

REFERENCES

Al-Ayed, O.S., Matouq, M., Anbar, Z., Khaleel, A.M., and Abu-Nameh, E. Oil shale pyrolysis kinetics and variable activation energy principle. *Applied Energy* 87 (2010): 1269–1272.

Akahira,T. and Sunose, T. Paper No. 246, 1969 research report, Trans. joint convention of four electrical institutes. Chiba Inst Technol (Sci. Technol.) vol. 16, pp. 22–31 (1971).

Amati, L., Campanella, L., Dragone, R., Nuccilli, A., Tomassetti, M., and Vecchio, S. New investigation of the isothermal oxidation of extra virgin olive oil: Determination of free radicals, total polyphenols, total antioxidant capacity, and kinetic data. *Journal of Agricultural and Food Chemistry* 56 (2008): 8287–8295.

Balkose, D., Egbuchunam, T.O., and Okieimen, F.E. Thermal behaviour of metal soaps from biodegradable rubber seed oil. *Journal of Thermal Analysis and Calorimetry* 101 (2010): 795–799.

Barkia, H., Belkbir, L., and Jayaweera, S.A.A. Oxidation kinetics of Timahdit and Tarfaya Moroccan oil shales. *Journal of Thermal Analysis and Calorimetry* 71 (2003): 97–106.

Brown, M.E., Dollimore, D., and Galwey, A.K. *Reactions in the Solid State. Comprehensive Chemical Kinetics*, vol. 22, Amsterdam: Elsevier (1980).

Brown, M.E., Maciejewski, M., and Vyazovkin, S. et al. Computational aspects of kinetic analysis: Part A: The ICTAC kinetics project-data, methods and results. *Thermochimica Acta* 355(1–2) (2000): 125–143.

Buzás, I. and Kurucz, É. Study of thermooxidative behavior of edible oils by thermal analysis. *Journal of the American Oil Chemists' Society* 56 (1979): 685–688.

Campanella, L., Nuccilli, A., Tomassetti, M., and Vecchio, S. Biosensor analysis for the kinetic study of polyphenols deterioration during the forced thermal oxidation of extra-virgin olive oil. *Talanta* 74 (2008a): 1287–1298.

Campanella, L., Nuccilli, A., Tomassetti, M., and Vecchio, S. Studio cinetico mediante termoanalisi della degradazione termossidativa e della resistenza all'ossidazione di alcuni dei principali trigliceridi e dell'olio extravergine di oliva. *Rivista Italiana delle Sostanze Grasse* 85 (2008b): 26–34.

Cerretani, L., Bendini, A., Barbieri, S., and Lercker, G. Preliminary observations on the change of some chemical characteristics of virgin olive oils subjected to a "soft deodorization" process. *Rivista Italiana delle Sostanze Grasse* 85 (2008): 75–82.

Doyle, C.D. Estimating isothermal life from thermogravimetric data. *Journal of Applied Polymer Science* 6(24) (1962): 639–642.

Dweck, J. and Sampaio, C.M.S. Analysis of the thermal decomposition of commercial vegetable oils in air by simultaneous TG/DTA. *Journal of Thermal Analysis and Calorimetry* 75 (2004): 385–391.

Felsner, M.L. and Matos, J.R. Análise da Estabilidade Térmica e Temperatura de Oxidação de Óleos Comestíveis Comerciais por Termogravimetria. *Anais da Associação Brasileira de Química* 47 (1998): 308–312.

Flynn, J.H. and Wall, L.A. A quick direct method for the determination of activation energy from thermogravimetric data. *Journal of Polymer Science B: Polymer Letters* 4 (1966): 323–328.

Font, R. and Rey, M.D. Kinetics of olive oil pyrolysis. *Journal of Analytical Applied Pyrolysis* 103 (2013): 181–188.

García, A.V., Beltrán, S.A., and Garrigós Selva, M.C. Characterization and classification of almond cultivars by using spectroscopic and thermal techniques. *Journal of Food Science* 78 (2013): C138–C144.

Garcia, C.C., Franco, P.I.B.M., Zuppa, T.O., Antoniosi Filho, N.R., and Leles, M.I.G. Thermal stability studies of some cerrado plant oils. *Journal of Thermal Analysis and Calorimetry* 87 (2007): 645–648.

Gennaro, L., Piccioli Bocca, A., Modesti, D., Masella, R., and Coni, E. Effect of biophenols on olive oil stability evaluated by thermogravimetric analysis. *Journal of Agricultural and Food Chemistry* 46 (1998): 4465–4469.

Javidialesaadi, A. and Raeissi, S. Biodiesel production from high free fatty acid-content oils: Experimental investigation of the pretreatment step. *APCBEE Procedia* 5 (2013): 474–478.

Jurconi, B., Feher, L., Doca, N. et al. Evaluation of oily soil biodegradability by means of thermoanalytical methods. *Journal of Thermal Analysis and Calorimetry* 88 (2007): 373–375.

Kulkarni, M.G. and Dalai, A.K. Waste cooking oils an economical source for biodiesel: A review. *Industrial Engineering Chemical Research* 45 (2006): 2901–2913.

Lee, S. and Min, D.B. Effects, quenching, mechanisms and kinetics of carotenoids in chlorophyll-sensitized photooxidation of soybean oil. *Journal of Agricultural and Food Chemistry* 38 (1990): 1630–1634.

Maher, K.D. and Bressler, D.C. Pyrolysis of triglyceride materials for the production of renewable fuels and chemicals. *Bioresource Technology* 98 (2007): 2351–2368.

Moretto, E. and Fett, R. *Tecnologia de Óleos e Gorduras na Indústria de Alimentos.* São Paulo: Varela Editora e Livraria Ltda (1998).

Ozawa, T. A new method of analyzing thermogravimetric data. *Bulletin of Chemical Society of Japan* 38 (1965): 1881–1886.

Ramalho, E.F.S.M., Santos, I.M.G., Maia, A.S., Souza, A.L., and Souza, A.G. Thermal characterization of the poultry fat biodiesel. *Journal of Thermal Analysis and Calorimetry* 106 (2011): 825–829.

Ren, X., Meng, J., Moore, A.M., Chang, J., Gou, J., and Park, S. Thermogravimetric investigation on the degradation properties and combustion performance of bio-oils. *Bioresource Technology* 152 (2014): 267–274.

Roos, J.H. Thermal analysis, state transitions and food quality. *Journal of Thermal Analysis and Calorimetry* 71 (2003): 197–203.

Santos, C.S.P., Cruz, R., Cunha, S.C., and Casal, S. Effect of cooking on olive oil quality attributes. *Food Research International* 54 (2013): 2016–2024.

Santos, J.C.O., Santos, A.V., and Souza, A.G. Thermal analysis in quality control of the olive oil. *European Journal of Pharmaceutical Science* 13(Suppl. 1) (2001): S23–S24.

Santos, J.C.O., Santos, A.V., Souza, A.G., Prasad, S., and Santos, I.M.G. Thermal stability and kinetic study on thermal decomposition of commercial edible oils by thermogravimetry. *Journal of Food Science* 67 (2002): 1393–1398.

Santos, J.C.O., Santos, I.M.G., Conceicão, M.M., Porto, S.L., Trindade, M.F.S., Souza, A.G., Prasad, S., Fernandez, V.J. Jr., and Araújo, A.S. Thermoanalytical, kinetic and rheological parameters of commercial edible oils. *Journal of Thermal Analysis and Calorimetry* 75 (2004): 419–428.

Silva, H.K.T.A., Chellappa, T., Carvalho, F.C. et al. Thermal stability evaluation of methylic biodiesel obtained for different oilseeds. *Journal of Thermal Analysis and Calorimetry* 106 (2011): 731–733.

Starink, M.J. The determination of activation energy from linear heating rate experiments: A comparison of the accuracy of isoconversion methods. *Thermochimica Acta* 404 (2003): 163–176.

Syed, S., Qudaih, R., Talab, I., and Janajreh, I. Kinetics of pyrolysis and combustion of oil shale sample from thermogravimetric data. *Fuel* 90 (2011): 1631–1637.

Tomassetti, M., Favero, G., and Campanella, L. Kinetic thermal analytical study of saturated mono-, di- and tri-glycerides. *Journal of Thermal Analysis and Calorimetry* 112(1) (2013a): 519–527.

Tomassetti, M., Vecchio, S., Campanella, L., and Dragone, R. Biosensors for monitoring the isothermal breakdown kinetics of peanut oil heated at 180°C. Comparison with results obtained for extra virgin olive oil. *Food Chemistry* 140 (2013b): 700–710.

Turri, B., Vicini, S., Margutti, S., and Pedemonte, E. Calorimetric analysis of the polymerisation process of linseed oil. *Journal of Thermal Analysis and Calorimetry* 66 (2001): 343–348.

Vecchio, S., Campanella, L., Nuccilli, A., and Tomassetti, M. Kinetic study of thermal breakdown of triglycerides contained in extra-virgin olive oil. *Journal of Thermal Analysis and Calorimetry* 91(2008): 51–56.

Vecchio, S., Cerretani, L., Bendini, A., and Chiavaro, E. Thermal decomposition study of monovarietal extra virgin olive oil by simultaneous thermogravimetry/differential scanning calorimetry: Relation with chemical composition. *Journal of Agricultural and Food Chemistry* 57 (2009): 4793–4800.

Vyazovkin, S., Burnham, A.K., Criado, J.M., Pérez-Maqueda, L.A., Popescu, C., and Sbirrazzuoli, N. ICTAC kinetics committee recommendations for performing kinetic computations on thermal analysis data. *Thermochimica Acta* 520 (2011): 1–19.

Wang, Z., Deng, S., Gu, Q., Zhang, Y., Cui, X., and Wang, H. Pyrolysis kinetic study of Huadian oil shale, spent oil shale and their mixtures by thermogravimetric analysis. *Fuel Processing Technology* 110 (2013): 103–108.

Zhang, L., Luo, Y., Hou, Z., He, Z., and Eli, W. Synthesis of carbonated cotton seed oil and its application as lubricating base oil. *Journal of American Oil Chemists' Society* 91 (2014a): 143–150.

Zhang, Q., Li, Q., Zhang, L., Fang, Y., and Wang, Z. Experimental and kinetic investigation of the pyrolysis, combustion, and gasification of deoiled asphalt. *Journal of Thermal Analysis and Calorimetry* 115 (2014b): 1929–1938.

6 Application of DSC-XRD Coupled Techniques for the Evaluation of Phase Transition in Oils and Fats and Related Polymorphic Forms

Sonia Calligaris, Luisa Barba, Gianmichele Arrighetti, and Maria Cristina Nicoli

CONTENTS

6.1 INTRODUCTION

The phase transition behavior of food lipids has been widely studied due to their multifunctional properties. They are irreplaceable in giving quality, sensory, and health characteristics to food. From a technological perspective, they function as plasticizing and structuring agents. All these properties are related to the lipid chemical composition, which affects their physical properties, health characteristics, and stability, as shown in Figure 6.1.

It is common practice to divide food lipids into fats and oils, depending on their appearance at room temperature: semisolid or liquid, respectively. Fats are mainly made from saturated-rich triacylglycerol (TAG), which can form crystals that interact in the matrix, forming a macroscopic network that finally affects fat functionalities (Acevedo et al., 2011). The control of the physical properties of plastic lipids is of primary importance in the attempt to obtain the desired structure of fat-structured foods (e.g., bakery products, snacks, spreads, ice creams). On the other hand, vegetable oils are the primary source of dietary unsaturated and polyunsaturated TAGs. They are mainly used by consumers as flavoring and cooking lipids, but they are also among the ingredients of a wide range of ambient-stable foods. For instance, besides their use as a filling medium for canned products, oils represent the lipid phase of a number of formulated foods, such as salad dressings, mayonnaise, sauces, and ready-to-eat products. By definition, they are consumed as liquids, since they are liquid at ambient temperature. However, storage in chilled (4°C) and frozen (−18°C) conditions causes the occurrence of crystallization phenomena. The latter could definitely affect the product's visual appearance and structure.

The study of fat structure in foods has received great interest due to their practical relevance. As reviewed by Acevedo et al. (2011), the solid fat structure of TAGs (macro scale) is determined by the formation of a network made of polycrystalline aggregates (micro scale) composed of several single-crystal structures (nano scale).

A typical characteristic of lipid structure is that lipids with the same molecular moiety can form different nanoscale primary crystals with the ability to pack their long hydrocarbon chains into various crystal lattices (polymorphism). Full details of the structure of lipid crystals have already been accurately described in the literature

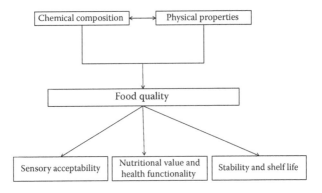

FIGURE 6.1 Schematic representation of food lipid functionalities.

(Sato, 1999; Lawler and Dimick, 1998; Larsson et al., 2006; Marangoni and Wesdorp, 2013; Marangoni et al., 2012). There are three main organizations of crystalline sub-cells, named by Larsson (1966): the hexagonal α, the orthorhombic-perpendicular β', and the triclinic-parallel β. These subcells can stack longitudinally, with a chain length structure that can vary, depending on the fatty acid moieties, from double to triple, quadruple, or sextuple structure (Larsson et al., 2006). Since polymorphs have different thermodynamic stability, polymorphic transformations are frequently observed moving from the α form—the least stable—to either the β' or the β form. Formation of fat crystal network and crystalline morphology are modified by factors associated with the TAG composition of the system and fatty acid positional distri-bution, as well as by external factors such as thermal treatments (cooling/heating rates, tempering), agitation, storage time, and so on (Sato et al., 2013). Thus, different structures, or better architectures, could be obtained by applying different process-ing conditions to the same lipid-based material. On the basis of these considerations, it is clear that it is possible to engineer the most appropriate lipid "architecture" for the specific application.

In real fat-structured products, the α form could be present in frozen foods and ice creams, since it can form directly from liquid as a consequence of rapid cool-ing. On the contrary, β' and β can be obtained directly from liquid by using dif-ferent processing conditions as well as through polymorphic transformations. It is interesting to remember that the β' form is highly desired in margarines and shortening, and in nontempered types of confectionery, where it allows good palat-ability. Polymorphic transformations from β' to β in such products are undesirable, causing product defects such as sandiness due to the large and plate-like crystal shapes. Finally, β is the most desired form in tempered fats such as chocolate-based products.

With the aim of engineering the most appropriate lipid "architecture" for the spe-cific application, it is essential to know in-depth the lipid structure at different length scales. Table 6.1 shows some examples of methodologies that could be applied for the characterization of lipid structure. Besides techniques that could give a picture of the fat structure under the actual studied conditions, the study of crystallization

TABLE 6.1
Examples of Methodologies That Could Be Applied for the Characterization of Lipid Structure

Scale Length	Characteristics	Methodology
Macro	Rheological properties, mechanical properties	Rheological analysis, texture analysis
Micro	Microstructure	Microscopy techniques (e.g., polarized light microscopy, electron microscopy, confocal scanning light microscopy, multiple photon microscopy)
Nano	Crystal morphology	X-ray diffraction (XRD)

kinetics as well as thermal behavior, other techniques, such as differential scanning calorimetry (DSC), have to be applied to deeply understand lipid phase transition.

Among different techniques that could be applied to study lipid physical properties, this chapter will focus on the application of x-ray diffraction methodology (XRD) in combination with DSC analysis. Besides a brief description of XRD techniques, with particular focus on synchrotron x-ray methodologies, examples of applications of these techniques will be described. Finally, possible practical implications in the food industry will be discussed.

6.2   XRD ANALYSIS

XRD techniques are the most direct methodologies for studying lipid crystal structure, and most current knowledge of fat crystalline structure comes from XRD studies. This is due to the fact that the two levels of organization of lipid crystals can be identified from their characteristic XRD pattern. The signals observed in the wide angle x-ray scattering (WAXS) region can be used to define the type of polymorphic form. As extensively reported in the literature, a single peak in the WAXS region around 4.15 Å is characteristic of the hexagonal α form; two lines around 3.8 and 4.2 Å are associated with the orthorhombic-perpendicular β′ subcell; and, finally, a sharp signal at about 4.6 Å along with other peaks from about 3.6 and 5.2 Å is typical of the triclinic-parallel subcell of the β polymorph. On the other hand, the peak position in the small-angle x-ray scattering (SAXS) region can be useful to deduce the chain length structure in longitudinal stacking.

As mentioned above, conventional XRD diffraction techniques have been widely applied to study lipid phase transition behavior. More recent is the use of high-brilliance x-ray sources (synchrotron) radiation instead of conventional methods. Since it provides greater x-ray brilliance than conventional sources, this technique has been indicated as the best method for the characterization of the dynamic phase behavior and polymorphism of lipids (Keller et al., 1996; Lopez et al., 2001; Sato and Ueno, 2011). Due to the excellent brightness and collimation properties of the synchrotron beam, it is possible to record weak structures generated by small amounts of crystallites dispersed in the amorphous matrix, which is difficult to observe in the laboratory. For these reasons, the use of synchrotron radiation instead of conventional radiations is becoming more frequent.

6.2.1   SYNCHROTRON XRD EXPERIMENTAL SETUP

The synchrotron x-ray setup used at the Synchrotron Radiation Facility located in Trieste (Italy) is shown in Figure 6.2. The x-ray beam emitted by the wiggler source on the Elettra 2 GeV electron storage ring is monochromatized by a Si(111) double crystal monochromator, focused on the sample and collimated by a double set of slits giving a spot size of 0.2 × 0.2 mm. The sample is positioned between the synchrotron XRD beam exit pinhole and the detector.

Different sample holders can be used for experiments. Traditionally, glass capillaries are used for powder and fluid samples. Additionally, samples can be located in a nylon loop for crystallographic experiments, as shown in Figure 6.3. This allows

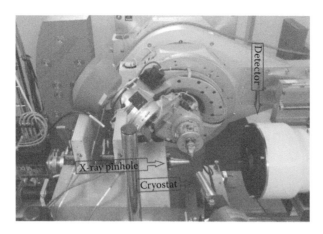

FIGURE 6.2   Typical setup used at Synchrotron Radiation Facility Elettra located in Trieste (Italy) to study phase behavior of lipids. (Courtesy of Sincrotrone Trieste SCpA.)

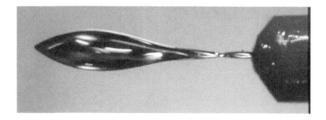

FIGURE 6.3   Nylon loop filled with liquid oil.

the collection of a transmission diffraction pattern free from the background introduced by the diffusion of x-rays from glass capillary walls. A second advantage of the nylon loop holder is that the control of sample temperature can be performed with higher precision, with a cryostat reducing temperature variations. Independently of the sample holder used, the lipid matrix has to be completely melted before mounting on the holder to allow an easier and uniform sample deposition.

The incoming x-ray beam hits the sample, being diffused or, when crystals are formed in the sample, diffracted. It is known that the difference between the two cases consists in the constructive interference of x-rays diffused by spatially ordered molecules.

Finally, rays diffracted by the sample are collected by means of a bidimensional x-ray detector. Dynamic range (span between maximum and minimum measurable x-ray intensities) and acquisition times are the most critical properties of detectors. The first affects the signal-to-noise ratio of the diffraction pattern, influencing the possibility of observing weak signals; the second accounts for the capability to follow the phase transition with the proper time resolution. An example of a bidimensional image obtained from the detector is shown in Figure 6.4a. Clearly visible rings emerge from the amorphous background at both low and high angle due to the presence of TAG crystals in sunflower oil cooled to $-30°C$. An approximate Q scale

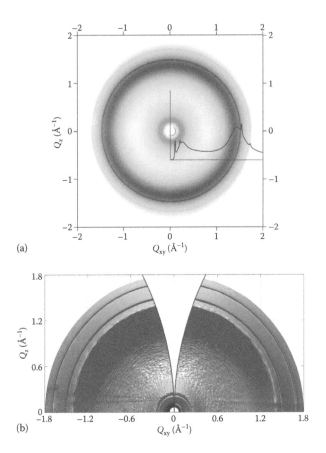

(a)

(b)

FIGURE 6.4 (a) Bidimensional synchrotron XRD pattern of sunflower oil measured at −30°C. Approximate Q scale is here reported only for indicative purposes. (b) Data are represented in real Q scale by means of functional representation.

is reported here only for indicative purposes. After elaboration, the same data represented in the real Q scale by means of functional representation are also reported in Figure 6.4b.

6.2.2 DATA ANALYSIS

The bidimensional patterns can be converted into plots of radial intensity as a function of various parameters. The reflection angle (2θ) is widely used to visualize WAXS and SAXS results at the same time. Alternatively, θ values can be converted into *interplanar* spacings by means of Bragg's law:

$$n\lambda = 2d \sin\theta$$

where the parameter n refers to the diffraction order ($n = 1$ is the first-order reflection), λ is the incident wavelength, and d is the minimum distance between two atomic planes of the same crystallographic family. When performing synchrotron experiments, results are generally plotted as intensity versus exchanged momentum Q:

$$Q = \frac{4\pi \sin \theta}{\lambda} = n \frac{2\pi}{d}$$

A plot of diffracted intensity as a function of exchanged momentum will have evenly spaced peaks for every family of crystallographic planes for which diffraction conditions occurred during the experiment. Thus, the spacing value being $2\pi/d$, data representation as a Q value allows the easier recognition of peaks belonging to the same crystallographic family.

Figure 6.5 shows, as an example, a synchrotron XRD pattern reporting intensity as a function of Q value obtained by cooling tripalmitin from 80°C to 20°C at 2°C/min (data obtained at Synchrotron Radiation Facility Elettra, Trieste, Italy). The presence of an α form is well evidenced ($d = 4.13$ Å). Equidistance peaks due to successive crystallographic orders are clearly visible on the left part of the figure.

6.2.3 SYNCHROTRON XRD ANALYSIS AS A FUNCTION OF TEMPERATURE

As previously stated and recently reviewed by Sato et al. (2013), phase transition behavior and polymorphic transformation of pure TAG as well as complex lipids are greatly affected by the thermal treatment applied. Fast cooling rates force nucleation and growth events to take place in a short time. This leads to the formation of numerous crystals in the less stable conformation (mainly α). On the other hand, slow cooling rates lead to the formation of the more thermostable polymorphs (β′ or β) associated with a reduced number of bigger crystals. After crystallization, polymorphic transformation can take place, always looking for a greater stability. Upon

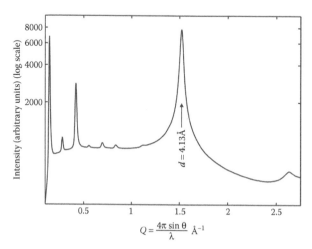

FIGURE 6.5 Intensity as a function of Q value in tripalmitin (PPP) cooled from 80°C to 20°C at 2°C/min. The d spacing of the wide angle peak associated with the α polymorph is also reported.

heating, two types of polymorphic modification can occur: melt-mediated or solid-state transformations, whose kinetics are determined by activation energies for the transformation and heating rates (Sato et al., 2013). In the first case, a less thermostable polymorph is rearranged on melting and recrystallized into a more stable polymorph. Conversely, solid-to-solid transformations can occur without melting events.

The intense flux of synchrotron radiation allows recording to be performed, exposing the sample to x-rays for a few seconds, permitting the continuous monitoring of crystallization/polymorphic transformation during sample cooling or heating. This generates a picture of the dynamic phase behavior occurring during the experiment. During synchrotron XRD analysis, the lipid sample is cooled or heated at a defined scan rate by fluxing a cryogenic fluid (e.g., nitrogen). In this way it is possible to simulate not only DSC analysis conditions but also—from a practical point of view—the processing conditions that could be applied during technological treatments of lipids.

A clear example of polymorphic transformation recorded during the heating at 2°C/min of the pure TAG triolein (OOO), previously cooled at the same scan rate, is shown in Figure 6.6. This TAG, made of three oleic acid groups (C18:1) linked to a glycerol group, is the main triunsaturated TAG that can be found in olive oil. Its polymorphism has been studied by other authors, who identified the different polymorphs that could form as a consequence of different thermal treatments (Ferguson and Lutton, 1947; Akita et al., 2006; Bayés-García et al., 2013). The XRD results obtained at Synchrotron Radiation Facility Elettra (Trieste, Italy) and shown in Figure 6.6 clearly evidence the presence of a β′ phase formed as a consequence of

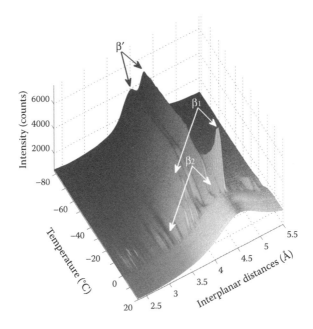

FIGURE 6.6 WAXS diffraction patterns of triolein (OOO) as a function of temperature recorded during heating at 2°C/min from −80°C to 20°C.

cooling performed at 2°C/min, in good agreement with data previously reported by Bayés-Garcia et al. (2013), who worked at the same cooling rate. Further heating highlighted the formation of a first β phase (β$_1$) followed by a second phase (β$_2$), evidencing a β$_1$→β$_2$ transformation.

6.2.4 MACROBEAM VERSUS MICROBEAM SYNCHROTRON X-RAY DIFFRACTION ANALYSIS

A further implementation of the previously described synchrotron x-ray diffraction technique (macrobeam) is termed the microbeam methodology. The basic principle relies on x-ray focusing optics and the synchrotron radiation x-ray source, enabling an intense x-ray microbeam to be obtained. The latter is scanned on a thin section of the sample with steps on the order of the beam size. Then, by collecting each bidimensional XRD pattern with a sensitive detector, it is possible to build 2-D micrometer-dimension images in real space. Using this technique, polymorphic structure and lamellar planes of fat crystals could be observed with a local resolution of 5–20 μm. The use of microbeam x-ray diffraction analysis instead of macrobeam systems can provide microscopic information about crystallization on the order of micrometer to submicrometer dimensions. This technique has been applied for the characterization of fat crystallization especially by Sato and coworkers (Shinohara et al., 2008; Arima et al., 2009; Tanaka et al., 2009; Ueno et al., 2008; Wassell et al., 2012). By using this experimental setup, for instance, Tanaka et al. (2009) clarified the structure and crystallization mechanisms of granular crystals in palm-based margarine, highlighting the presence of different polymorphs in different areas of granular crystals. Furthermore, Wassell et al. (2012) studied the interfacial crystallization of fats in water-in-oil (W/O) emulsions.

6.3 EXAMPLES OF APPLICATION OF DSC-SYNCHROTRON XRD COUPLED TECHNIQUES

The phase transition behavior of lipids has been widely studied by using thermoanalytical techniques, among which the most widely applied is DSC. As extensively reported in other chapters of this book, this technique allows the heat changes associated with phase transitions to be measured, highlighting the crystallization and melting temperatures and eventual polymorphic transformations as well as the enthalpy associated with these transitions. The monitoring of heat changes associated with temperature variation begets endothermic peaks, exothermic peaks, or both, whose area is proportional to the enthalpy gained or lost by the material undergoing phase transition. Unfortunately, since the DSC thermograms of natural fats and oils result in a complicated overlapping of crystallization and melting peaks, the interpretation of DSC curves could be complex and often impossible. Moreover, the detection of phase transitions with very low enthalpies is difficult. Due to these technical problems, data obtained from DSC analysis should be combined with data gained by XRD techniques in an attempt to associate thermal events with the polymorph involved. For these reasons, information provided by XRD and DSC data has been frequently combined to obtain an in-depth description of the structural behavior of lipids. Focusing on application of synchrotron radiation, and thus on more recent literature data, Sato and coworkers reported extensive information on the crystallization kinetics of pure triacylglycerols having different fatty acid moieties measured by means of results deriving from synchrotron XRD and DSC analysis (Boubekri et al., 1999; Zhang et al., 2007, 2009; Bayés-Garcia et al., 2011). Ollivon and coworkers developed an apparatus permitting the synchronous recording of synchrotron XRD and DSC data (Lopez et al., 2001, 2007). They applied this technique especially to milk fat, as described in Chapter 10. Other authors also applied these techniques to characterize phase transitions of different plastic fats at room temperature. For instance, literature data report the phase transition behavior of palm stearin in bulk state and in emulsion (Sonoda et al., 2004), fat blends made of hydrogenated and unhydrogenated vegetable oils (Szydlowska-Czerniak et al., 2005), cocoa butter (Loisel et al., 1998), aerated food emulsions containing interesterified palm oil (Kalnin et al., 2002), lard (Kalnin et al., 2005), and fat from Capuassu (Silva et al., 2009).

Among natural lipids, the physical properties of vegetable oils have been studied to a lesser extent, probably because they are consumed as liquids at room temperature. However, since they may be partly crystalline in chilled and frozen food formulations, the understanding of their phase transition behavior could be crucial to evaluate the feasibility of their use as ingredients in complex foods, which have to be stored under defined conditions.

By measuring the XRD pattern of liquid oils, clear signals are generally detected, indicating that a certain degree of order is retained in the liquid state. Figure 6.7 shows the diffraction results of extra-virgin olive oil recorded at 20°C. Two clear diffraction signals can be noted at about 4.5 and 23.0 Å. These bumps are associated

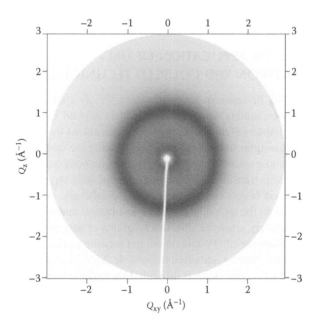

FIGURE 6.7   Bidimensional pattern of liquid extra-virgin olive oil recorded at 20°C.

with the short-range organization of the TAG molecules in the liquid phase (Larsson, 1997). Similar results were previously reported by other authors considering other lipid matrixes (Loisel et al., 1998; Lopez et al., 2001, 2007; Barba et al., 2013).

Upon cooling of vegetable liquid oils, phase transitions occur, evidencing clear, intense diffraction peaks in the XRD pattern as well as exothermic signals in the DSC curve.

As an example of a "simple" phase transition behavior, data on coffee oil are described. The lipid fraction of roasted coffee is an interesting food ingredient that could be used in a wide number of food formulations. Coffee oil has peculiar flavoring properties as well as nutraceutical characteristics. The feasibility of the use of coffee oil as an ingredient greatly depends not only on its chemical characteristics but also on its physical properties. The main fatty acid is linoleic acid (L) (about 46%), followed by palmitic acid (P) (about 34%). Significant amounts of stearic (S) (6.5%) and oleic acids (O) (8.5%) are also present, while the percentage of linolenic and arachidonic acids is about 1%–2%. Finally, gadoleic and behenic acids are present only in trace amounts (Speer and Kolling-Speer, 2001).

Figures 6.8 and 6.9 report XRD and DSC results obtained in studying the phase transition of coffee oil upon cooling the sample at 2°C/min. As reported by Calligaris et al. (2009), at about 6.5°C, in agreement with DSC data, the crystallization of coffee oil starts, with the appearance of a number of wide angle diffraction peaks (4.17, 3.73 Å) as well as small-angle peaks (54.80, 27.60, 18.41, 13.81, and 9.22 Å). The interplanar distances detected in the wide angle region are those typical of organization of the triacylglycerol chains in orthorhombic-perpendicular β' subcells. At the same time, the small-angle diffraction peaks correspond to a double chain length organization (2L) with a cell parameter c of 55.29 Å. It should be noted that the same crystalline structure was detected even by using different scanning rates (5, 10°C/min) and after a flash freezing (cooling at −30°C at 20°C/min) of the coffee oil. As is well known, flash freezing of the lipid matrix is generally applied to induce the formation of the less thermostable α crystal form (Sato, 1999).

The occurrence of the β' polymorph can be expected when one of the three fatty acid chains of TAG is somehow different from the other two (Sato, 1999). Thus, the formation of β' crystal in coffee oil could be related to the predominance of this type of triacylglycerol (i.e., PLP, PLL, SLL, POP, SLL). Since 18 can be considered the mean number of carbon atoms in the fatty acid chain and 1.52 Å is the average carbon–carbon distance in the zigzag plane of the acylglycerol chain (van Langevelde et al., 1999), the mean length of a fatty acid chain is 27.36 Å. Thus, the length corresponding to a double layer is 54.72 Å, in good agreement with experimental data.

The study of phase transitions of sunflower oil, one of the most widely used oils in frozen products, is more complex. As is well known, this oil is rich in unsaturated fatty acids, especially oleic acid (C18:1) and linoleic acid (C18:2). The cooling DSC curve of sunflower oil recorded at 2°C/min presents two exothermic peaks (A, B) associated with the appearance of wide and small-angle diffraction peaks (Figure 6.10). Associated with these events, in the XRD pattern at ~−17°C, the first peaks to emerge are identifiable as reflections of the hexagonal subcell α with a reticular parameter of 61.87 Å. Upon further cooling, the appearance of a new phase transition event at ~−54°C can be noted, associated with the formation of a β' form

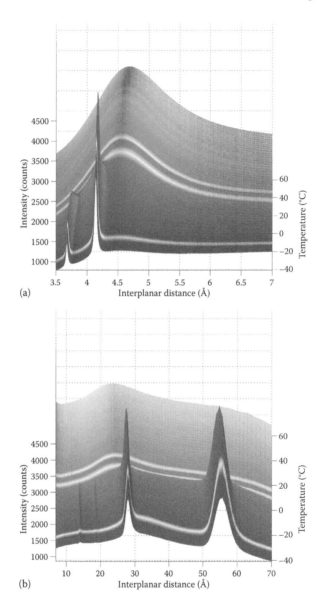

(a)

(b)

FIGURE 6.8   (a) WAXS and (b) SAXS as a function of temperature recorded during cooling of coffee oil at scanning rate of 2°C/min. (Modified from Calligaris, S., Munari, M., Arrighetti, G., and Barba, L., *European Journal of Lipid Science and Technology*, 111, 1270–1277, 2009.)

with a reticular parameter of 82.89 Å (Calligaris et al., 2008). From a technological point of view, these structures coexist at the temperature normally used for the frozen storage of foods (−18°C). However, in such conditions the oil is not completely crystallized, and part of the oil remains in an amorphous state, as evidenced by the presence of bumps during the entire XRD experiment.

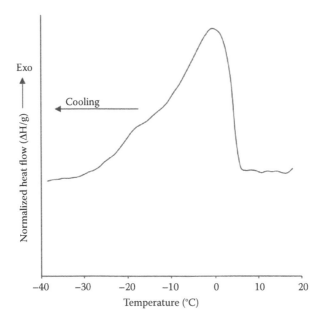

FIGURE 6.9 DSC crystallization and melting curve obtained by cooling the sample at 2°C/ min. (Modified from Calligaris, S., Munari, M., Arrighetti, G., and Barba, L., *European Journal of Lipid Science and Technology*, 111, 1270–1277, 2009.)

Moving to the heating phase, the DSC results highlight not only melting events (C, E) but also a clear crystallization event (D) (Figure 6.10). This kind of behavior can often be observed during phase transition of TAG and fats (Kalnin et al., 2002; Tan and Che Man, 2002; MacNaughtan et al., 2006). The presence of an exothermic peak—occurring before an endothermic one—in DSC heating curves is linked to the occurrence of polymorphic transformation. As the sample is heated, some of the less thermostable polymorph melts and is rearranged and recrystallized into a more stable polymorph. On this point, it should be remembered that solid-to-solid transformations can also occur, highlighted by an exothermic peak without a preceding melting event (Sato et al., 2013). As stated by Tan and Che Man (2002), polymorphs in vegetable oils, unlike pure TAG, cannot be deduced conclusively from DSC analysis but can only be determined by also performing XRD analysis. In addition, the scanning rate could greatly affect the transitions and thus the calorimetric curve. The use of slow scan rates (<5°C/min) is advisable to ensure thermal equilibrium of the sample during data collection (Tan and Che Man, 2002). Similar considerations must be applied to XRD analysis.

In the case of sunflower oil, shown in Figure 6.10, combining DSC with synchrotron XRD results, it can be stated that during heating the α form slowly melts (peak C), while the β′ form continues to organize, probably incorporating into the crystal network molecules previously melted from the α form, subtracted from the amorphous phase, or both.

The application of synchrotron radiation in combination with DSC analysis can also be useful to study the crystallization and polymorphic transformations of

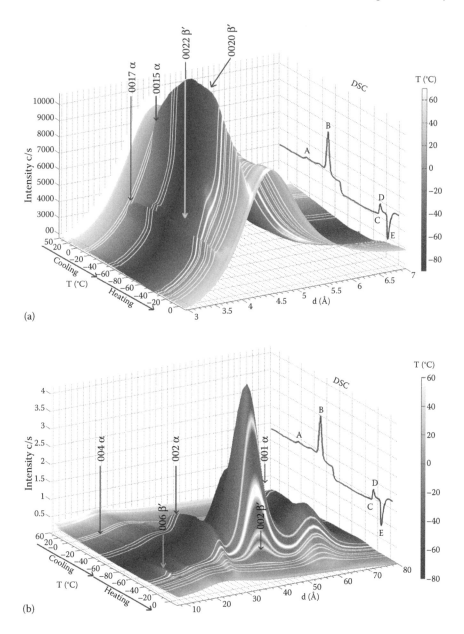

FIGURE 6.10 SAXS (a) synchrotron x-ray diffraction patterns as a function of temperature recorded during cooling of sunflower oil at scanning rate of 2°C/min from 20°C to −80°C (b) and WAXS. (With kind permission from Springer: *Journal of the American Oil Chemists' Society*, Phase transition of sunflower oil as affected by the oxidation level, 85, 591–598, Calligaris, S., Arrighetti, G., Barba, L., and Nicoli, M.C., 2008.)

complex systems, such as emulsions. Investigation of the crystallization of TAG in emulsion droplets is much more challenging than in the anhydrous state due to the presence of nonlipid components. The literature data available on the crystallization of lipids in food emulsions and fat-containing food products have shown that crystallization behavior, expressed as crystallization temperature, crystal polymorphism, and solid fat content, is influenced by numerous factors, such as the nature of lipid and emulsifier, the composition of the water phase, temperature, and emulsification conditions (Relkin et al., 2003; Higami et al., 2003; Truong et al., 2014).

Synchrotron XRD studies in combination with DSC analysis have been performed to observe the structural and thermal properties of crystalline nanoparticles of TAG in emulsions (Bunjes et al., 1996). The correlation between line widths of small-angle diffraction spectra with DSC thermopeaks indicated that the reduction of melting temperature of TAG crystals was related to the reduction of particle size of emulsion droplets. In accordance with these results, Higami et al. (2003) observed not only a reduction of melting/crystallization temperature but also an increase in transformation rate of $\alpha \to \beta' \to \beta$ polymorphs. Moreover, Sonoda et al. (2004) reported interesting results on differences in phase transition behavior of palm stearin in bulk state and in O/W emulsion. By applying the same cooling and heating cycle to both samples, they highlighted that palm stearin forms only α polymorphs in emulsions, whereas in bulk state the β' polymorph was also noted. In bulk phase as well as in emulsion, polymorphic transformations were evidenced during further heating ($\alpha \to \beta' \to \beta$). As stated by these authors, conventional laboratory XRD techniques cannot clarify the rapid and complex polymorphic behavior of palm stearin.

Finally, the application of DSC-XRD coupled techniques could also be used to study the crystallization behavior of crystallizing non-TAG emulsifiers (e.g., monoglycerides, phospholipids). The structuring behavior of these molecules is different from that of triacylglycerols: they are able to self-assemble into a wide number of structures, which include lamellar, micellar, cubic, and hexagonal mesophases, due to their amphiphilic character (Sagalowicz et al., 2006). In this context, we should remember the growing interest in the peculiar structuring properties of saturated monoglycerides. They have the capacity to self-assemble in both hydrophobic and hydrophilic domains, forming different mesophases (Goldstein et al., 2012). Once introduced into oils, these molecules are able to self-assemble into inverse bilayers, leading to the formation of a continuous three-dimensional network, called an organogel, preventing oil from flowing (Da Pieve et al., 2010; Co and Marangoni, 2012). Similarly, when introduced into water under specific physicochemical conditions, they can organize by themselves to obtain lamellar phases forming a gel network able to encapsulate large amounts of oil (Batte et al., 2007). The crystallizing behavior of saturated monoglycerides in both systems has been described by different authors using both XRD and DSC analysis (Da Pieve et al., 2011; Calligaris et al., 2010). Recently Mao et al. (2014) described the monoglyceride self-assembled structure in O/W emulsion. The formation and characterization of monoglyceride mesophases as a function of temperature was studied by using DSC/XRD combined data, allowing the observation of polymorphic transformation also in systems containing very small quantities of monoglycerides (1% w/w).

6.4 FUTURE PERSPECTIVES FOR APPLIED PURPOSES IN FOOD TECHNOLOGY

In recent years much effort has been directed toward the development of foods with improved or novel functional properties by applying structural design principles (Betoret et al., 2011; McClements et al., 2009; Co and Marangoni, 2012). Macromolecules, such as lipids, have been proposed as efficient structuring agents (McClements et al., 2009; Simo et al., 2012; Co and Marangoni, 2012). For this reason, there is a pressing need to perform in-depth studies of lipid physical properties in complex food systems. It is now well accepted that the creation of novel foods and the improvement of existing ones depend on a better understanding of thermodynamic and kinetic factors affecting the complex interrelationships between the structure, at nano, micro, and macro level, and food performance (Aguilera, 2006). Figure 6.11 shows a schematic representation of the cognitive steps for the understanding of biopolymer structural hierarchies.

Due to the versatility of lipids in creating different structures by changing the molecular features and processing conditions, lipid structure engineering could be exploited to improve the functionality of existing foods as well as to design novel foods. In our opinion, the possibility of engineering lipid physical properties could be exploited for the following research topics.

6.4.1 DEVELOPMENT OF FOODS LOW IN FAT, IN SATURATED AND TRANS FATTY ACIDS, OR BOTH

The simple replacement of trans fatty acids, saturated fatty acids, or both, or reduction of the total fat content of foods, does not allow the achievement of foods that match the quality characteristics expected by consumers. It is a matter of fact that the final structure of the product, as well as its sensory properties, is highly related

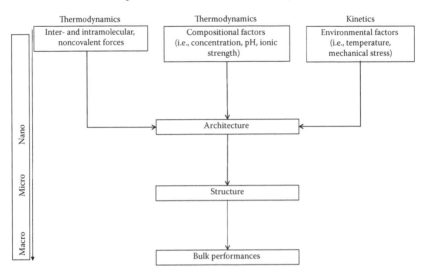

FIGURE 6.11 Cognitive steps for the understanding of biopolymer structural hierarchies.

to the presence of lipid. A winning strategy for the production of low-fat or low-saturated-fat products seems to be the substitution of plastic fats with unsaturated oils structured by the addition of molecules that form self-assembly nanostructures. In this context, we should remember the growing interest in the structuring properties of saturated monoglycerides due to their excellent capacity to self-assemble in both hydrophobic and hydrophilic domains (Goldstein et al., 2012; Co and Marangoni, 2012; Manzocco et al., 2012; Calligaris et al., 2013).

6.4.2 Design of Delivery Systems for Lipophilic Bioactive Compounds in Foods

Lipid structures have been proposed as systems to introduce the desired guest molecules into a nanoscaled domain that creates a compartmentalization between the components of the food matrix. Successful application of self-assembly structures as delivery systems depends on the ability to incorporate the potential guest molecules within the various phases in appropriate amounts, and to disperse them inside the food matrix without losing their expected functionality. The physical properties of the lipids used to obtain these nanostructures greatly affect the chemical stability as well as the rate of release of the compounds delivered. At the current stage of knowledge, many aspects necessary for a rational and reliable design of lipid architectures as delivery systems remain to be deeply understood.

6.4.3 Understanding the Effect of Lipid Structure on the Bioavailability of Lipophilic Bioactives and Lipid Nutrients

The rate of release of nutrients from food materials and their bioavailability are both important considerations in the development of value-added food products. When lipids pass through the gastrointestinal tract they are subjected to changes in their physical structure and chemical profile due to the occurrence of a complex set of chemical, enzymatic, and physical events. These events are affected by factors not only related to molecular characteristics but also to food structure. Potentially, the physical state of a lipid may alter the ability of enzymes to digest the lipid or alter the absorption of the digestion product. Some studies have shown that fat crystallization within lipid droplets (emulsified solid lipid particles) may reduce the rate of lipid digestion and thus the release of the encapsulated lipophilic bioactive (McClements et al., 2009). Besides this information, very few studies report links between bioavailability and lipid/food structure.

6.4.4 Understanding the Effect of Lipid Structure on Oxidation Rate

Although great efforts have been made in the past to understand the "chemistry" of lipid oxidation in foods, only recently has increasing attention been paid to the effect of food structure on oxidation. In particular, a key topic is the role of lipid physical state. Some authors (Calligaris et al., 2004, 2006, 2007; Okuda et al., 2005; Helgason et al., 2009) reported that the physical state of food components could greatly affect the oxidative reaction kinetics in foods. The consequences are positive deviations from the temperature dependence of oxidation rate as described by the Arrhenius equation. The reasons seem to be related to the occurrence of crystallization. The latter could induce a cascade of temperature-dependent events, such as solute concentration and changes in physicochemical properties in the liquid phases surrounding crystals, affecting lipid stability. These compositional modifications might be able to counterbalance or even oppose the direct effect of temperature on the oxidation rate, giving a reason for the observed deviations (Calligaris et al., 2004). In addition, Helgason et al. (2009) found that the inclusion of carotenoids within a solid lipid core could provide additional protection against oxidative attack by reducing the mobility of reactants. In this case, the structural arrangement of triacylglycerol molecules and the polymorphic transitions could affect the oxidative stability. Besides this fragmentary information, a rational study of the effect of lipid structure organization at nano, micro, and macro levels on oxidation kinetics is lacking.

In the light of this view, the application of sophisticated and expensive techniques, such as synchrotron XRD analysis in combination with other analytical methods (e.g., DSC, polarized light microscopy, low-intensity ultrasound), is justified by the evidence that only a deep understanding of lipid physical structure in foods and its relation with other system components will allow the effective exploitation of lipid structure in the design of foods with new and improved functionalities.

REFERENCES

Acevedo, N., Peryronerl, F., and Marangoni, A. 2011. Nanoscale structure intercrystalline interactions in fat crystal network. *Current Opinion in Colloid and Interface Science* 16:374–383.

Aguilera, J. M. 2006. Food product engineering: Building the right structures. *Journal of the Science of Food and Agriculture* 86:1147–1155.

Akita, C., Kawaguchi, T., and Kaneko, F. 2006. Structural study on polymorphism of *cis*-unsaturated triacylglycerol: Triolein. *The Journal of Physical Chemistry B* 110:4346–4353.

Arima, S., Ueno, S., Ogawa, A., and Sato, K. 2009. Scanning microbeam small-angle X-ray diffraction study of interfacial heterogeneous crystallization of fat crystals in oil-in-water emulsion droplets. *Langmuir* 25:9777–9784.

Barba, L., Arrighetti, G., and Calligaris, S. 2013. Crystallization and melting properties of extra virgin olive oil studied by synchrotron XRD and DSC. *European Journal of Lipid Science and Technology* 115:322–329.

Batte, H. D., Wright, A. J., Rush, J. W., Idziak, S. H. J., and Marangoni, A. G. 2007. Phase behaviour, stability, and mesomorphism of monostearin-oil-water gels. *Food Biophysics* 2:29–37.

Bayés-Garcia, L., Calvet, T., Cuevas-Diarte, M. A., Ueno, S., and Sato, K. 2011. *In situ* synchrotron X-ray diffraction study of crystallization of polymorphs of 1,3-dioleoyl-2-palmitoyl glycerol (OPO). *CrystEngComm* 13:3592–3599.

Bayés-Garcia, L., Calvet, T., Cuevas-Diarte, M. A., Ueno, S., and Sato, K. 2013. Crystallization and transformation of polymorphic forms of trioleoyl glycerol and 1,2-dioleoyl-3-rac-linoleoyl glycerol. *The Journal of Physical Chemistry B* 117:9170–9181.

Betoret, E., Betoret, N., Vidal, D., and Fito, P. 2011. Functional foods development: Trends and technologies. *Trends in Food Science and Technology* 22:498–508.

Boubekri, K., Yano, J., Ueno, S., and Sato, K. 1999. Polymorphic transformations in sn-1,3-distearoyl-2-ricinoleyl-glycerol. *Journal of the American Oil Chemists' Society* 76:949–955.

Bunjes, H., Westesen, K., and Koch, M. H. J. 1996. Crystallization tendency and polymorphic transitions in triglyceride nanoparticles. *International Journal of Pharmaceutics* 129:159–173.

Calligaris, S., Arrighetti, G., Barba, L., and Nicoli, M. C. 2008. Phase transition of sunflower oil as affected by the oxidation level. *Journal of the American Oil Chemists' Society* 85:591–598.

Calligaris, S., Da Pieve, S., Arrighetti, G., and Barba, L. 2010. Effect of the structure of monoglyceride-oil-water gels on aroma partition. *Food Research International* 43:671–677.

Calligaris, S., Manzocco, L., Conte, L. S., and Nicoli, M. C. 2004. Application of a modified Arrhenius equation for the evaluation of oxidation rate of sunflower oil at sub-zero temperature. *Journal of Food Science* 69:E361–E366.

Calligaris, S., Manzocco, L., and Nicoli, M. C. 2006. Influence of crystallization on the oxidative stability of extra virgin olive oil. *Journal of Agricultural and Food Chemistry* 54:529–535.

Calligaris, S., Manzocco, L., and Nicoli, M. C. 2007. Modeling the temperature dependence of oxidation rate in water-in-oil emulsions stored at sub-zero temperatures. *Food Chemistry* 101:1019–1024.

Calligaris, S., Manzocco, L., Valoppi, F., and Nicoli, M. C. 2013. Effect of palm oil replacement with monoglyceride organogel and hydrogel on sweet bread properties. *Food Research International* 51:596–602.

Calligaris, S., Munari, M., Arrighetti, G., and Barba, L. 2009. An insight into physicochemical properties of coffee oil. *European Journal of Lipid Science and Technology* 111:1270–1277.

Co, E. D. and Marangoni, A. G. 2012. Organogels: An alternative edible oil-structuring method. *Journal of the American Oil Chemists' Society* 89:749–780.

Da Pieve, S., Calligaris, S., Co, E., Nicoli, M. C., and Marangoni, A. G. 2010. Shear nanostructuring of monoglyceride organogel. *Food Biophysics* 5:211–217.

Da Pieve, S., Calligaris, S., Panozzo, A., Arrighetti, G., and Nicoli, M. C. 2011. Effect of monoglyceride organogel structure on cod liver oil stability. *Food Research International* 44:2978–2983.

Ferguson, R. H. and Lutton, E. S. 1947. The polymorphism of triolein. *Journal of the American Oil Chemists' Society* 132:687–699.

Goldstein, A., Marangoni, A., and Seetharaman, K. 2012. Monoglyceride stabilized oil in water emulsions: An investigation of structuring and shear history on phase behavior. *Food Biophysics* 7:227–235.

Helgason, T., Awad, T. S., Kristbergsson, K., Decker, E. A., McClements, D. J., and Weiss, J. 2009. Impact of surfactant properties on oxidative stability of β-carotene encapsulated within solid lipid nanoparticles. *Journal of Agriculture and Food Chemistry* 57:8033–8040.

Higami, M., Ueno, S., Segawa, T., Iwanami, K., and Sato, K. 2003. Simultaneous synchrotron radiation X-ray diffraction-DSC analysis of melting and crystallization behavior of trilauroylglycerol in nanoparticles of oil-in-water emulsion. *Journal of the American Oil Chemists' Society* 80:731–739.

Kalnin, D., Garnaud, H., Ametisch, H., and Ollivon, M. 2002. Monitoring fat crystallization in aerated food emulsions by combined DSC and time-resolved synchrotron X-ray diffraction. *Food Research International* 35:927–934.

Kalnin, D., Lesier, P., Artzner, F., Keller, G., and Ollivon, M. 2005. Systematic investigation of lard polymorphism using combined DSC and time resolved synchrotron X-ray diffraction. *European Journal of Lipid Science and Technology* 107:594–606.

Keller, G., Lavigne, F., Loisel, L., Ollivon, M., and Bourgaux, C. 1996. Investigation of the complex thermal behaviour of fats: Combined DSC and X-ray diffraction techniques. *Journal of Thermal Analysis* 47:1545–1565.

Larsson, K. 1966. Classification of glyceride crystal form. *Acta Chemica Scandinavica* 20:2255–2260.

Larsson, K. 1997. Molecular organization in lipids. In Friberg, S. E. and Larsson, K. (eds), *Food Emulsions*. New York: Marcel Dekker.

Larsson, K., Quinn, P., Sato, K., and Tiberg, F. 2006. *Lipids: Structure, Physical Properties and Functionality*. Bridgwater: The Oil Press.

Lawler, J. L. and Dimick, P. S. 1998. Crystallization and polymorphism of fats. In Akoh, C. C. and Min, D. B. (eds), *Food Lipids*, pp. 229–250. New York: Marcel Dekker.

Loisel, C., Keller, G., Lecq, G., Bourgaux, C., and Ollivon, M. 1998. Phase transition and polymorphism of cocoa butter. *Journal of the American Oil Chemists' Society* 75:425–439.

Lopez, C., Bourgaux, C., Lesieur, P., and Ollivon, M. 2007. Coupling of time-resolved synchrotron X-ray diffraction and DSC to elucidate the crystallisation properties and polymorphism of triglycerides in milk fat globules. *Lait* 87:459–480.

Lopez, C., Lavigne, F., Lesieur, P., Keller, G., and Ollivon, M. 2001. Thermal and structural behavior of anhydrous milk fat. 2. Crystalline forms obtained by slow cooling. *Journal of Dairy Science* 84:2402–2421.

MacNaughtan, W., Farhat, I. A., Himawan, C., Starov, V. M., and Stapley, A. G. F. 2006. A differential scanning calorimetry study of the crystallization kinetics of tristearin-tripalmitin mixtures. *Journal of the American Oil Chemists' Society* 83:1–9.

Manzocco, L., Calligaris, S., Da Pieve, S., Marzona, S., and Nicoli, M. C. 2012. Effect of monoglyceride-oil-water gels on white bread properties. *Food Research International* 49:778–782.

Mao, L., Calligaris, S., Barba L., and Miao, S. 2014. Monoglyceride self-assembled structure in O/W emulsion: Formation, characterization and its effect on emulsion properties. *Food Research International* 58:81–88.

Marangoni, A. G., Acevedo, N., Malaky, F., Co, E., Peyronel, F., Mazzanti, G., Quinn, B., and Pink, D. 2012. Structure and functionality of edible fats. *Soft Matter* 8:1275–1300.

Marangoni, A. G. and Wesdorp, L. H. 2013. *Structure and Properties of Fat Crystal Network*, 2nd edn. Boca Raton: Taylor and Francis.

McClements, D. J., Decker, E. A., Park, Y., and Weiss, J. 2009. Structural design principles for delivery of bioactive components in nutraceuticals and functional foods. *Critical Reviews in Food Science and Nutrition* 49:577–606.

Okuda, S., McClements, D. J., and Decker, E. A. 2005. Impact of lipid physical state on the oxidation of methyllinoleate in oil-in-water emulsions. *Journal of Agriculture and Food Chemistry* 53:9624–9628.

Relkin, P., Sourdet, S., and Fosseux, P. Y. 2003. Fat crystallization in complex food emulsions—Effects of adsorbed milk proteins and of a whipping process. *Journal of Thermal Analysis and Calorimetry* 71:187–195.

Sagalowicz, L., Leser, M. E., Watzke, H. J., and Michel, M. 2006. Monoglyceride self-assembly structures as delivery vehicles. *Trends in Food Sciences and Technology* 17:204–214.

Sato, K. 1999. Solidification and phase transformation behavior of fats: A review. *Lipid* 12:467–474.

Sato, K., Bayés-Garcia, L., Calvet, T., Cuevas-Diarte, M., and Ueno, S. 2013. External factors affecting polymorphic crystallization of lipids. *European Journal of Lipid Science and Technology* 115:1224–1238.

Sato, K. and Ueno, S. 2011. Crystallization, transformation and microstructure of polymorphic fats in colloidal dispersion states. *Current Opinion in Colloids and Interface Science* 16:384–390.

Shinohara, Y., Takamizawa, T., Ueno, S., Sato, K., Kobayashi, M., Nakajima, M., and Amemiya, Y. 2008. Microbeam X-ray diffraction analysis of interfacial heterogeneous nucleation of *n*-hexadecane inside oil-in-water emulsion droplets. *Crystal Growth Design* 8:3123–3126.

Silva, J. C., Plivelic, T. S., Herrera, M. L., Ruscheinsky, N., Kieckbuck, T. G., Luccas, V., and Torriani, I. L. 2009. Polymorphic phase of natural fat from Capuassu (*Theobroma grandiflorum*) beans: A WaXS/SAXS/DSC study. *Crystal Growth Design* 9:5155–5163.

Simo, O. K., Mao, Y., Tokle, T., Decker, E. A., and McClements, D. J. 2012. Novel strategies for fabricating reduced fat foods: Heteroaggregation of lipid droplets with polysaccharides. *Food Research International* 48:337–345.

Sonoda, T., Takata, Y., Ueno, S., and Sato, K. 2004. DSC and synchrotron-radiation X-ray diffraction studies on crystallization and polymorphic behaviour of palm stearin in bulk and oil in water emulsion states. *Journal of the American Oil Chemists' Society* 81:365–373.

Speer, K. and Kolling-Speer, I. 2001. Lipids. In Clarke, R. J. and Vitzthum, O. G. (eds), *Coffee: Recent Developments*, pp. 33–49. Oxford: Blackwell Science.

Szydlowska-Czerniak, A., Karlovits, G., Lanch, M., and Szlyk, E. 2005. X-ray diffraction and differential scanning calorimetry studies of β'-β transition in fat mixtures. *Food Chemistry* 92:133–141.

Tan, C. P. and Che Man, Y. B. 2002. Comparative differential scanning calorimetry analysis of vegetable oils: Effect of heating rate variation. *Phytochemical Analysis* 13:129–141.

Tanaka, L., Tanaka, K., Yamato, S., Ueno, S., and Sato, K. 2009. Microbeam X-ray diffraction study of granular crystals formed in water-in-oil emulsion. *Food Biophysics* 4:331–339.

Truong, T., Bansal, N., Sharma, R., Palmer, M., and Bhandari, B. 2014. Effects of emulsion droplet sizes on the crystallisation of milk fat. *Food Chemistry* 145:725–735.

Ueno, S., Nishida, T., and Sato, K. 2008. Synchrotron radiation microbeam X-ray analysis of microstructures and polymorphic transformation of spherulite crystals of trilaurin. *Crystal Growth Design* 8:751–754.

van Langevelde, A., van Malssen, K., Sonneveld, E., Peschar, R., and Schenk, H. 1999. Crystal packing of a homologous series of β'-stable triacylglycerols. *Journal of the American Oil Chemists' Society* 76:603–609.

Wassell, P., Okumara, A., Young, N. W. G., Bonwick, G., Smith, C., Sato, K., and Ueno, S. 2012. Synchrotron radiation macrobeam and microbeam X-ray diffraction studies on interfacial crystallization of fats in water-in-oil emulsions. *Langmuir* 28:5539–5547.

Zhang, L., Ueno, S., Miura, S., and Sato, K. 2007. Binary phase behaviour of 1,3-dipalmitoyl-2-oleoyl-*sn*-glycerol and 1,2-dioleoyl-3-palmitoyl-*rac*-glycerol. *Journal of the American Oil Chemists' Society* 84:219–227.

Zhang, L., Ueno, S., Sato, K., Adlof, R. O., and List, G. R. 2009. Thermal and structural properties of binary mixtures of 1,3-distearoyl-2-oleoyl-glycerol (SOS) and 1,2-dioleoyl-3-stearoyl-*sn*-glycerol (*sn*-OOS). *Journal of Thermal Analysis and Calorimetry* 98:105–111.

7 Application of DSC, Pulsed NMR, and Other Analytical Techniques to Study the Crystallization Kinetics of Lipid Models, Oils, Fats, and Their Blends in the Field of Food Technology

Silvana Martini

CONTENTS

7.1   INTRODUCTION

Lipids, together with water, carbohydrates, and proteins, are the major components of foods. Lipids are an excellent source of energy, provide nutrition to humans, and deliver appropriate flavor, texture, and mouthfeel to foods. The chemical composition of the lipids is responsible for providing appropriate flavor to foods, while the crystallization behavior is responsible for providing appropriate texture and mouthfeel. Lipids undergo phase transitions when they are stored at different temperatures. Lower temperatures induce the crystallization of lipids with the formation of a crystalline network. The characteristics of the crystalline network formed on cooling define the functional properties obtained. For example, a crystalline network formed by small and interconnected crystals results in a material with a smooth mouthfeel and texture, while a crystalline network formed by large crystals provides a coarse and grainy texture. The characteristics of the crystalline network obtained can be tailored to any food application by changing the chemical composition of the lipid and processing parameters such as cooling rate, crystallization temperature, agitation rate, and the use of additives.

Controlling and monitoring lipid crystallization in foods is essential for maintaining food product quality. A thorough understanding of the effect of processing conditions on lipid crystallization is needed to obtain crystalline networks with desired functional properties. Several techniques can be used to monitor the crystallization behavior of lipids, such as low-resolution nuclear magnetic resonance (p-NMR), x-ray diffraction (XRD), low-intensity ultrasound, polarized light microscopy (PLM), polarized light turbidimetry (PLT), and differential scanning calorimetry (DSC). DSC is one of the most commonly used techniques in industry and academia due to its ease of use, relatively low cost, accuracy, and reproducibility.

This chapter will discuss the use of DSC to study phase transitions in lipids with an emphasis on lipid crystallization kinetics. The use of other analytical techniques often coupled to DSC experiments will also be discussed. The advantages and disadvantages of each technique will be presented. The overall objective of this chapter is to describe the use of DSC and other instrumental techniques for evaluating phase behavior in lipids affected by processing conditions.

7.2   USE OF DSC TO STUDY PHASE TRANSITIONS IN LIPIDS

DSC is commonly used to study phase transitions of a wide range of materials, and lipids are no exception. In general, edible lipids can undergo two types of phase transitions when subjected to a change in temperature: liquid–solid transitions (crystallization) and solid–liquid transitions (melting). During processing of lipid-based foods, a decrease in temperature is usually applied to the system to induce lipid crystallization. The objective of inducing crystallization is to form a crystalline network that will provide appropriate texture and mouthfeel to the food. When the product is consumed, changes in temperature can induce melting of the lipid phase, which contributes to mouthfeel and texture. In addition, changes in temperature during transportation and storage can induce phase transitions of lipids that will ultimately affect the quality of the final product. In summary, phase transitions in lipids must be controlled during manufacture, storage, and transportation to maintain food product quality.

7.2.1 Liquid–Solid Transitions (Crystallization)

Liquid–solid transitions occur when a lipid is cooled from the molten state to a temperature below the melting point of the material. In the liquid state, lipid molecules (triacylglycerols [TAGs]) are randomly oriented and have high mobility. During cooling, TAGs rearrange to form a highly organized crystalline lattice in which the molecular mobility is reduced. This crystallization process can be easily induced and characterized using DSC. Experiments designed to study crystallization processes in lipids must begin with a sample that is completely melted. Therefore, fats are melted at temperatures above their melting point, usually 20°C–40°C above the melting point, and kept at that temperature for at least 30 min to erase any crystal memory in the system. This is an important step, since the presence of crystals or nuclei will affect the crystallization behavior of the system. After ensuring that the material is completely melted, the system is cooled to the desired crystallization temperature (T_c). The difference between T_c and the melting point (T_m) of the sample is defined as the supercooling $(\Delta T = T_m - T_c)$ of the system. In general, supercooling is the driving force for crystallization in complex lipids found in nature such as cocoa butter, palm kernel oil, palm oil, milk fat, lard, and tallow. This means that, for a specific lipid system (constant T_m), the higher the supercooling (lower T_c) the faster the crystallization. For example, if a lipid has a melting point of 30°C, the sample must be placed at a temperature below 30°C to induce crystallization. If a $T_c = 20$°C is chosen, a $\Delta T = 10$°C is generated. This is a simplistic description of driving force of crystallization in lipids, since this definition can change depending on the chemical composition of the system and on the T_c used. For more detail on driving force of crystallization in different lipids, please refer to Hartel (2001a). Once the appropriate supercooling is achieved, the formation of a solid phase occurs and a certain amount of heat is released from the system. DSC can be used to quantify the amount of heat released during the crystallization process, which is recorded as an exothermic peak in the DSC curves. It is important to note that the exothermic peak can be either an upward or a downward peak depending on the instrument setup. In general, an arrow is included in DSC plots to denote the direction of an exothermic process. DSC is a very versatile technique used to study crystallization events, since several parameters can be accurately controlled, such as T_c, time, and cooling rate. Crystallization temperatures can be set as low as −60°C, depending on the instrument, and cooling rates can range from 20°C/min to 0.5°C/min.

Crystallization processes can be classified into two different types: (i) *isothermal crystallization* and (ii) *nonisothermal crystallization*. Isothermal crystallization occurs at very low ΔT values. In these cases, the driving force for crystallization is not high enough to induce the formation of a new nucleus, and therefore spontaneous nucleation does not occur. However, nucleation and crystal growth might occur after a certain amount of time. The time elapsed between the moment the sample reaches crystallization temperature (the system reaches the appropriate driving force for crystallization) and the moment it starts to crystallize is called the induction time of crystallization. Note that crystallization occurs after the sample reaches crystallization temperature, and therefore crystallization (nucleation and growth) occurs isothermally. In the previous example of a sample with melting point of 30°C, isothermal crystallization processes would occur at low supercoolings ($\Delta T \approx 2°C$–$3°C$), for example, $T_c \approx 28°C$. The lower the T_c, the higher the ΔT, and therefore isothermal crystallization is less likely to occur. When ΔT values are very high, the driving force for crystallization is also high and nucleation occurs spontaneously. In these cases, crystallization might occur during the cooling step (before T_c is reached), and therefore crystallization occurs under nonisothermal conditions. In general, $\Delta T \approx 10°C$ is needed to induce nonisothermal crystallization. Isothermal and nonisothermal crystallization events studied by DSC are described in the next sections.

7.2.1.1 Isothermal Crystallization Studies by DSC

Isothermal crystallization is usually achieved when samples are crystallized at T_c close to the melting point of the sample (low supercoolings). In general, supercoolings between 0°C and 4°C are desired, although this will vary depending on the type of sample and on the processing conditions used. It is important to emphasize that, during isothermal crystallization experiments, exothermic crystallization peaks should not occur during the cooling. A typical temperature protocol for an isothermal crystallization experiment should contain the following steps:

1. Heat sample to 60°C–80°C (depending on the melting point of the sample).
2. Isothermal for 15–30 min (to ensure complete melting of the sample and the absence of nuclei).
3. Cool at a constant cooling rate to T_c (usually 0°C–4°C below the melting point of the sample).
4. Isothermal at T_c for 60–90 min.
5. Heat at a specific heating rate to 60°C–80°C (this step is optional).

Steps 1 and 2 allow the complete melting of the sample and the elimination of any crystals or nuclei in the material. The temperature is decreased to T_c in step 3 and crystallization (nucleation and growth) is observed in step 4. An exothermic crystallization peak should be detected during this step (step 4) and not during step 3. The experiment can be terminated after step 4, or a new step in the method can be included to evaluate the melting behavior of the crystalline network generated during the isothermal step. This additional step (step 5) consists of increasing the temperature from T_c to 20°C–30°C above the sample's melting point.

Isothermal crystallization experiments are commonly used to calculate the induction time of crystallization (τ), which is defined as the time it takes for the crystallization peak to occur during the isothermal step (step 4). Several authors have reported induction times of crystallization for different fats, such as palm oil (Saberi et al., 2011b; Verstringe et al., 2013), palm stearin and palm kernel oil (Rashid et al., 2012), cocoa butter (Ray et al., 2012; Toro-Vazquez et al., 2005), and anhydrous milk fat (AMF) (Martini et al., 2008) as affected by the addition of different additives such as diacylglycerols (Saberi et al., 2011b), monopalmitin (Verstringe et al., 2013), limonene (Ray et al., 2012), polar lipids (Toro-Vazquez et al., 2005), and waxes (Martini et al., 2008). These studies are excellent examples of how different processing conditions such as the use of additives can alter the crystallization behavior of lipids. Figure 7.1 shows typical DSC profiles obtained during isothermal crystallization.

Figure 7.1a shows the DSC profiles obtained for tripalmitin (PPP) at different crystallization temperatures, while Figure 7.1b shows the crystallization profiles of tristearin (SSS) (MacNaughtan et al., 2006). It is very clear from Figure 7.1a that PPP has a longer induction time of crystallization as T_c increases due to the lower supercooling. These authors used XRD to identify the different polymorphic forms obtained during the isothermal crystallization and reported that PPP crystallizes in different polymorphic forms depending on the T_c chosen. For example, both β and β' polymorphic forms are present at low T_c (48.5°C and 50.5°C), while only β forms

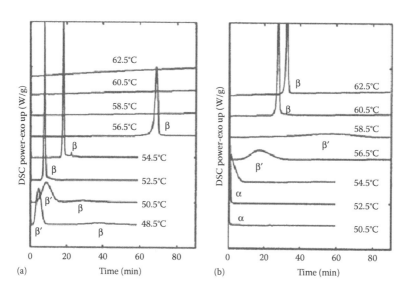

FIGURE 7.1   (a) DSC crystallization profiles obtained for tripalmitin at different crystallization temperatures (48.5°C, 50.5°C, 52.5°C, 54.5°C, 56.5°C, 58.5°C, 60.5°C, and 62.5°C); (b) DSC crystallization profiles of tristearin at different crystallization temperatures (50.5°C, 52.5°C, 54.5°C, 56.5°C, 58.5°C, 60.5°C, and 62.5°C). (With kind permission from Springer Science and Business Media: *Journal of the American Oil Chemists' Society*, A differential scanning calorimetry study of the crystallization kinetics of tristearin-tripalmitin mixtures, 83, 2006, 1–9, MacNaughtan, W., Farhat, I.A., Himawan, C., Starov, V.M., and Satpley, A.G.F.)

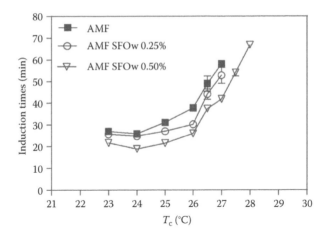

FIGURE 7.2 Induction times of crystallization of anhydrous milk fat (AMF) as affected by the addition of sunflower waxes (SFOw) at different crystallization temperatures. (With kind permission from Springer Science and Business Media: *Journal of the American Oil Chemists' Society*, Effect of the addition of waxes on the crystallization behavior of anhydrous milk fat, 85, 2008, 1097–1104, Martini, S., Carelli, A., and Lee, J.)

are present at high T_c ($T_c \geq 52.5°C$). Similar results are shown in Figure 7.1b for SSS, where α crystals are observed at low T_c ($T_c \leq 54.5°C$) and β' crystals are obtained when the sample is crystallized at 56.5°C and 58.5°C, while β crystals are obtained at higher T_c ($T_c \geq 60.5°C$). It is very clear from Figure 7.1b that only β and β' polymorphs crystallize under true isothermal conditions at $T_c \geq 56.5°C$ (note that the α crystallization peaks occur before the isothermal step begins, $t < 0$).

Figure 7.2 shows a different way of reporting results obtained from isothermal crystallization experiments performed using DSC. In this case, the induction times of crystallization determined by DSC are plotted as a function of T_c. As expected, an exponential increase in the induction times is observed as a function of T_c. The higher the T_c, the lower the supercooling, and therefore the longer the induction times of crystallization. In this particular study the effect of adding sunflower oil waxes (SFOw) to AMF was reported. The induction of crystallization was significantly reduced for a constant supercooling when SFOw was added at 0.25% and 0.5% (Martini et al., 2008).

DSC isothermal crystallization experiments can also be used to quantify the kinetics of crystallization of lipids. Kinetics parameters such as rate constant and the index of the crystallization reaction can be quantified. Most of the research in the literature uses the Avrami equation (Equation 7.1) to model the kinetics of crystallization under isothermal conditions (Avrami, 1940).

$$-\ln\left(1-f\right) = k_n t^n \qquad (7.1)$$

where:
  t = time
  k_n = rate constant

f = fractional extent of crystallization at any time t
n = index of reaction

DSC isothermal crystallization experiments can be used to calculate f using different approaches. Some authors have used partial integration of the crystallization peak at different timepoints during isothermal crystallization (Rashid et al., 2012; Toro-Vazquez et al., 2000, 2005; Saberi et al., 2011b; MacNaughtan et al., 2006; Chaleepa et al., 2010) to calculate f using Equation 7.2:

$$f(t) = \frac{\Delta H_t}{\Delta H_{total}} \times 100 \tag{7.2}$$

where:
ΔH_t = partial area under the DSC crystallization curve peak at time t
ΔH_{total} = total area under the crystallization peak

The fractional extent of crystallization (f) is therefore expressed as the amount of solids present at a specific timepoint and represents the cumulative solid fraction as a function of time. The shape of f curves as a function of time will depend on the crystallization kinetics of the system as affected by the chemical composition of the sample and by processing conditions such as T_c, the use of emulsifiers, and agitation. The fractional extent of crystallization (f) can also be determined using the "stop-and-return" method, where the sample is cooled to T_c and kept at T_c for different periods of time to achieve partial crystallization. When these specific periods of time elapse, the sample is heated to 80°C to induce melting and quantify the amount of crystalline material formed during that time at T_c. The fractional extent of crystallization (f) is calculated from the area under the curve of the endothermic melting peak as a function of time (Foubert et al., 2008; Verstringe et al., 2013; Fernandes et al., 2013; Wang et al., 2011). Independently of the method used to calculate f, the change in f as a function of time is plotted and a sigmoidal curve is usually obtained, which can be fitted to Equation 7.1 to quantify the kinetics of crystallization in terms of k and n values (Equation 7.1). While n values provide information about the dimensionality of crystal growth (one, two, or three dimensions), k is the rate constant that considers nucleation and growth events. In addition to k and n, $t_{1/2}$ is commonly used to describe the time taken to achieve half of the overall crystallization and is calculated using Equation 7.3 (MacNaughtan et al., 2006; Wang et al., 2011; Martini et al., 2006b; Herrera et al., 1999; Rincón-Cardona et al., 2013):

$$t_{1/2} = \left(\frac{\ln 2}{k} \right)^{1/n} \tag{7.3}$$

Figure 7.3 shows a very clear representation of the type of data that can be collected from isothermal experiments for the calculation of crystallization kinetics (Saberi et al., 2011b). It can be seen that the crystallization kinetics is affected by

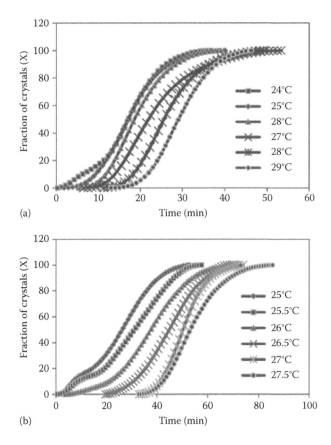

(a)

(b)

FIGURE 7.3 Cumulative solid fractions as a function for time for palm oil (a) and palm oil with 5% of a palm-based diacylglycerol (b). (Reprinted from *Food Research International*, 44, Saberi, A.H., Lai, O., and Toro-Vazquez, J.F., Crystallization kinetics of palm oil in blends with palm-based diacylglycerol, 425–435, Copyright [2011], with kind permission from Elsevier.)

crystallization temperature, with faster crystallization occurring at lower T_c, and by the addition of diacylglycerols, which causes a significant delay in crystallization, especially at high T_c.

7.2.1.2 Nonisothermal Crystallization Studies by DSC

DSC can also be used to evaluate crystallization of lipids under nonisothermal conditions. In this case crystallization occurs during the cooling step of the process, which means that nucleation, growth, or both occur before the sample reaches the final temperature set in the DSC. This type of crystallization usually occurs when a sample is cooled from the melt to a very low temperature (usually between −20°C and −60°C); the supercooling generated is so high that the sample crystallizes before it reaches the final temperature. The onset of crystallization can be quantified as the temperature at which the sample starts to crystallize during cooling. The majority of the isothermal studies described in the previous section are supplemented by nonisothermal experiments. Such experiments are usually performed to evaluate how different additives such as emulsifiers affect the crystallization behavior of the sample or to evaluate how changes in chemical composition of the lipid (fatty acid and TAG composition) affect onset temperature of crystallization (T_{on}).

Figure 7.4 shows typical nonisothermal crystallization thermograms obtained from palm oil crystallized in the presence of monopalmitin (Verstringe et al., 2013). It can be observed that, when palm oil is crystallized without the monoacylglycerol, $T_{on} \approx 20°C$. When 8% of monoacylglyerol is added, T_{on} increases significantly to approximately 45°C, showing a clear induction in the crystallization of palm oil.

Several researchers have reported nonisothermal crystallization data in various lipid systems, such as palm oil (Saberi et al., 2011b; Verstringe et al., 2013; Basso et al., 2010); lard (Miklos et al., 2013); diacylglycerol-based oils (Saitou et al., 2012); palm oil and palm stearin blends (Saadi et al., 2012); palm kernel oil, palm olein, palm midfraction, and palm stearin (Saberi et al., 2011a); cocoa butter (Toro-Vazquez et al., 2005; Ray et al., 2012); partially hydrogenated soybean oil blended with coconut oil and palm stearin (Reyes-Hernandez et al., 2007); and interesterified lipids (Adhikari et al., 2012; Claro da Silva et al., 2013), to name a few. These studies show that the crystallization behavior of lipids under nonisothermal conditions is affected by the chemical composition of the sample and by processing conditions such as cooling rate and the use of additives, including mono- and diacylglycerols and emulsifiers.

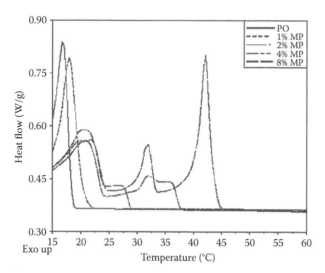

FIGURE 7.4 Example of nonisothermal crystallization thermograms of palm oil and palm oil with different content of monoacylglycerols. (Reprinted from *Food Research International*, 51, Verstringe, S., Danthine, S., Blecker, C., Depypere, F., and Dewenttinck, K., Influence of monopalmitin on the isothermal crystallization mechanism of palm oil, 344–353, Copyright [2013], with kind permission from Elsevier.)

7.2.1.3   Advantages and Disadvantages of Using DSC for Crystallization Studies

It is important to discuss some advantages and disadvantages of using DSC to evaluate the crystallization of lipids. The key advantages of using DSC over other techniques include speed, reliability, reproducibility, and tight control of experimental conditions, including temperature and cooling rate. The primary limitations of DSC in the food industry are that it is an offline technique, only very small sample sizes can be used (10–15 mg), and samples are crystallized without agitation. In addition, DSC does not provide data about morphology, size, and size distribution of crystals formed during the crystallization process. These are important characteristics that determine the functional properties of the crystalline network formed. It is also important to note that it is not recommended to use DSC techniques to evaluate early stages of crystallization such as nucleation events. DSC has relatively low sensitivity toward detecting crystalline and solid fat. DSC crystallization exotherms are detected only when approximately 1% of crystals are present. This must be taken into account when interpreting DSC crystallization results, especially when dealing with onset of crystallization. DSC crystallization exotherms are detected when crystals have undergone a significant amount of growth, and therefore these curves should not be used to study nucleation events. Induction times calculated from DSC crystallization curves must be defined as induction times of crystallization and not as induction times of nucleation. If nucleation processes must be measured, the use of more sensitive techniques, such as PLM and laser scattering techniques, is more appropriate (Cerdeira et al., 2004).

7.2.2   Solid–Liquid Phase Transitions (Melting)

Solid–liquid phase transitions occur when the temperature of a solid or semisolid fat increases above its melting temperature. DSC can be used to monitor the amount of heat absorbed during this transformation and the temperatures at which this transformation occurs. The most common use of DSC in melting experiments is to quantify the melting point of lipids. In this particular application samples are cooled nonisothermally in the DSC to a very low temperature (usually −20°C to −60°C), kept at this temperature for at least 90 min to allow the formation of the most stable crystalline structure, and then heated to 80°C. The peak temperature obtained from the melting endotherm is used to quantify the melting point of the sample. For highly polymorphic fats the procedure for melting point determination using DSC is slightly different. The melted fat is placed in a DSC pan and then kept at −20°C for many weeks to generate the most stable polymorphic form. After this period of time the DSC pan is placed in the DSC cell and heated to evaluate the melting profile.

In general, lipids are composed of several hundred species of TAGs, and therefore their melting behavior is characterized by a broad melting peak that covers a wide range of temperatures. A typical example is the melting behavior of AMF, in which a number of different peaks are observed. The melting of AMF can be divided into three endothermic peaks: (i) a high-melting-point peak observed at temperatures above 50°C, (ii) a medium-melting peak observed at temperatures between 35°C and 40°C, and (iii) a low-melting peak observed at temperatures between 15°C and 35°C

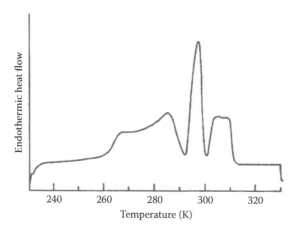

FIGURE 7.5   DSC melting profile of anhydrous milk fat (AMF) after nonisothermal crystallization to 13°C and storage at 13°C for four weeks. (Reprinted from *LWT—Food Science and Technology*, 13, Timms, R.E., The phase behavior of mixtures of cocoa butter and milk fat, 61–65, Copyright [1980], with kind permission from Elsevier.)

(Timms, 1980). Figure 7.5 shows a typical melting profile of AMF obtained after cooling the sample to 13°C and leaving it at this temperature for four weeks.

The characteristic broad melting profile of AMF can be observed in this figure (Timms, 1980). The melting behavior of a lipid sample with a more uniform composition of TAGs, such as cocoa butter, will result in a sharper melting peak that will cover a narrow range of temperatures (Hartel, 2001b).

Typical information retrieved from melting profiles includes (a) melting enthalpy (ΔH), expressed as joules per gram, (b) onset temperature (T_{on}), and (c) peak temperature (T_p) of the transition. Peak temperatures represent the temperature at which the melting peak reaches its maximum (or minimum, depending on the DSC settings). Onset temperatures are defined as the temperature at which the first deviation from the baseline is observed, corresponding to the first melting processes in the sample. Melting enthalpies are calculated by integrating the area under the melting peak and are used to quantify the amount of solid fat present in the crystalline network. Note that the latent heat of fusion also depends on the chemical composition of the crystallizing species and on the type of polymorphic form generated. For example, higher latent heat of fusion is obtained by more stable polymorphic forms (Hartel, 2001b).

7.2.2.1   Using DSC to Evaluate the Effect of Processing Conditions on the Melting Profile of Solid and Semisolid Lipids

When a lipid is cooled a crystalline network is formed. As previously discussed, the characteristics of this crystalline network determine some of the functional properties of the material. The melting profile of a crystalline network is one the most important characteristics, since it is responsible for mouthfeel and texture of the sample. The DSC technique has some flexibility to evaluate the melting behavior of lipids. It can be used to evaluate the melting behavior of samples crystallized under controlled conditions in the DSC, or of samples crystallized elsewhere, such as in a beaker, in a crystallization cell, or on an industrial processing line. This is an advantage, since other processing conditions, such as agitation and storage time under real environmental conditions, can be evaluated. The type of melting profile obtained through DSC can provide information about the structure of the crystalline network of the semisolid material formed during crystallization. Examples include the formation of different polymorphic forms and the evaluation of interactions or intersolubility between TAG molecular species. The use of DSC to identify different polymorphic forms is discussed in detail in Chapter 6 and will be only briefly described in Section 7.2.2.2.

The effect of processing conditions on the melting behavior of lipids has been studied extensively by a number of researchers. For example, DSC was used to evaluate the melting behavior of shortenings processed with different emulsifiers and stored under different environmental conditions (Martini and Herrera, 2008). In this study, the use of DSC was very helpful to evaluate how the crystalline network of the material changed as a function of storage conditions. Storage at different time and temperature combinations resulted in the melting and recrystallization of TAG species, which were detected by DSC. Other uses of DSC include the characterization of crystalline networks obtained as a consequence of sonication (Suzuki et al., 2010; Ye et al., 2011; Wagh et al., 2013), wax addition (Kerr et al., 2011), and emulsifier addition (Cerdeira et al., 2003; Wagh et al., 2013). DSC was also used to evaluate the melting behavior of several fats, such as fractions of high-oleic high-stearic sunflower oil (Rincón-Cardona et al., 2013; Martini et al., 2013), interesterified palm kernel oil, palm stearin oils, and their physical blends (Norizzah et al., 2004), palm olein and tripalmitin blends (Calliauw et al., 2010), and palm kernel oil blends with canola oil (Martini et al., 2006a,b). The effect of cooling rate on the type of crystalline network formed was evaluated for several systems, such as waxes (Martini and Añón, 2003), high-melting milk fat fractions and sunflower oil blends (Martini et al., 2001, 2002), lard and AMF (Campos et al., 2002), and milk fat and rapeseed oil blends (Kaufmann et al., 2012). These studies report that slow-cooled samples have a broader melting peak compared with fast-cooled samples, suggesting a crystalline network formed by a more heterogeneous population of TAGs. Figure 7.6 shows an example of the melting behavior of the crystals obtained after crystallizing a mixture of 40% sunflower oil in high-melting milk fat fraction crystallized at 36°C using slow (0.1°C/min) and fast (5.5°C/min) cooling rates (Martini et al., 2001). This figure clearly shows a broader melting peak in the samples crystallized at a slow cooling rate.

Depending on the intersolubility of the different TAGs present in the sample, either solid solutions, eutectics, or compound crystals can be formed during the crystallization process. Interactions that occur at the molecular level might also be affected by processing conditions. Results reported in the studies mentioned above suggest that during slow cooling the intersolubility or interaction among TAG species is increased, enabling cocrystallization. The slow cooling rate provides enough time for molecular species to reorganize and to interact. As a consequence, different populations of crystals are obtained: one population of crystals melts at lower temperatures (see shoulder at approximately 32°C) and another population of crystals melts at higher temperatures (main peak at 46°C). When the sample is crystallized

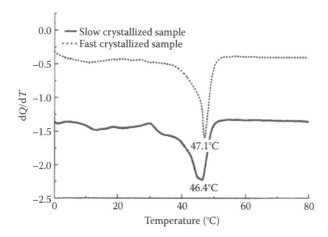

FIGURE 7.6 Effect of cooling rate on the melting behavior of 40% sunflower oil in high-melting fraction of milk fat after crystallizing at slow (0.1°C/min; solid line) and fast (5.5°C/min; dotted line) cooling rates. (Reprinted with permission from Martini, S., Herrera, M.L., and Hartel, R.W., *Journal of Agricultural and Food Chemistry*, 49, 3223–3229. Copyright [2001] American Chemical Society.)

rapidly, only TAGs with high melting points crystallize and the intersolubility or interaction among TAGs is less likely to occur. The lack of interaction is represented by a sharper melting peak with no evident shoulders (Figure 7.6).

7.2.2.2 Use of DSC to Identify Different Polymorphic Forms

Certain lipids, such as cocoa butter, are very polymorphic in nature. The type of polymorph formed is affected by processing and storage conditions and will ultimately impact the functional properties of the material. Therefore, characterization of the different polymorphic forms in lipids is critical to ensure product quality and stability. The presence of polymorphic forms in lipids can be quantified using DSC, since different polymorphic forms will result in melting peaks with different melting peak temperatures (T_p). However, it is important to note that the presence of melting peaks with different T_p does not necessarily mean that different polymorphic forms exist. XRD experiments must be performed to identify polymorphic forms associated with specific melting peaks. The use of XRD in combination with DSC to identify different polymorphic forms will be discussed in Chapter 6, and only a brief discussion will be provided in this chapter. Wille and Lutton (1966) reported for the first time the melting points of the six different polymorphic forms of cocoa butter at approximately 17°C, 23°C, 25°C, 27°C, 34°C, and 36°C for β_3' (sub-α), α-2, β_2'-2, β_1'-2 , β_2-3, and β_1-3, respectively. Figure 7.7 shows typical examples of DSC profiles for PPP crystallized in an α polymorphic form, which transforms into different polymorphic forms during heating in the DSC. This figure also shows the synchrotron XRD (SXRD) patterns that are used to identify each type of polymorphic form (Sato et al., 1999). It is clear from this figure that PPP crystallizes in an α polymorph (Figure 7.7: exothermic peak—dotted line) and that during heating the α form melts (Figure 7.7: endothermic peak—solid line) followed by the crystallization of a β form

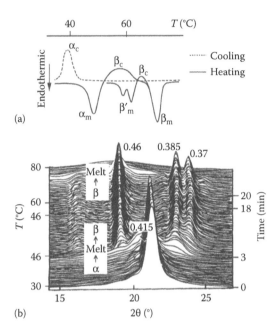

FIGURE 7.7 (a) DSC cooling and heating profiles of tripalmitin and (b) XRD of tripalmitin used to identify each polymorphic form. (Reprinted from *Progress in Lipid Research*, 38, Sato, K., Ueno, S., and Yano, J., Molecular interactions and kinetic properties of fats, 91–116, Copyright [1999], with kind permission from Elsevier.)

(Figure 7.7: exothermic peak—solid line) and the subsequent melting of the β form (Figure 7.7: endothermic peak—solid line). In addition, the presence of the β' form was observed for a very short period of time, and its melting peaks are also observed in Figure 7.7, followed by the crystallization and melting of the β form.

Other researchers used DSC to identify melting peaks of different polymorphic forms in hard and soft fractions of high-oleic high-stearic sunflower oil (Rincón-Cardona et al., 2013). These authors reported T_p values of approximately 24°C, 25°C, and 32°C for $β'_2$, $β'_1$, and β polymorphs, respectively, for soft stearins and T_p values of approximately 30°C, 32°C, and 37°C for the $β'_2$, $β'_1$, and β polymorphs, respectively, for hard stearins. Similarly, Takeuchi et al. (2002b) reported different T_p values for polymorphic forms of a 50% blend of SOS/SSO with values of 27.5°C, 34°C, and 41°C for the α, β', and β form, respectively. The identification of melting peak temperatures of polymorphic forms of other pure TAG systems such as SOS/SLS (Takeuchi et al., 2002a) and POP/OPO and PPP/POP (Sato 2001) was also reported, where S = stearic acid, O = oleic acid, P = palmitic acid, and L = linoleic acid.

7.2.2.3 Calculation of Percentage of Liquid at a Specific Temperature

Melting profiles obtained using DSC can be used to calculate the percentage of solids (or liquid) present at a specific temperature. This information is useful to evaluate the melting behavior of a sample, which can be affected by the type of molecular organization formed during the crystallization process. Similarly to the discussion presented in Section 7.2.1.1, where the fractional extent of crystallization (f) is calculated as a ratio of enthalpy values as a function of time, the melting behavior or fractional extent of melting [$f(T)$] can be calculated as a ratio between enthalpy values obtained at a specific temperature and the total enthalpy of the melting peak (Equation 7.4).

$$f(T) = \frac{\Delta H_T}{\Delta H_{\text{total}}} \times 100 \tag{7.4}$$

where:

$f(T)$ = amount of solid at a specific temperature T

ΔH_T = melting enthalpy calculated at the temperature T

ΔH_{total} = total melting enthalpy

The melting profile of a crystalline network can therefore be evaluated by plotting the cumulative amount of solids as a function of temperature $f(T)$. Melting thermograms obtained using DSC were used to describe the sharper melting profile of AMF, palm kernel oil, and an all-purpose shortening when crystallized in the presence of ultrasound waves (Suzuki et al., 2010; Ye et al., 2011) and to evaluate the melting profile of different fractions of high-oleic high-stearic sunflower oil (Bootello et al., 2011); palm midfraction, cocoa butter, and their blends (Ali and Dimick, 1994); and coconut oil (Marikkar et al., 2013), to name a few. Figure 7.8 shows the melting profile of interesterified soybean oil crystallized at 32°C for 90 min with and without the use of power ultrasound (Ye et al., 2011). A sharper melting profile is observed in the sonicated samples (10/20 and 10/10) when compared with the nonsonicated ones (10/wo).

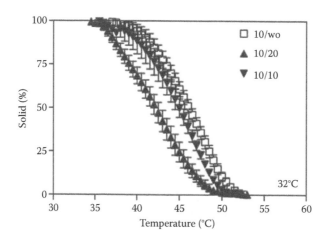

FIGURE 7.8 Melting behavior expressed as the percentage of solid material remaining at specific temperatures for interesterified soybean oil crystallized at 32°C for 90 min with and without the use of power ultrasound. (Reprinted with permission from Ye, Y., Wagh, A., and Martini, S., *Journal of Agricultural and Food Chemistry*, 59, 10712–10722. Copyright [2011] American Chemical Society.)

Sonication is more efficient when applied in the presence of crystals (i.e., 10/20 conditions).

Understanding the melting profile of a lipid is very important to establish the functional properties of the sample and therefore its potential for use in food products. Lipids can have a wide plastic range, in which the amount of solids remains constant over a wide range of temperatures, or can have a very steep melting profile. The functional characteristics of these two types of products are very different. Wide plastic-range lipids are usually used as all-purpose shortenings, since they can be processed over a wide range of temperatures without affecting the amount of solids in the system. Lipids with sharper melting profiles are usually employed in applications where a brittle consistency is needed at room temperature while complete melting of the sample must be achieved at higher temperatures (O'Brien, 2009). This is the case with cocoa butter, which completely melts at body temperature (37°C), providing a smooth and fast melting that results in the typical cooling sensation experienced with chocolates (Beckett, 2009).

7.2.2.4 Phase Diagrams

Melting parameters extracted from DSC melting profiles can also be used to build phase diagrams. These diagrams represent temperature and concentration conditions in which two phases are in equilibrium. In particular, when dealing with lipids, the equilibrium between liquid and solid phases is evaluated. Phase diagrams can be either binary or ternary. In binary phase diagrams, two lipids are mixed in different proportions, then crystallized using a slow cooling rate to promote crystallization under equilibrium conditions, and then melted at a slow cooling rate. The melting T_{on} and T_p of each blend are quantified and plotted as a function of sample composition. Figure 7.9 shows examples of the different phase diagrams that describe the formation of solid solutions and eutectic, monotectic, and peritectic systems (Timms, 1984; Hartel, 2001a; Craven and Lencki, 2011).

Timms (1984) was one of the first researchers to use DSC to construct phase diagrams of lipids and reported the formation of solid solutions, eutectics, and monotectic systems, depending on the type of TAGs tested. This author also constructed pseudo-phase diagrams using mixtures of two fats instead of two pure molecular entities. He reported, for example, the phase behavior of mixtures of different fractions of milk fat and cocoa butter (Timms, 1980) and cocoa butter and cocoa butter replacers (CBRs) (Timms, 1984). Timms was also one of the first to use isosolid diagrams (see Section 7.3.2). DSC has been used to build phase diagrams of several lipid systems, such as 1,3-diacylglycerols (Craven and Lencki, 2011), rice bran wax (Dassanayake et al., 2009), POS/SOS (Rousset et al., 1998), SOS/OSO (Koyano et al., 1992), POP/OPO (Zhang et al., 2007; Sato et al., 1999), PPO-POP (Sato et al., 1999), PPP/POP (Sato, 2001), SOS/SSO (Takeuchi et al., 2002b), and SOS/SLS (Takeuchi et al., 2002a); where P = palmitic acid, O = oleic acid, S = stearic acid, and L = linoleic acid.

Constructing phase diagrams helps us to understand molecular interactions that occur between the different TAGs present in lipids and how these interactions affect the functional properties of the crystalline network formed. For example, it is important to understand how the addition of a CBR affects the crystallization behavior and therefore the functional properties of the crystalline network formed.

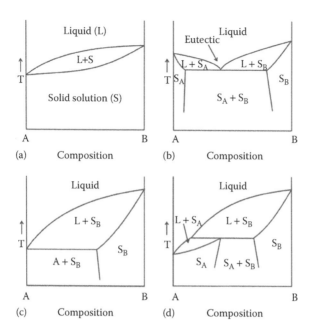

FIGURE 7.9 The four main types of phase diagram observed with binary mixtures of triglycerides: (a) monotectic, continuous solid solution; (b) eutectic; (c) monotectic, partial solid solution; (d) peritectic. (Reprinted from *Progress in Lipid Research*, 23, Timms, R.E., Phase behavior of fats and their mixtures, 1–38, Copyright [1984], with kind permission from Elsevier.)

7.2.2.5 Ideal Solution Behavior

Lipids are composed of several hundred species of TAGs of different chemical compositions and melting points. This means that, when the system is placed at a specific temperature, high-melting-point TAGs crystallize while low-melting-point TAGs remain in a liquid state. In these cases the crystallization behavior of the system can be treated as a solution, with high-melting-point TAGs being the solute and low-melting-point TAGs being the solvent. This behavior only occurs for certain TAGs and for specific crystallization temperatures. It is therefore important to establish the intersolubility of different TAGs present in natural lipids. The intersolubility of different TAGs can be evaluated using the Hildebrand equation (Equation 7.5):

$$\ln X = \frac{\Delta H_f}{R}\left(\frac{1}{T_m} - \frac{1}{T_b}\right) \tag{7.5}$$

where:

X = mole fraction of the high-melting lipid
ΔH_f = the enthalpy of melting for the high-melting lipid (J/mol)
R = universal gas constant (8.314 J×mol^{-1}×K^{-1})
T_m = melting temperature (K) of the high-melting lipid
T_b = melting temperature of the blend

The Hildebrand equation is used to evaluate whether two components behave as an ideal solution. In the specific case of TAGs, the Hildebrand equation can be used to evaluate whether or not high-melting TAGs form ideal solutions with low-melting TAGs. Based on Equation 7.5, DSC can be used to measure ΔH_f, T_m, and T_p of different TAG molecules and evaluate whether these behave as ideal solutions. When plotting $\ln X$ versus $1/T_b$, a straight line with a negative slope is expected for systems that behave as ideal solutions. Sometimes this behavior is observed over a certain range of concentrations and not over the entire range of concentrations, suggesting a partial ideal solution behavior.

The Hildebrand equation has been used by several researchers to evaluate whether binary blends of edible lipids behave as ideal solutions. Some of the research includes evaluating solubility of high-melting fats such as tripalmitin, cocoa butter, and palm oil stearin in low-melting fats such as tricaprylin, canola oils, sunflower oil, lard olein, and palm oil olein (Zhou and Hartel, 2006); fully hydrogenated palm oil in sunflower oil (Kloek et al., 2000); and high-melting milk fat fraction and middle-melting milk fat fraction in low-melting milk fat fraction and canola oil (Wright et al., 2000). Evaluating ideal solution behavior and intersolubility of TAGs is important for predicting how these molecular interactions affect polymorphism and functional properties such as texture.

7.3 FUNDAMENTALS OF p-NMR

Solid fat content (SFC) measures the amount of solid fat in a lipid and is expressed as a percentage. As previously mentioned, lipids are composed of hundreds of different TAGs with a wide range of melting points. Therefore, when lipids are placed at a specific temperature, high-melting-point TAGs crystallize while low-melting-point TAGs remain in the liquid state. The amount of solid formed as a consequence of TAG crystallization can be accurately quantified using p-NMR and expressed as SFC. The amount of solids present in a crystalline network is usually related to the functional properties of the lipid. As expected, SFC values are affected by the chemical composition of the sample and by processing conditions such as crystallization temperature, cooling rate, agitation, and the presence of additives. Therefore, it is common to express SFC values as a function of temperature. SFC can either be relatively constant as a function of temperature or can decrease with high temperatures. An example of the first case is an all-purpose shortening for which SFC values do not change over a wide range of temperatures. Cocoa butter is a typical example of the second case, in which SFC values decrease rapidly over a narrow range of temperatures. This sharp melting behavior is required in lipids to provide snap and optimum mouthfeel in food products. The p-NMR technique uses an external continuous magnetic field and a radio-frequency pulse that interact with the sample's natural magnetic moments. Relaxation times of hydrogen nuclei associated with these events are then measured as a function of time. The calculation of SFC values is based on the different relaxation times observed in nuclei present in the liquid phase versus those present in the solid phase. Nuclei in the liquid phase relax more slowly (70 μs) than those in the solid phase (<10 μs). A detailed description of the fundamentals of SFC measurement by p-NMR techniques can be found in Peyronel and Campos (2012), and falls beyond the scope of this chapter.

The American Oil Chemists' Society has issued two official methods to measure SFC in fats using p-NMR (AOCS Official Method Cd 16-81, 1999; AOCS Official Method Cd 16b-93, 1999). One of the methods (Cd 16-81) is called the indirect method, in which the SFC is calculated using Equation 7.6:

$$\mathrm{SFC} = 100 - \left(\frac{R60 \times ST}{S60 \times RT} \right) \times 100 \qquad (7.6)$$

where:

$R60$ = signal intensity of the reference oil at 60°C
$S60$ = signal intensity of the sample at 60°C
RT = signal intensity of reference oil at a specific temperature T
ST = signal intensity of the sample at a specific temperature T

As detailed in Equation 7.6, this method measures the NMR signal from the sample at a certain temperature (T) and compares it with the signal of the same sample when it is fully melted (60°C). This method is based on the measurement of NMR signals of the liquid phase only, and the measurements are compared with a reference oil at all temperatures to eliminate temperature effects on the NMR signal. The ratio $R60/RT$ shown in Equation 7.6 represents the correction factor for the temperature dependence of the NMR signal at a certain temperature T. The second AOCS Official Method (Cd 16b-93) describes the use of NMR using direct measurement, in which the NMR signals from both the solid and liquid phases are measured (Equation 7.7).

$$\mathrm{SFC} = \left(\frac{E11 - E70}{E70 + \left[(E11 - E70) \times F \right] + D} \right) \times F \times 100 \qquad (7.7)$$

where:

$E11$ = NMR signal measured at 11 μs after the pulse
$E70$ = NMR signal measured at 70 μs after the pulse
F = empirical factor that corrects for the dead time of the receiver
D = digital offset factor to correct for the non-linearity or offset of the detector

Both F and D are calculated from specific calibration standards provided with the instrument. The signal measured at 11 μs ($E11$) corresponds to the relaxation of protons present in the solid and liquid phases, while the signal measured at 70 μs ($E70$) corresponds to the relaxation of the protons present in the liquid phase. This technique is very easy to use, since Equation 7.7 is usually programmed into the instrument data system software, permitting the SFC to be reported directly.

As previously discussed, the amount of solids in lipid samples can be calculated using DSC (Equations 7.2 and 7.4). Some differences in the values obtained between the solid contents determined by DSC and p-NMR are observed. In general, SFC values obtained by DSC are higher than the ones obtained by p-NMR. Marquez et al. (2013) state that these differences are attributed to an increase in consumed energy per unit of melted mass as a function of increasing temperature. These authors provide a correction to the DSC method that results in SFC values similar to those obtained by p-NMR. Lambelet (1983a) made a similar observation in mixtures of cocoa butter and milk fat and attributed this difference to the presence of an amorphous phase in the fats (Lambelet et al., 1986). Lambelet (1983b) also noticed that these differences are a function of the chemical composition of the lipid sample, indicating that greater differences between DSC and p-NMR values were observed when a higher amount of lower-melting components was present in the lipid blend.

7.3.1 Use of p-NMR to Study Crystallization in Lipids

AOCS Official Methods provide a standardized temperature–time protocol specific to stabilizing and nonstabilizing fats. A specific temperature–time protocol was developed for stabilizing fats to generate the most stable polymorphic form and therefore obtain reproducible SFC values. These methods have been described and discussed in detail by Peyronel and Campos (2012) and will not be discussed in this chapter. SFC measurements obtained by the AOCS Official Methods are excellent resources to compare the SFC of different fats under very controlled crystallization conditions (temperature–time protocols). However, they do not provide information about SFC values of samples processed under real-world processing of lipids. For example, the SFC of a sample crystallized in an industrial or lab setting, either in a flow setup or in batch, cannot be determined using the temperature–time protocols described in the AOCS Official Methods. In these cases the direct and indirect methods described by the AOCS Official Methods can still be used without the temperature–time protocols; instead, SFC can be measured directly from the crystallizing material in a flow cell or in a batch system. The measurement of SFC using p-NMR is very versatile, since it can be used to measure the SFC of a sample crystallized in a flow cell or in a crystallization reactor. These setups allow the effect of several processing conditions on SFC values to be explored. In addition, SFC can be measured in static samples (without agitation); samples can be crystallized from the melt in the NMR tube and placed at different T_c to evaluate and quantify the crystallization kinetics of the sample as a function of time. Even though these samples are crystallized under static conditions, the effect of certain processing conditions on SFC values such as cooling rate and the use of additives can be explored. The SFC can be measured as a function of time, and kinetic parameters such as those derived from the Avrami equation (Equation 7.1) can be calculated.

NMR has been used to measure the SFC as a function of time and T_c of several lipids and lipid blends, such as stearic/oleic mixtures (Morselli Ribeiro et al., 2012); palm oil/palm stearin (Saadi et al., 2012); palm kernel oil, palm oil, palm midfraction, palm olein, and palm stearin (Saberi et al., 2011b); tripalmin/palm olein (Calliauw et al., 2010); partially hydrogenated soybean oil and *trans*-free soybean oil blended with coconut oil and palm stearin (Reyes-Hernandez et al., 2007); high-melting milk fat fractions in low-melting milk fat fraction (Herrera et al., 1999);

palm oil crystallized with and without the addition of diacylglycerols and palm oil fatty acids (Timms, 1985); and interesterified palm kernel and palm stearin oils (Norizzah et al., 2004), and to describe differences between chemically and enzymatically interesterified tristearin and triolein (Ahmadi et al., 2008). p-NMR has also been used to evaluate the effect of different processing conditions on SFC values of different lipid systems. For example, samples crystallized at slow cooling rates usually result in lower SFC values (Kaufmann et al., 2012; Herrera and Hartel, 2000; Martini et al., 2002a), especially for high SFC values (>30%). However, no effect of agitation on SFC values was reported (Herrera and Hartel, 2000).

The evaluation of SFC values is important to predict functional properties of the crystalline network obtained as a consequence of crystallization. Higher SFC values indicate that a higher amount of solid lipid is present in the system and, for example, a harder material is obtained. However, this is not always the case, since other physical properties of the crystalline network, such as crystal size, might affect its functional properties. Braipson-Danthine and Deroanne (2004) showed that the hardness of lipid networks with the same SFC is strongly affected by the microstructure of the sample. Crystalline networks with smaller crystals and the same amount of solids are generally harder and more elastic than samples with bigger crystals (Suzuki et al., 2010). In addition, the presence of different polymorphic forms can affect the functional properties of lipid materials with similar SFC values (Braipson-Danthine and Deroanne, 2004).

7.3.2 Isosolid Diagrams

SFC measurements can be used to build isosolid phase diagrams. These diagrams are constructed by plotting the temperature at which different mixtures of two fats have the same SFC. Similarly to the description of phase diagrams in Section 7.2.2.4, these isosolid phase diagrams are very helpful for evaluating whether two fats are compatible. Timms (1984) was one of the first researchers to build isosolid diagrams to test compatibility of two fats. Most of the early isosolid diagrams were built between milk fat and cocoa butter and between cocoa butter and CBRs, including palm kernel stearins, palm oil fractions, milk fat fractions, and lard (Gordon et al., 1979; Timms, 1980; Hartel, 1996; Kaylegian et al., 1993; Ali and Dimick, 1994). Figure 7.10 shows typical isosolid diagrams for fats that are completely miscible over the entire range of concentrations or compositions (Figure 7.10a), in which cocoa butter was mixed with a CBR rich in SOS, SOP, and POP triacylglycerols, where S = stearic acid, O = oleic acid, and P = palmitic acid (Gordon et al., 1979). Any mixture of cocoa butter with the CBR can be used for confectionery products, since these two fats are completely miscible. Figure 7.10b shows a typical example of partial miscibility of fats, such as cocoa butter and palm kernel stearin, where only mixtures close to the left and right sides of the diagram can be used for confectionery formulations.

The use of isosolid diagrams has been extended to applications other than confections. For example, isosolid diagrams have been built for mixtures of palm oil and AMF (Danthine, 2012); different fractions of milk fat (Marangoni and Lencki, 1998); palm kernel oil, palm stearin and palm olein (Zhou et al., 2010); and stearic/oleic mixtures (Morselli Ribeiro et al., 2012), to name a few.

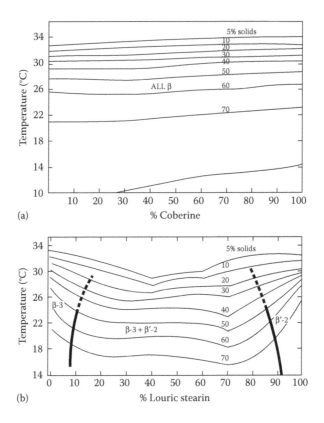

FIGURE 7.10 Isosolid diagrams of cocoa butter with CBRs: (a) CBR with symmetric TAGs; (b) palm kernel stearin. (From Gordon, M.H., Padley, F.B., and Timms, R.E., *Fette, Seifen, Anstrichmittel*, 3, 116–121, 1979.)

7.3.3 CRYSTALLIZATION KINETICS

As mentioned, SFC measured with p-NMR can be used to monitor the crystallization behavior of lipids and their blends. As discussed in Section 7.2.1.1, the kinetics of crystallization can be quantified using the Avrami equation (Equation 7.1). Figure 7.11 shows an example of SFC data collected as a function of time for rapidly cooled samples (5.5°C/min) of high-melting fraction (HMF) of milk fat blended with different amounts of sunflower oil (10%, 20%, and 40%) (Martini et al., 2002a).

Table 7.1 shows the kinetic parameters obtained from fitting the Avrami equation to these data. It is evident from these data that SFC is affected by the chemical composition of the sample and by processing conditions such as crystallization temperature (T_c). For example, for a constant chemical composition, it is evident that slower crystallization occurs at higher temperatures due to the low supercooling generated in the sample. A decrease of five orders of magnitude is observed in k_n values when T_c in HMF is increased from 5°C to 30°C, while a decrease of two orders of magnitude is observed for k_n values for HMF crystallized at 25°C when 40% of SFO is added (Table 7.1). The crystallization of several other lipid systems, such as palm

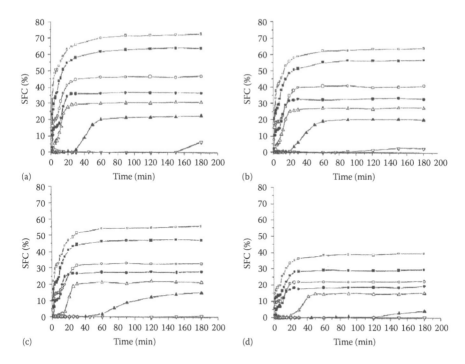

FIGURE 7.11 SFC as a function of time and temperature for high-melting milk fat fraction (a) blended with 10% sunflower oil (b), 20% sunflower oil (c), and 40% sunflower oil (d). Symbols ∇, □, ■, ○, ●, △, and ▲ represent crystallization performed at 5°C, 10°C, 15°C, 20°C, 25°C, 30°C, and 35°C, respectively. (With kind permission from Springer Science and Business Media: *Journal of the American Oil Chemists' Society*, Effect of cooling rate on crystallization behavior of milk fat fraction/sunflower oil blends, 79, 2002a, 1055–1062, Martini, S., Herrera, M.L., and Hartel, R.W.)

kernel oil and canola oil blends (Martini et al., 2006b), low-*trans* fats with and without the addition of emulsifiers (Cerdeira et al., 2006), high-melting milk fat fractions and sunflower oil blends with and without the addition of emulsifiers (Puppo et al., 2002), and hard stearin fractions of high-oleic high-stearic sunflower oil (Bootello et al., 2013), has also been studied using SFC measurements.

7.4 OTHER TECHNIQUES TO STUDY CRYSTALLIZATION OF LIPIDS

The crystallization of lipids can also be monitored using other techniques besides p-NMR and DSC. These include low-intensity ultrasound, PLM, and PLT. These techniques will be discussed briefly in the following sections.

7.4.1 LOW-INTENSITY ULTRASOUND

Low-intensity ultrasound refers to acoustic techniques that use frequencies between 0.1 and 10 MHz. This technique has recently been used to monitor the crystallization of lipids. The technique is based on the measurement of an increase in the ultrasonic

TABLE 7.1

Kinetic Parameters Obtained from Fitting the Data Reported in Figure 7.11 to the Avrami Model

Lipid Blend	T_c (°C)	k_n (min^{-n})	n
HMF	5	2.5×10^{-1}	0.605
	10	1.7×10^{-1}	0.523
	15	1.0×10^{-1}	0.607
	20	5.0×10^{-2}	0.902
	25	2.0×10^{-3}	2.118
	30	4.6×10^{-6}	2.663
10% SFO	5	2.0×10^{-1}	0.587
	10	1.7×10^{-1}	0.406
	15	1.1×10^{-1}	0.411
	20	5.9×10^{-2}	0.580
	25	5.6×10^{-3}	1.330
	30	1.1×10^{-4}	1.872
20% SFO	5	1.6×10^{-1}	0.514
	10	1.2×10^{-1}	0.445
	15	8.0×10^{-2}	0.411
	20	2.7×10^{-2}	0.807
	25	1.8×10^{-4}	2.139
	30	4.2×10^{-5}	1.654
40% SFO	5	1.1×10^{-1}	0.416
	10	7.8×10^{-2}	0.412
	15	3.4×10^{-2}	0.749
	20	8.1×10^{-3}	1.350
	25	1.6×10^{-5}	2.375

Source: Martini, S., Herrera, M.L. and Hartel, R.W., *Journal of the American Oil Chemists' Society*, 79, 1055–1062, 2002a.

HMF, high-melting milk fat fraction; SFO, sunflower oil; T_c, crystallization temperature; k_n, rate constant; n, Avrami exponent.

velocity as a consequence of the presence of solids in the system for a constant temperature. Ultrasonic velocity values were first used by McClements and Povey (1987, 1988) to calculate SFC values using density and compressibility values of the continuous (liquid oil) and dispersed (solid lipid) phases. Other research expanded on these results and used low-intensity ultrasound to monitor the crystallization of lipids, with possible application in an inline system (Martini et al., 2005a–c), and to monitor changes in the crystallization of lipids during storage (Santacatalina et al., 2011). The advantage of using low-intensity ultrasound is to monitor crystallization in real time without the need of transferring the sample to a tube (p-NMR) or a pan

(DSC). However, the use of this technique is limited to an intermediate amount of solids. A high attenuation of the ultrasound signal is observed when high amounts of solids (>30%) are present in the medium. In addition, further research in this area must be performed to evaluate the effect of crystal size on ultrasound velocity and attenuation (Martini et al., 2005a).

7.4.2 POLARIZED LIGHT MICROSCOPY

The microstructure of the crystalline network formed during crystallization plays an important role in the functional properties of the materials. This means that lipids with similar SFC or melting behavior but with different crystal sizes and morphology might behave differently in terms of texture, viscoelasticity, and mouthfeel. It is therefore important to complement crystallization experiments with evaluation of the microstructure of the lipid crystals. PLM is an optical microscopy technique commonly used to evaluate the microstructure of lipid crystals or clusters obtained during crystallization. The microstructure of crystalline networks formed during lipid crystallization is highly dependent on the processing conditions used. For example, high supercoolings usually result in very small, ill-defined crystals, while lower supercoolings result in bigger crystals that arrange themselves into spherulites. Other processing conditions that affect crystal morphology and size are cooling rate, agitation, and the presence of emulsifiers (Martini et al., 2002b). Agitation and fast cooling rate usually generate smaller crystals, while emulsifiers might generate either bigger or smaller crystals. This is usually a consequence of either induction or delay in the crystallization behavior of the fat due to the presence of emulsifiers. Emulsifiers can affect the crystallization behavior of fats differently depending on their chemical composition. When the chemical structure of the emulsifier is similar to that of the lipid, the emulsifier can cocrystallize with the fat and the structural dissimilarities between TAG and the emulsifier can delay nucleation and inhibit growth. If the structure or chemical composition of the emulsifier is different from that of the lipid, the emulsifier can act as a heteronucleus, accelerating the nucleation process. However, during crystal growth, the emulsifier might be adsorbed at steps or kinks on the surface of the growing fat and therefore inhibit crystal growth and modify crystal morphology.

PLM is very useful to evaluate early stages of crystallization, since it can detect very small crystals (1–3 µm). The morphology of crystals formed during crystallization can be observed either by crystallizing the sample directly in a slide and cover-slide using a heating stage to control temperature, or by crystallizing the material using agitation in a crystallization cell. When the sample is crystallized in a crystallization cell, aliquots of the crystallizing lipid can be taken from the crystallization cell and placed in a previously tempered slide and cover-slide to avoid secondary crystallization in the slide. Needless to say, the microstructure of crystals obtained under these two conditions (directly in the slide and in a crystallization cell) is significantly different (Martini et al., 2002b, 2013). Figure 7.12 shows differences in crystal morphology of samples crystallized in a crystallization cell under agitation and in a slide and cover-slide (Martini et al., 2002b).

The advantage of doing crystallization experiments directly on a microscope slide is that crystals are well defined, and therefore crystal sizes and distributions

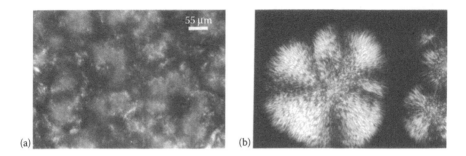

FIGURE 7.12 PLM pictures of a mixture of 40% sunflower oil in high-melting milk fat fraction crystallized at 30°C at 0.1°C/min and kept at 30°C for 120 min when samples are crystallized in a crystallization cell under agitation (a) and directly in a cover-slide (b). White bar represents 55 μm. (With kind permission from Springer: *Journal of the American Oil Chemists' Society*, Effect of processing conditions on microstructure of milk fat fraction/ sunflower oil blends, 79, 2002b, 1063–1068, Martini, S., Herrera, M.L., and Hartel, R.W.)

are easy to quantify. The disadvantage is that it may not accurately represent industrial processing conditions that involve agitation or shear. Overall, PLM allows us to evaluate changes in crystal sizes and shapes as a consequence of processing conditions and as a function of time. In addition, PLM can be used to identify different polymorphic forms (Martini et al., 2013); however, these experiments must always be supported by XRD measurements.

7.4.3 POLARIZED LIGHT TURBIDIMETRY (PLT)

The crystallization behavior of lipids can also be evaluated using PLT. This technique is very sensitive and can detect very early stages of the crystallization process. It is very common to use this technique to calculate induction times of nucleation, which in turn can be used to calculate activation free energies of nucleation. This technique has several advantages, such as the evaluation of very early stages of the crystallization process and the ability to assess the effect on crystallization behavior of different processing conditions, such as cooling rate, agitation, and the use of additives. These processing conditions can be controlled by using a crystallization cell that allows temperature control using an external water bath and agitation using, for example, a magnetic stirrer. This piece of equipment is not available commercially, but it can be easily built in the laboratory. A schematic figure of a piece of PLT equipment can be found in Herrera (1994). In short, PLT consists of a helium–neon laser light source and a polarized lens placed between the laser and the sample. The laser light is transmitted through the sample, reaching a second analyzer that is placed at the Cross-Nicholls position with the first analyzer. After crossing the second analyzer, the laser beam is detected by a photodiode. This arrangement allows the detection of optically anisotropic crystals.

Herrera (1994) used a PLT to measure induction times of nucleation of hydrogenated sunflower oil as affected by cooling rate, crystallization temperature, and emulsifier addition (Herrera and Marquez Rocha, 1996). Kerr et al. (2011) and Wagh et al. (2013) used a PLT to measure the induction times of nucleation of AMF with and without the addition of sunflower waxes (Kerr et al., 2011), emulsifiers (Wagh et al., 2013), and high-intensity ultrasound (Wagh et al., 2013).

The induction times of nucleation calculated with this technique were used by several researchers (Herrera et al., 1998; Martini et al., 2001; Martini and Añón, 2003; Cerdeira et al., 2003, 2005) to calculate activation free energies of nucleation (ΔG_c) using the Fisher–Turnbull equation (Equation 7.8; Strickland-Constable, 1968):

$$J = \left(\frac{NkT}{h}\right) \times \exp\left(\frac{-\Delta G_d}{kT}\right) \times \exp\left(\frac{-\Delta G_c}{kT}\right) \tag{7.8}$$

where:

J = rate of nucleation
ΔG_d = activation free energy of diffusion
k = gas constant per molecule
T = temperature
N = number of molecules per cubic centimeter in the liquid phase
h = Planck's constant

J can be taken as being proportional to the inverse of the induction time (τ) of nucleation. For a spherical nucleus, the activation free energy of nucleation (ΔG_c) is related to the surface free energy of the crystal/melt interface, σ, and the supercooling (melting point − crystallization temperature) $\Delta T = (T_m - T_c)$ by the following equation:

$$\Delta G_c = \left(\frac{16}{3}\right)\frac{\pi \sigma^3 T_m^2}{(\Delta H)^2 (\Delta T)^2} \tag{7.9}$$

where:
ΔH is the enthalpy of nucleation

For a TAG system, the main barrier to diffusion is the molecular conformation, and, therefore, the first exponential in Equation 7.8 is equal to:

$$\frac{-\alpha \Delta S}{R} \tag{7.10}$$

where:
α is the fraction of molecules that should be in the right conformation for incorporation in a nucleus
ΔS is the decrease of entropy on crystallization of one mole of TAG
R is the gas constant

Combining Equations 7.8 through 7.10 and rearranging, the following equation is obtained:

$$\tau \times T = \frac{h}{Nk} \times \exp\left(\frac{\alpha \Delta S}{k}\right) \times \exp\left[\left(\frac{16}{3}\right)\frac{\pi\sigma^3 T_m^2}{kT\left(\Delta H\right)^2\left(\Delta T\right)^2}\right] \qquad (7.11)$$

From a plot of $\ln(\tau T)$ versus $1/T(\Delta T)^2$ a slope (s) can be evaluated, which allows calculation of the activation free energy of nucleation from:

$$\Delta G_c = \frac{sk}{\left(T_m - T_c\right)^2} \qquad (7.12)$$

Although the slope obtained from this analysis is a constant, the activation free energy is a function of supercooling (Ng, 1990; Herrera et al., 1998).

As previously mentioned, the Fisher–Turnbull equation was used to calculate activation free energies of nucleation of palm oil and palm stearin (Ng, 1990), hydrogenated sunflower oil (Herrera et al., 1998), sunflower waxes (Martini and Añón, 2003), and milk fat/sunflower oil blends as affected by cooling rate and sunflower oil addition (Martini et al., 2001) and emulsifier addition (Cerdeira et al., 2003, 2005). In general, for the same supercooling, activation free energies increased with the addition of a liquid oil, such as sunflower oil (Martini et al., 2001), and with the addition of emulsifiers (Cerdeira et al., 2003, 2005). Increases in activation free energy values suggest a delay in the nucleation process in the system. Calculation of activation free energies helps in the understanding of the molecular events that drive the crystallization of lipids. Processing conditions can affect crystallization by either inducing or delaying nucleation or growth processes or both. Changes in activation free energy values suggest that these effects on crystallization occur at the nucleation step of the crystallization process.

REFERENCES

Adhikari, P., Shin, J., Lee, J., et al. Crystallization, physicochemical properties, and oxidative stability of the interesterified hard fat from rice bran oil, fully hydrogenated soybean oil, and coconut oil through lipase-catalyzed reaction. *Food Bioprocess Technology* 5 (2012): 2474–2487.

Ahmadi, L., Wright, A. J., and Marangoni, A. G. Chemical and enzymatic interesterification of tristerain/triolein-rich blends: Chemical composition, solid fat content and thermal properties. *European Journal of Lipid Science and Technology* 110 (2008): 1014–1024.

Ali, A. R. M. and Dimick, P. S. Thermal analysis of palm mid-fraction, cocoa butter and milk fat blends by differential scanning calorimetry. *Journal of the American Oil Chemists' Society* 71 (1994): 299–302.

AOCS (American Oil Chemists' Society Official Methods: Cd 16-81). *Solid Fat Content (SFC) by Low Resolution Nuclear Magnetic Resonance—The Indirect Method*. Urbana, IL: AOCS Press (1999).

AOCS (American Oil Chemists' Society Official Methods: Cd 16b-93). *Solid Fat Content (SFC) by Low Resolution Nuclear Magnetic Resonance—The Direct Method*. Urbana, IL: AOCS Press (1999).

Avrami, M. Kinetics of phase change: II. Transformation-time relations for random distribution of nuclei. *Journal of Chemical Physics* 8 (1940): 212–224.

Basso, R. C., Badan Ribeiro, A. P., Masuchi, M. H., et al. Tripalmitin and monoacylglycerols as modifiers in the crystallization of palm oil. *Food Chemistry* 122 (2010): 1185–1192.

Beckett, S. T. Chapter 19: Vegetable fats. In S. T. Beckett (ed.), *Industrial Chocolate Manufacture and Use*, 4th edn., pp. 415–433. Ames, IA: Wiley-Blackwell (2009).

Bootello, M. A., Garces, R., Matrinez-Force, E., and Salas, J. J. Dry fractionation and crystallization kinetics of high-oleic high-stearic sunflower oil. *Journal of the American Oil Chemists' Society* 88 (2011): 1511–1519.

Bootello, M. A., Hartel, R. W., Levin, M., et al. Studies of isothermal crystallization kinetics of sunflower hard stearin-based confectionery fats. *Food Chemistry* 139 (2013): 184–195.

Braipson-Danthine, S. and Deroanne, C. Influence of SFC, microstructure and polymorphism on texture (hardness) of binary blends of fats involved in the preparation of industrial shortenings. *Food Research International* 37 (2004): 941–948.

Calliauw, G., Fredrick, E., Gibon, V., et al. On the fractional crystallization of palm olein: Solid solutions and eutectic solidification. *Food Research International* 43 (2010): 972–981.

Campos, R., Narine, S. S., and Marangoni, A. G. Effect of cooling rate on the structure and mechanical properties of milk fat and lard. *Food Research International* 35 (2002): 971–981.

Cerdeira, M., Candal, R. J., and Herrera, M. L. Analytical techniques for nucleation studies in lipids: Advantages and disadvantages. *Journal of Food Science* 69 (2004): R185–R191.

Cerdeira, M., Martini, S., Candal, R. J., and Herrera, M. L. Polymorphism and growth behavior of low-trans fatty acids blends formulated with and without emulsifiers. *Journal of the American Oil Chemists' Society* 83 (2006): 489–496.

Cerdeira, M., Martini, S., Hartel, R. W., and Herrera, M. L. Effect of sucrose ester addition on nucleation and growth behavior of milk fat-sunflower oil blends. *Journal of Agricultural and Food Chemistry* 51 (2003): 6550–6557.

Cerdeira, M., Pastore, V., Vera, L., et al. Nucleation behavior of blended high-melting fractions of milk fat as affected by emulsifiers. *European Journal of Lipid Science and Technology* 107(12) (2005): 877–885.

Chaleepa, K., Szepes, A., and Ulrich, J. Effect of additives on isothermal crystallization kinetics and physical characteristics of coconut oil. *Chemistry and Physics of Lipids* 163 (2010): 390–396.

Claro da Silva, R., Badan Ribeiro, A. P., Schafer de Martini Soares, F. A., et al. Microstructure and thermal profile of structured lipids produced by continuous enzymatic interesterification. *Journal of the American Oil Chemists' Society* 90 (2013): 631–639.

Craven, R. J. and Lencki, R. W. Binary phase behavior of diacid 1,3-diacylglycerols. *Journal of the American Oil Chemists' Society* 88 (2011): 1125–1134.

Danthine, S. Physicochemical and structural properties of compound dairy fat blends. *Food Research International* 48 (2012): 187–195.

Dassanayake, L. S. K., Kodali, D. R., Ueno, S., and Sato, K. Physical properties of rice bran wax in bulk and organogels. *Journal of the American Oil Chemists' Society* 86 (2009): 1163–1173.

Fernandes, V. A., Muller, A. J., and Sandoval, A. J. Thermal, structural and rheological characteristics of dark chocolate with different compositions. *Journal of Food Engineering* 116 (2013): 97–108.

Foubert, I., Fredrick, E., Vereecken, J., Sichien, M., and Dewettinck, K. Stop-and-return DSC method to study fat crystallization. *Thermochimica Acta* 47 (2008): 7–13.

Gordon, M. H., Padley, F. B., and Timms, R. E. Factors influencing the use of vegetable fats in chocolate. *Fette, Seifen, Anstrichmittel* 3 (1979): 116–121.

Hartel, R. W. Applications of milk-fat fractions in confectionery products. *Journal of the American Oil Chemists' Society* 73 (1996): 945–953.

Hartel, R. W. Chapter 4: Solution characteristics and glass transition. In *Crystallization in Foods*, pp. 91–144. Maryland: Aspen (2001a).

Hartel, R. W. Chapter 3: Measurement of crystalline structure in foods. In *Crystallization in Foods*, pp. 34–90. Maryland: Aspen (2001b).

Herrera, M. L. Crystallization behavior of hydrogenated sunflower seed oil: Kinetics and polymorphism. *Journal of the American Oil Chemists' Society* 71 (1994): 1255–1260.

Herrera, M. L., de Leon Gatti, M., and Hartel, R. W. A kinetic analysis of crystallization of a milk fat model system. *Food Research International* 32 (1999): 289–298.

Herrera, M. L., Falabella, C., Melgarejo, M., and Añón, M. C. Isothermal crystallization of hydrogenated sunflower oil: I nucleation. *Journal of the American Oil Chemists' Society* 75 (1998): 1273–1280.

Herrera, M. L. and Hartel, R. W. Effect of processing conditions on crystallization kinetics of a milk fat model system. *Journal of the American Oil Chemists' Society* 77 (2000): 1177–1187.

Herrera, M. L. and Marquez Rocha, F. J. Effects of sucrose ester on the kinetics of polymorphic transition in hydrogenated sunflower oil. *Journal of the American Oil Chemists' Society* 73 (1996): 321–326.

Kaufmann, N., Andersen, U., and Wiking, L. The effect of cooling rate and rapeseed oil addition on the texture and microstructure of anhydrous milk fat. *International Dairy Journal* 25 (2012): 73–79.

Kaylegian, K. E., Hartel, R. W., and Lindsay, R. C. Applications of modified milk-fat in food products. *Journal of Dairy Science* 76 (1993): 1782–1796.

Kerr, R., Tombokan, X., Ghosh, S., and Martini, S. Crystallization behavior of anhydrous milk fat-sunflower oil wax blends. *Journal of Agricultural and Food Chemistry* 59 (2011): 2689–2695.

Kloek, W., Walstra, P., and van Vliet, T. Crystallization kinetics of fully hydrogenated palm oil in sunflower oil mixtures. *Journal of the American Oil Chemists' Society* 77 (2000): 389–398.

Koyano, T., Hachiya, I., and Sato, K. Phase behavior of mixed systems of SOS and OSO. *The Journal of Physical Chemistry* 96 (1992): 10514–10520.

Lambelet, P. Comparison of NMR and DSC methods for determining solid content of fats. Application to cocoa butter and its admixtures with milk fat. *LWT—Food Science and Technology* 16 (1983a): 200–202.

Lambelet, P. Comparison of NMR and DSC methods for determining solid content of fats. Application to milk fat and its fractions. *LWT—Food Science and Technology* 16 (1983b): 90–95.

Lambelet, P., Desarzens, C., and Raemy, A. Comparison of NMR and DSC methods for determining the solid content of fats III: Protons transverse relaxation times in cocoa butter and edible oils. *LWT—Food Science and Technology* 19 (1986): 77–81.

MacNaughtan, W., Farhat, I. A., Himawan, C., Starov, V. M., and Stapley, A. G. F. A differential scanning calorimetry study of the crystallization kinetics of tristearin-tripalmitin mixtures. *Journal of the American Oil Chemists' Society* 83 (2006): 1–9.

Marangoni, A. G. and Lencki, R. W. Ternary phase behavior of milk fat fractions. *Journal of Agricultural and Food Chemistry* 46 (1998): 3879–3884.

Marikkar, J. M. N., Saraf, D., and Dzulkifly, M. H. Effect of fractional crystallization on composition and thermal behavior of coconut oil. *International Journal of Food Properties* 16 (2013): 1284–1292.

Marquez, A. L., Perez, M. P., and Wagner, J. R. Solid fat content estimate by differential scanning calorimetry: Prior treatment and proposed correction. *Journal of the American Oil Chemists' Society* 90 (2013): 467–473.

Martini, S. and Añón, M. C. Crystallization of sunflower oil waxes. *Journal of the American Oil Chemists' Society* 80 (2003): 525–532.

Martini, S., Bertoli, C., Herrera, M. L., Neeson, I., and Marangoni, A. G. Attenuation of ultrasonic waves: Influence of microstructure and solid fat content. *Journal of the American Oil Chemists' Society* 82 (2005a): 319–328.

Martini, S., Bertoli, C., Herrera, M. L., Neeson, I., and Marangoni, A. G. In-situ monitoring of solid fat content by means of p-NMR and ultrasonics. *Journal of the American Oil Chemists' Society* 82 (2005b): 305–312.

Martini, S., Carelli, A., and Lee, J. Effect of the addition of waxes on the crystallization behavior of anhydrous milk fat. *Journal of the American Oil Chemists' Society* 85 (2008): 1097–1104.

Martini, S. and Herrera, M. L. Physical properties of low-trans shortenings as affected by emulsifiers and storage conditions. *European Journal of Lipid Science and Technology* 110 (2008): 172–182.

Martini, S., Herrera, M. L., and Hartel, R. W. Effect of cooling rate on nucleation behavior of milk fat-sunflower oil blends. *Journal of Agricultural and Food Chemistry* 49 (2001): 3223–3229.

Martini, S., Herrera, M. L., and Hartel, R. W. Effect of cooling rate on crystallization behavior of milk fat fraction/sunflower oil blends. *Journal of the American Oil Chemists' Society* 79 (2002a): 1055–1062.

Martini, S., Herrera, M. L., and Hartel, R. W. Effect of processing conditions on microstructure of milk fat fraction/sunflower oil blends. *Journal of the American Oil Chemists' Society* 79 (2002b): 1063–1068.

Martini, S., Herrera, M. L., and Marangoni, A. G. New technologies to determine solid fat content on-line. *Journal of the American Oil Chemists' Society* 82 (2005c): 313–317.

Martini, S., Kim, D., Ollivon, M., and Marangoni, A. G. The water vapor permeability of polycrystalline fat barrier films. *Journal of Agricultural and Food Chemistry* 54 (2006a): 1880–1886.

Martini, S., Kim, D., Ollivon, M., and Marangoni, A. G. Structural factors responsible for the permeability of water vapor through fat barrier films. *Food Research International* 39 (2006b): 550–558.

Martini, S., Rincón Cardona, J. A., Ye, Y., et al. Crystallization behavior of high stearic high oleic sunflower oil stearins. *Journal of the American Oil Chemists' Society* 90 (2013): 1773–1786.

McClements, D. J. and Povey, J. W. Solid fat content determination using ultrasonic velocity measurements. *International Journal of Food Science and Technology* 22 (1987): 491–499.

McClements, D. J. and Povey, J. W. Comparison of pulsed NMR and ultrasonic velocity techniques for determining solid fat content. *International Journal of Food Science and Technology* 23 (1988): 159–170.

Miklos, R., Zhang, H., Lametsch, R., and Xu, X. Physicochemical properties of lard-based diacylglycerols in blends with lard. *Food Chemistry* 138 (2013): 608–614.

Morselli Ribeiro, M. D. M., Barrera Arellano, D., and Ferreria Grosso, C. R. The effect of adding oleic acid in the production of stearic acid lipid microparticles with a hydrophilic core by spray-cooling process. *Food Research International* 47 (2012): 38–44.

Ng, W. L. A study of the kinetics of nucleation in a palm oil melt. *Journal of the American Oil Chemists' Society* 67 (1990): 879–882.

Norizzah, A. R., Chong, C. L., Cheow, C. S., and Zaliha, O. Effects of chemical interesterification on physicochemical properties of palm stearin and palm kernel olein blends. *Food Chemistry* 86 (2004): 229–235.

O'Brien, R. D. Chapter 4: Fats and oils formulations. In *Fats and Oils: Formulating and Processing for Applications*, pp. 263–345. New York: CRC Press (2009).

Peyronel, F. and Campos, R. Methods to study the physical properties of fats. In A. G. Marangoni (ed.), *Structure-Function Analysis of Edible Fats*, pp. 231–294. Urbana, IL: AOCS Press (2012).

Puppo, M. C., Martini, S., Hartel, R. W., and Herrera, M. L. Effect of sucrose esters on isothermal crystallization and rheological behaviors of blends of high-melting milk fat fraction and sunflower oil. *Journal of Food Science* 67 (2002): 3419–3426.

Rashid, N. A., Let, C. C., Seng, C. C., and Omar, Z. Crystallization kinetics of palm stearin, palm kernel olein and their blends. *LWT—Food Science and Technology* 46 (2012): 571–573.

Ray, J., MacNaughtan, W., Chong, P. S., Vieira, J., and Wolf, B. The effect of limonene on the crystallization of cocoa butter. *Journal of the American Oil Chemists' Society* 89 (2012): 437–445.

Reyes-Hernandez, J., Dibilidox-Alvarado, E., Charo-Alonso, M. A., and Toro-Vazquez, J. F. Physicochemical and rheological properties of crystallized blends containing trans-free and partially hydrogenated soybean oil. *Journal of the American Oil Chemists' Society* 84 (2007): 1081–1093.

Rincón-Cardona, J. A., Martini, S., Candal, R. J., and Herrera, M. L. Polymorphic behavior during isothermal crystallization of high stearic high oleic sunflower oil stearins. *Food Research International* 51 (2013): 86–97.

Rousset, Ph., Rappz, M., and Minner, E. Polymorphism and solidification kinetics of the binary system POS-SOS. *Journal of the American Oil Chemists' Society* 75 (1998): 857–864.

Saadi, S., Ariffin, A. A., Ghazali, H. M., et al. Application of differential scanning calorimetry (DSC), HPLC and pNMR for interpretation primary crystallization caused by combined low and high melting TAGs. *Food Chemistry* 132 (2012): 603–612.

Saberi, A. H., Kee, B. B., Oi-Ming, L., and Miskandar, M. S. Physicochemical properties of various palm-based diacylglycerols oils in comparison with their corresponding palm-based oils. *Food Chemistry* 127 (2011a): 1031–1038.

Saberi, A. H., Lai, O., and Toro-Vazquez, J. F. Crystallization kinetics of palm oil in blends with palm-based diacylglycerol. *Food Research International* 44 (2011b): 425–435.

Saitou, K., Mitsui, Y., Shimizu, M., et al. Crystallization behavior of diacylglycerols-rich oils produced from rapeseed oil. *Journal of the American Oil Chemists' Society* 89 (2012): 1231–1239.

Santacatalina, J. V., Garcia-Perez, J. V., Corona, E., and Benedito, J. Ultrasonic monitoring of lard crystallization during storage. *Food Research International* 44 (2011): 146–155.

Sato, K. Crystallization behavior of fats and lipids—A review. *Chemical Engineering Science* 56 (2001): 2255–2265.

Sato, K., Ueno, S., and Yano, J. Molecular interactions and kinetic properties of fats. *Progress in Lipid Research* 38 (1999): 91–116.

Strickland-Constable, R. F. Nucleation of solids. In *Kinetics and Mechanism of Crystallization*, pp. 74–129. London: Academic Press (1968).

Suzuki, A., Lee, J., Padilla, S., and Martini, S. Altering functional properties of fats using power ultrasound. *Journal of Food Science* 75 (2010): E208–E214.

Takeuchi, M., Ueno, S., Flöter, E., and Sato, K. Binary phase behavior of 1,3-distearoyl-2-oleoyl-sn-glycerol (SOS) and 1,3-distearoyl-2-linoleoyl-sn-glycerol (SLS). *Journal of the American Oil Chemists' Society* 79 (2002a): 627–632.

Takeuchi, M., Ueno, S., and Sato, K. Crystallization kinetics of polymorphic forms of a molecular compound constructed by SOS (1,3-distearoyl-2-oleoyl-sn-glycerol) and SSO (1,2,-distearoyl-3-oleoyl-rac-glycerol). *Food Research International* 35 (2002b): 919–926.

Timms, R. E. The phase behavior of mixtures of cocoa butter and milk fat. *LWT—Food Science and Technology* 13 (1980): 61–65.

Timms, R. E. Phase behavior of fats and their mixtures. *Progress in Lipid Research* 23 (1984): 1–38.

Timms, R. E. Physical properties of oils and mixtures of oils. *Journal of the American Oil Chemists' Society* 62 (1985): 241–249.

Toro-Vazquez, J. F., Briceño-Montelongo, M., Dibilidox-Alvarado, E., Charo-Alonso, M., and Reyes-Hernandez, J. Crystallization kinetics of palm stearin in blends with sesame seed oil. *Journal of the American Oil Chemists' Society* 77 (2000): 297–310.

Toro-Vazquez, J. F., Rangel-Vargas, E., Dibilidox-Albarado, E., and Charo-Alonso, M. A. Crystallization of cocoa butter with and without polar lipids evaluated by rheometry, calorimetry and polarized light microscopy. *European Journal of Lipid Science and Technology* 107 (2005): 641–655.

Verstringe, S., Danthine, S., Blecker, C., Depypere, F., and Dewenttinck, K. Influence of monopalmitin on the isothermal crystallization mechanism of palm oil. *Food Research International* 51 (2013): 344–353.

Wagh, A., Walsh, M., and Martini, S. Effect of lactose monolaurate and high intensity ultrasound on crystallization behavior of anhydrous milk fat. *Journal of the American Oil Chemists' Society* 90 (2013): 977–987.

Wang, F., Liu, Y., Jin, Q., et al. Kinetic analysis of isothermal crystallization in hydrogenated palm kernel stearin with emulsifier mixtures. *Food Research International* 44 (2011): 3021–3025.

Wille, R. L. and Lutton, E. S. Polymorphism of cocoa butter. *Journal of the American Oil Chemists' Society* 43 (1966): 491–496.

Wright, A. J., McGauley, S. E., Narine, S. S., et al. Solvent effects on the crystallization behavior of milk fat fractions. *Journal of Agricultural and Food Chemistry* 48 (2000): 1033–1040.

Ye, Y., Wagh, A., and Martini, S. Using high intensity ultrasound as a tool to change the functional properties of interesterified soybean oil. *Journal of Agricultural and Food Chemistry* 59 (2011): 10712–10722.

Zhang, L., Ueno, S., Miura, S., and Sato, K. Binary phase behavior of 1,3-dipalmitoyl-2-oleoyl-*sn*-glycerol and 1,2-dioleyl-3-palmitoyl-*rac*-glycerol. *Journal of the American Oil Chemists' Society* 84 (2007): 219–227.

Zhou, Y. and Hartel, R. W. Phase behavior of model lipid systems: Solubility of high-melting fats in low-melting fats. *Journal of the American Oil Chemists' Society* 83 (2006): 505–511.

Zhou, S., Zhang, F., Jin, Q., et al. Characterization of palm kernel oil, palm stearin, and palm olein blends in isosolid diagrams. *European Journal of Lipid Science and Technology* 112 (2010): 1041–1047.

Section III

DSC in Food Technology: Palm Products, Lipid Modification, Emulsion Stability

8 Application of DSC Analysis in Palm Oil, Palm Kernel Oil, and Coconut Oil
From Thermal Behaviors to Quality Parameters

Chin Ping Tan, Siou Pei Ng, and Hong Kwong Lim

CONTENTS

8.1 OILS AND FATS

Oils and fats, basic components of man's caloric diet, have been used since time immemorial as food, as food ingredients, and in preparation of other foods. Oil-bearing fruits, nuts, and seeds have been grown and used for food for many centuries. Some 250,000 plant species are known, with 4500 species being examined for oil (Orthoefer,

1996). One major source of vegetable oils is the oil-bearing fruits and nuts of perennial trees such as coconut, palm, and palm kernel (Padley, 1994). The perennial oil-bearing trees are grown in the tropical regions. Today, palm-based oils and coconut oil are raw materials for cooking oil, margarine, shortening, and other specialty or tailored products that have become essential ingredients in food products prepared in the home, in restaurants, and by the food processing industry. Spurred by income and population growth in developing countries, as well as the rapidly expanding food industry in Asia and other regions, the global growth in consumption and trade of vegetable oils is outpacing that of most other agricultural commodities (Morgan and Sanford, 1996).

8.1.1 PALM OIL PRODUCTS

Today, palm oil is the world's largest edible oil in terms of production and trade. Palm oil has surpassed soybean oil as the most important vegetable oil in the world. Global production of palm oil was 46 million tons in 2009–2010, which accounted for 31% of the total global production of 139 million tons (Gunstone, 2010). Palm oil, the principal palmitic acid oil, is derived from the fruit of the oil palm tree, *Elaes guineensis*, which has the appearance of a date palm with a large head of pinnate feathery fronds growing from a sturdy perennial trunk (O'Brien, 1998). Crude palm oil has a deep orange-red color due to its high carotene content: 0.03%–0.15%, of which 90% consists of α- and β-carotene. It has a characteristic "nutty" or "fruity" flavor. The oxidative stability of palm oil is affected by the presence of high levels of carotene, which acts as a prooxidant even in the presence of high tocopherol and tocotrienol concentrations. Therefore, it must be bleached (which eliminates the carotene) as well as refined and deodorized. Today, palm oil and its products are being used more often in foods such as cooking oil, margarines, shortenings, and confectionery products. The versatility of palm oil, and its adaptability to different food applications, are due to its chemical composition (Yap et al., 1989). Palm oil differs from many of the common vegetable oils in its high level of palmitic acid, about 44%. It has been found that addition of palm oil to fats destined for shortening and margarine production has a beneficial effect on their polymorphic stability (Berger, 1986).

Palm oil is commercially fractionated on a large scale, resulting in a low-melting and a high-melting fraction, known in the trade as palm olein and palm stearin, respectively (Salunkhe et al., 1992a). This fractionation process can be effected in the dry form in the presence of a detergent or solvent. The method employed, to a certain extent, determines some of the chemical and physical properties of the oleins and stearins produced, especially the stearins (Basiron, 1996). By varying the fractionation methods and conditions used, a range of stearins with differing chemical and physical properties can be produced, while keeping the chemical and physical properties of oleins within a very narrow range of values (Basiron, 1996). Palm olein is used as a liquid frying oil, and palm stearin as a solid component in margarine and shortening blends (Law and Thiagarajan, 1990).

8.1.2 PALM KERNEL OIL

Palm kernels are by-products from palm oil mills (Basiron, 1996). Palm kernel oil is obtained as a minor product during processing of oil palm fruit and represents about

2%–4% of the harvested palm fruit bunch (Gascon et al., 1989). It is extracted by mechanical pressure screw processing, solvent extraction, or preprocessing followed by solvent extraction (Tang and Teoh, 1985). Palm kernel oil is a lauric oil. The sharp melt, low melting point, and low level of unsaturates make palm kernel oil particularly suited for low-moisture food products in applications such as confectionery fats, candy centers, cookie fillers, nut roasting, coffee whiteners, and spray oils (Young, 1983; Traitler and Dieffenbacher, 1985).

8.1.3   COCONUT OIL

In the tropics, the coconut palm (*Cocos nucifera* L.), a member of the Palmae family, is one of the most useful trees (Salunkhe et al., 1992b). Coconut palms are traditionally found in coastal regions in Asia and the Pacific islands. Coconut oil obtained from the nuts of the coconut palm is a commercially important oil in the lauric oil group (Langstraat and Jurgens, 1976). The Philippines is the most important producing and exporting country, followed by Indonesia, India, Sri Lanka, Malaysia, and Oceania. More than 90% of the fatty acids of coconut oil are saturated (Graalmann, 1990). The saturated character of the oil imparts a strong resistance to oxidative rancidity (Canapi et al., 1996). Edible products derived from coconut oil rely upon high stability, high solids, and sharp melting characteristics (Orthoefer, 1996). In coconut-producing countries, coconut oil is used extensively as a frying medium. Physical blends and interesterified mixtures of coconut oil and hydrogenated coconut oil are processed into margarines and shortenings. Coconut oil is also widely used as cream fat and as a component in biscuit cream and confectionery oil (Canapi et al., 1996).

8.2   CHEMICAL COMPOSITION OF PALM-BASED OILS AND COCONUT OIL

The fatty acid (FA) composition of palm-based oils and coconut oil is shown in Table 8.1. The proportions of saturated (SFA), monounsaturated (MUFA), and polyunsaturated (PUFA) FA data are also tabulated in Table 8.1. Palm oil, palm superolein, red palm olein, and palm stearin are oil samples obtained from palm pulp. In these oil samples, the SFA accounted for more than 43%, while oleic acid made up about 20%–44% of the total FA. The PUFA content ranged from 4% to 12%. These oils can be distinguished from other oils by high levels (>35%) of palmitic acid. The fractionation of palm oil into olein and stearin fractions has a significant influence on fatty acid composition. As shown in Table 8.1, the palmitic acid tends to migrate into the palm stearin. Tan and Che Man (2000) studied two types of palm olein fractions. These two olein fractions (palm superolein and red palm olein) of palm oil show comparable FA composition. The FA composition of palm kernel oil is closer to that of coconut oil (Tan and Che Man, 2000). Although palm kernel oil and palm oil are derived from the same plant origin, both oils differ considerably in their characteristics and properties. Palm kernel oil is similar to palm oil in that they are both high in lauric and myristic acids. Nevertheless, palm kernel oil has a lower content of short-chain FA.

TABLE 8.1

Fatty Acid Composition of Palm-Based Oils and Coconut Oil

	Oil Sample					
Fatty Acid	Palm Oil	Palm Superolein	Red Palm Olein	Palm Stearin	Coconut Oil	Palm Kernel Oil
C6:0	–	–	–	–	1.4	0.5
C8:0	–	–	–	–	13.5	6.4
C10:0	–	–	–	–	8.7	5.2
C12:0	0.5	0.7	0.7	0.4	51.1	55.8
C14:0	1.7	1.5	1.7	2.1	14.5	14.7
C16:0	48.7	41.6	38.1	68.3	5.5	5.8
C18:0	3.9	3.8	3.4	4.0	1.4	1.3
C18:1	37.1	42.0	44.2	20.6	3.3	8.9
C18:2	8.1	10.4	12.0	4.6	0.7	1.5
SFA	54.7	47.6	43.8	74.8	96.0	89.7
MUFA	37.1	42.0	44.2	20.6	3.3	8.9
PUFA	8.1	10.4	12.0	4.6	0.7	1.5

Source: Tan, C.P. and Che Man, Y.B., *Journal of the American Oil Chemists' Society*, 77, 143–155, 2000.

Note: C6:0, caproic; C8:0, caprylic; C10:0, capric; C12:0, lauric; C14:0, myristic; C16:0, palmitic; C18:0, stearic; C18:1, oleic; C18:2, linoleic; C18:3, linolenic.

The triacylglycerol (TAG) composition of palm-based oils and coconut oil is presented in Table 8.2. For palm-based oils (palm oil, palm superolein, red palm olein, and palm stearin), POO and PPO (P, palmitic; O, oleic) account for up to 56% of the TAG (Tan and Che Man, 2000). Palm kernel and coconut oils are characterized as hard oils. There are many similarities between palm kernel and coconut oils in terms of their TAG compositions.

8.3   APPLICATIONS OF DSC IN OILS AND FATS

Differential scanning calorimetry (DSC) is the most widely used thermoanalytical technique for assessment of physical characteristics of oil and fats in terms of their melting and crystallization behavior (Che Man et al., 1999; Tan and Che Man, 2000; Kellens et al., 2007; Braipson-Danthine and Gibon, 2007). The interpretation of DSC thermal curves is not easy, especially the melting curves due to the polymorphic transitions that may occur during melting (Kellens et al., 2007; Chu et al., 2002). The melting and crystallization characteristics of an oil sample in a DSC scan can be indicated by various parameters. Figure 8.1 shows typical melting and crystallization curves of palm oil. The major parameters of a DSC thermal curve are thermal dynamic transition peaks: (T_{on}) (onset temperature) and (T_{off}) (offset temperature). The T_{on} of the crystallization curve is recorded at the beginning of oil/fat crystal formation, while the T_{off} of the melting properties is recorded when the oil/fat begins to melt.

TABLE 8.2

Triacylglycerol Composition of Palm-Based Oils and Coconut Oil

	Oil Sample					
Triacylglycerol	Palm Oil	Palm Superolein	Red Palm Olein	Palm Stearin	Coconut Oil	Palm Kernel Oil
CCLa	–	–	–	–	12.9	6.8
CLaLa	–	–	–	–	17.4	9.9
LaLaLa	–	–	–	–	21.2	21.2
LaLaM	–	–	–	–	18.0	17.0
LaLaO	–	–	–	–	3.1	5.3
LaMM	–	–	–	–	10.2	8.8
MMM	0.4	0.7	0.6	0.2	–	–
LaLaP	–	–	–	–	0.5	1.2
LaMO	–	–	–	–	2.4	4.6
MPL	2.4	3.2	3.7	1.0	–	–
LaMP	–	–	–	–	5.5	4.6
LaOO	–	–	–	–	1.1	3.8
LaPO	–	–	–	–	1.6	4.3
LaPP+MMO	–	–	–	–	2.1	1.9
OOL	0.7	0.7	0.8	0.1	–	–
MMP	1.8	2.3	2.6	0.8	0.2	0.7
MOO	–	–	–	–	0.8	2.0
MPO+POL	–	–	–	–	1.1	2.1
POL	10.1	12.8	15.8	5.3	–	–
PPL	9.8	10.7	11.2	7.8	–	0.6
MPP	0.6	–	–	2.3	–	–
OOO	4.1	4.9	5.6	1.8	0.6	1.4
POO	24.2	29.1	36.3	12.0	0.3	1.9
PPO	31.1	27.2	17.1	29.8	0.7	1.1
PPP	5.9	–	0.1	29.2	0.6	0.1
SOO	2.3	3.1	3.6	0.8	–	0.4
PSO	5.1	5.0	2.5	3.8	–	0.4
PPS	0.9	–	–	5.2	–	–
SSO	0.5	0.4	–	–	–	–

Source: Tan, C.P. and Che Man, Y.B., *Journal of the American Oil Chemists' Society*, 77, 143–155, 2000.

C, capric; L, linoleic; La, lauric; M, myristic; O, oleic; P, palmitic; S, stearic.

8.3.1 MELTING AND CRYSTALLIZATION BEHAVIORS OF PALM-BASED OILS AND COCONUT OIL BY DSC

It is well known that the thermal behaviors of oils and fats are profoundly influenced by physicochemical interactions, particularly those among TAG. These interactions of TAG are often complex, and a full understanding of their thermal behaviors

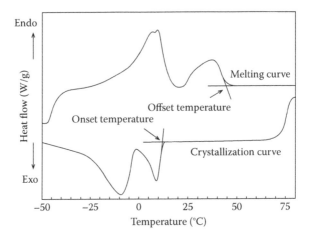

FIGURE 8.1 Typical DSC melting and crystallization curves of palm oil. The techniques used for determination of the DSC parameters are illustrated by constructed lines. The onset and offset temperatures corresponded closely to the intersection of the extrapolated baseline and the tangent line (leading edge) of the peak. (Reprinted with permission from Tan, C.P. and Che Man, Y.B., *Journal of the American Oil Chemists' Society*, 77, 143–155, 2000.)

requires examination of these interactions. Beyond these complex behaviors, TAG also shows temperature-dependent polymorphic behavior, complicating the interpretation of DSC thermal curves. Table 8.3 lists various studies to evaluate thermal behaviors of palm-based oils and coconut oil using DSC.

Melting and crystallization, two commonly used physical events to characterize thermal behavior of oil samples, require the intake or release of thermal enthalpy. DSC is eminently suitable to determine these physical properties of oil samples. Generally, in melting curves of oil samples, complex features that were not easily interpretable, such as shoulders not separable from peaks, were noticed. These results illustrated the complex nature of TAG in oil samples. This is a consequence of the known phenomenon of polymorphism of natural oils and fats, which has interested researchers for many years. The crystallization of oils and fats is a complex phenomenon, which has also interested researchers for many years (Smith et al., 1994). The process is complicated by the slow rate of crystal growth, caused by the polymorphic behavior of fats and the complex shape. Crystal lattice imperfections develop during crystallization and exert a major effect on oil and fat processing.

Thermal behavior of palm kernel oil products was studied by Rossell (1975). The author clearly stated that DSC cooling curves of these lauric fats are complicated and at times difficult to understand and explain. In contrast, Berger and Akehurst (1966) claimed that cooling curves are simpler and easier to interpret than heating curves. Lambelet and Ganguli (1983) and Coni et al. (1994) further confirmed the finding of Berger and Akehurst (1966). The difference between these two conclusions (Rossell vs. Berger and Akehurst) is probably due to the fact that Berger and Akehurst (1966) confined their attention to nonlauric oils, which do not display the rapid crystallization found in Rossell's work.

TABLE 8.3
Thermal Behavior of Palm-Based and Coconut Oils by DSC

Compound	Thermal Property	References
Failed-batch palm oil	M/C	Che Man and Swe (1995)
Palm kernel oil products	M/C	Rossell (1975)
Palm midfractions	M	Md.Ali and Dimick (1994)
Palm oil	M/C	Kawamura (1979, 1980), Jacobsberg and Ho (1976), Oh (1985), and Ng and Oh (1994)
Palm oil and its products	M	Yap et al. (1989), Ng (1990), and Che Man et al. (1999)
Palm olein	M/C	Swe et al. (1995) and Siew and Ng (1996)
Palm stearin and hydrogenated palm stearin	M	Busfield and Proschogo (1990)
Palm, palm kernel, and coconut oil	M/C	Dyszel and Baish (1992)
Palm, palm kernel, and coconut oil	M/C	Tan and Che Man (2002)
Palm oil and palm stearin	M/C	Abdul Azis et al. (2011)
Palm kernel oil	M/C	Anihouvi et al. (2013)

C, crystallization; M, melting.

The DSC melting curves of palm-based oils and coconut are presented in Figure 8.2 (palm oil, palm superolein, and red palm olein) and Figure 8.3 (palm stearin, coconut oil, and palm kernel oil). The melting curve of palm oil showed two major endothermic regions (Figure 8.2), corresponding to endothermic transitions of the olein (lower-temperature peak) and stearin (higher-temperature peak) fractions. Palm superolein and red palm olein showed only one major endotherm in the lower temperature region (olein fraction), and both oil samples have typical melting curves. Palm stearin shows both endothermic regions (Figure 8.3); the higher region is distinguished by a high peak (and two small fusion peaks) preceding the low-temperature region (consisting of four small merging peaks). The small low-temperature peak in the melting curve of palm stearin indicated that a small amount of olein fraction is still trapped in this oil sample after fractionation. For coconut oil, a major endothermic peak with a shoulder peak and a small distinct endothermic peak are observed (Figure 8.3), while a major endothermic peak with a shoulder peak and two small fusion peaks are noticed in palm kernel oil (Figure 8.3).

The crystallization curves of palm-based oils and coconut oil are illustrated in Figure 8.4 (palm oil, palm superolein, and red palm olein) and Figure 8.5 (palm stearin, coconut oil, and palm kernel oil). For palm oil, the crystallization curve displays two major exothermic regions. The higher-temperature region defines crystallization of the stearin fraction, while the lower-temperature region indicates crystallization of the olein fraction (Figure 8.4). The higher-temperature region is absent in the crystallization

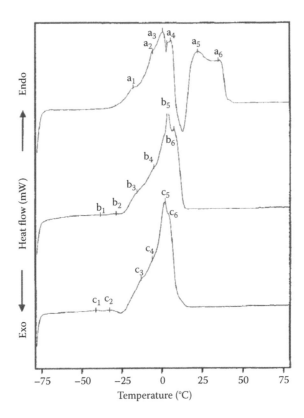

FIGURE 8.2 DSC melting curves of (a) palm oil, (b) palm superolein, and (c) red palm olein. (Reprinted with permission from Tan, C.P. and Che Man, Y.B., *Journal of the American Oil Chemists' Society*, 77, 143–155, 2000.)

curves of palm superolein and red palm olein. The crystallization curve of palm stearin has both regions, although the higher-temperature region is the major feature (Figure 8.5). The crystallization curve of coconut oil shows two distinct exothermic peaks, whereas palm kernel oil shows two overlapping exothermic peaks (Figure 8.5).

8.3.2 DSC as a Tool to Monitor Physical Processes in the Palm Oil Industry

Fractionation and blending are two widely used physical processes that help to alter the melting profile, physical properties, and chemical composition of the feed oil or fat, and the products so produced are in effect new ingredients suitable for use in applications in which the original oil or fat could never have been used or would have performed poorly (Krishnamurthy and Kellens, 1996; Ramli et al., 2014; Jahurul et al., 2014). Fractionation is a thermomechanical process, which eventually leads to two new products, olein and stearin. Meanwhile, blending is one of the cheapest processes for varying the oil composition to improve the physical and nutritional quality by admixture of two or more edible oils. Blending, therefore, helps to diversify this

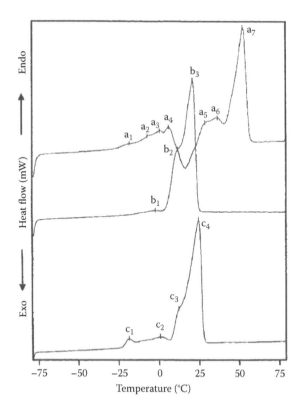

FIGURE 8.3 DSC melting curves of (a) palm stearin, (b) coconut oil, and (c) red palm kernel oil. (Reprinted with permission from Tan, C.P. and Che Man, Y.B., *Journal of the American Oil Chemists' Society*, 77, 143–155, 2000.)

oil's range of applications. Being able to predict changes in physical properties based on chemical property changes will be important in optimizing a blend or fractionated oil to suit a given application perfectly. These physical processes have enabled the introduction of many fat products, particularly palm oil, into new food applications (Talbot, 1995). DSC has been widely used to monitor these two processes (Herrera and Añón, 1991; D'Souza et al., 1991; Dimick et al., 1996; Che Man et al., 1999; Nor Hayati et al., 2009; Ramli et al., 2014).

DSC profiles resolved the issue of whether there are actually any substantive differences between the solid fractions obtained by simple thermal fractionation. Che Man et al. (1999) described the thermal profiles of palm oil and its products. The authors outlined the importance of thermal behavior of various palm oil products and concluded that the thermal profiles can be used as guidelines for fractionation of crude palm oil or refined, bleached, and deodorized (RBD) palm oil. The crystallization of palm oil is influenced by the presence of free FA, partial glycerides, and oxidation products (Jacobsberg and Ho, 1976). Ng (1990) studied the nucleation from a supercooled melt of palm oil by optical microscopy and DSC. The author confirmed that palm oil exhibits a rather simple cooling curve with high- and low-temperature

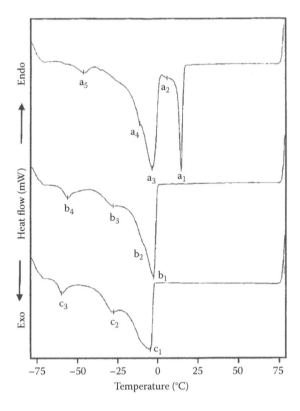

FIGURE 8.4 DSC crystallization curves of (a) palm oil, (b) palm superolein, and (c) red palm olein. (Reprinted with permission from Tan, C.P. and Che Man, Y.B., *Journal of the American Oil Chemists' Society*, 77, 143–155, 2000.)

exotherms, which is exclusively related to the "hard" and "soft" components of the oil. On the other hand, Che Man and Swe (1995) determined the thermal behavior of failed-batch palm oil. The authors concluded that a rapid and sudden surge of heat demand is observed for samples from failed crystallizers.

Previous reports (Nor Hayati et al., 2007, 2009) indicated that binary blends of palm kernel olein with soybean oil could improve the oxidative and physical stability of soybean oil–based emulsions to a certain extent. Figure 8.6 shows a typical series of DSC melting curves of oil blends consisting of palm kernel oil and soybean oil. Blending with fully hydrogenated soybean oil is primarily used to formulate products for bakery applications (NorAini et al., 2001). In these studies, DSC is one of the important tools to evaluate the developed fat/oil blends. In addition, several studies reported palm-based diacylglycerols as major components in blending with palm-based oil, palm kernel olein, and sunflower oil (Saberi et al., 2011, 2012) and blends of the stearin fraction of palm-based DAGs with sunflower oil, palm olein, and palm midfractions (Latip et al., 2013).

Most recently, Jahurul et al. (2014) produced hard cocoa butter replacers from palm stearin and mango seed fat blends. DSC has been used to monitor changes in

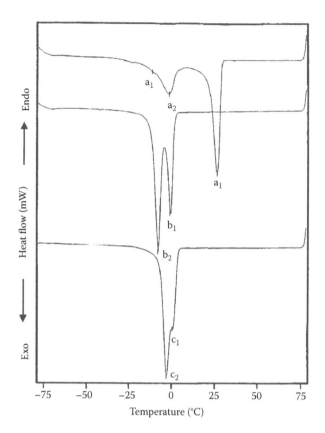

FIGURE 8.5 DSC crystallization curves of (a) palm stearin, (b) coconut oil, and (c) red palm kernel oil. (Reprinted with permission from Tan, C.P. and Che Man, Y.B., *Journal of the American Oil Chemists' Society*, 77, 143–155, 2000.)

melting behaviors of fat blends and subsequently determine fat blends that are suitable as cocoa butter replacers. This study revealed that cocoa butter replacers could be prepared by blending palm stearin and mango seed fat, and they could be utilized by chocolate manufacturers in tropical countries. In another study, blending of palm midfraction (PMF), palm stearin (POs), and olive oil has been shown to produce an alternative fat source for a cocoa butter substitute (Ramli et al., 2014). The ternary fat blend is found to exhibit better oxidative stability and may be suitable as a cocoa butter substitute.

Blending of palm olein with other vegetable oils provides an opportunity to reduce the melting point of palm olein and consequently enhance its market acceptability. Recently, DSC melting curves of the oil blends from palm olein and canola oil have been used to trace some new polymorphs of the TAG (Siddique et al., 2010). This study helps the oil industry to identify the most economically viable oil blends for cooking purposes, with maximum nutrition as well as desirable physicochemical properties. In another study, two ternary systems of fats consisting of low-erucic-acid rapeseed oil/hydrogenated low-erucic-acid rapeseed oil and palm

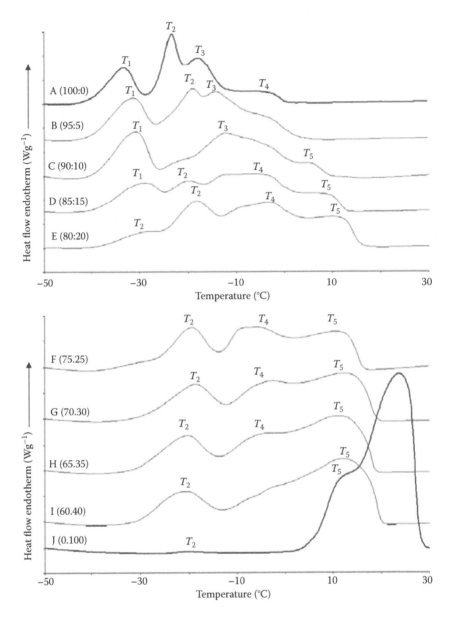

FIGURE 8.6 DSC melting curves for soybean oil (A), palm kernel olein (J), and their blends (soybean oil: palm kernel olein) (B–I). (Reprinted with permission from Nor Hayati, I., Che Man, Y.B., Tan, C.P., and Nor Aini, I., *International Journal of Food Science and Technology*, 44, 152–161, 2009.)

oil/hydrogenated palm oil have been produced (Danthine and Deroanne, 2003) and studied for their physical properties, such as solid fat content, DSC melting curves, and polymorphism.

8.3.3   DSC Characterization of Palm-Based Structured Oils and Fats

Besides fractionation and blending, fat and oil applications can be broadened by modifying the materials through interesterification to produce various functional structured oils and fats (Petrauskaite et al., 1998). The current popularity of structured oils and fats arises from the fact that they are technologically and nutritionally improved lipids. Structured oils and fats are TAG that have been modified to change the fatty acid composition and positional distribution of the glycerol backbone by chemically or enzymatically catalyzed reactions.

DSC has been frequently used to monitor the interesterification processes of various palm-based oils or oil blends. Changes in DSC melting curves of palm oil, sunflower oil, palm kernel olein, and their blends in various ratios have been evaluated by using a combination of blending and chemical interesterification techniques (Noor Lida et al., 2006). Grimaldi et al. (2001) also evaluated interesterified oil blends consisting of palm oil and palm kernel oil using DSC. In another study, production of cocoa butter equivalent through enzymic interesterification of palm oil midfraction with stearic acid was carried out by Undurraga et al. (2001). In a more recent study, Adhikari and Hu (2012) studied the effect of enzymatic interesterification of three different mixtures of rice bran oil, shea olein, and palm stearin using DSC as a tool to monitor crystallization behaviors of esterified oil blends.

Most recently, our group (Ng et al., 2014) studied the crystallization and melting curves obtained by DSC for palm olein-based diacylglycerol and palm superolein and various blends (0–100 wt.% palm olein-based diacylglycerol, at 10 wt.% increments). These DSC thermal curves are shown in Figure 8.7. The different characteristic curves of crystallization and melting profile indicated the composition of the diacylglycerol content in each of the oil blends. The crystallization profile of palm olein-based diacylglycerol displayed three exothermic transitions (Figure 8.7). The first crystallization peak appeared at 32.03°C, the second at 7.62°C, and the third at −9.54°C (Ng et al., 2014). In comparison to the palm olein-based diacylglycerol (Figure 8.7), the exothermic transition of palm superolein showed only one major sharp peak in the low-temperature region at −1.02°C. In terms of the binary blends, increasing palm olein-based diacylglycerol above 30 wt.% changed the crystallization profiles, and this was mainly dependent on the addition of palm olein-based diacylglycerol in the blends. The melting profiles of palm olein-based diacylglycerol and its binary blends with palm superolein showed melting in three fractions, resulting in three discrete melting peaks, while palm superolein showed only one endothermic transition (Figure 8.7).

8.3.4   Quantitative Chemical Analysis of Palm Oil–Based Products by DSC

DSC is a first-order instrument. Today, analytical chemists are busy using various tools from mathematical and statistical techniques to perform multianalyte

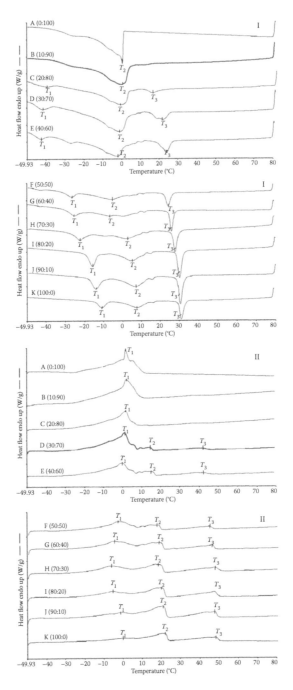

FIGURE 8.7 DSC crystallization (I) and melting (II) curves of palm olein-based diacylglycerol (K), palm superolein (A), and their binary blends (B–J). (Reprinted with permission from Ng, S.P., Lai, O.M., Abas, F., Lim, H.K., Beh, B.K., Ling, T.C., and Tan, C.P., *Food Research International*, 55, 62–69, 2014.)

quantitative analysis with data from first-order instruments (Booksh and Kowalski, 1994). Haryati et al. (1997) determined iodine value (IV) in palm oil products using DSC. On the other hand, Haryati (1999) also indicated that the energy of the cooling curve is sufficient to predict the IV of palm oil products. The author concluded that the DSC methods are more consistent than the standard methods and minimize the uses of solvents and labor. Chu et al. (2002) also determined the IV of various palm olein blends using DSC.

Tan (2001) also determined three important quality indices in the deep-fat frying industry, namely, total polar compounds (TPC), free fatty acid (FFA), and IV of heated palm olein, using the crystallization curves of DSC. In this study, heated oils exhibited a simple curve after cooling in the DSC with a well-defined single crystallization peak, as shown in Figure 8.8. It is clear that the DSC traces are affected in a systematic way by the increasing TPC level in palm olein (Tan and Che Man, 1999). In general, as the polar compounds increase, the peak of crystallization shifts to lower temperatures and becomes broader (an increase in the peak temperature range). This

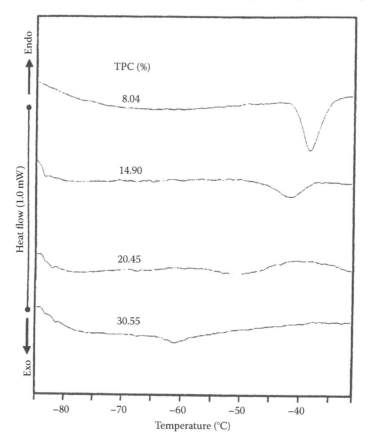

FIGURE 8.8 DSC crystallization curves of palm olein with different concentrations of total polar compounds. (Reprinted with permission from Tan, C.P. and Che Man, Y.B., *Journal of the American Oil Chemists' Society*, 76, 1047–1057, 1999.)

study revealed that a single DSC cooling curve could predict the TPC, FFA, and IV of palm olein using stepwise regression analysis. This finding showed that the DSC measurements were well correlated with the American Oil Chemists' Society official standard methods. This bodes well for the wider application of DSC in quality control. The speed and ease of data manipulation brought about by computer control during this study represent the further benefits of the DSC method. It is especially attractive to develop a DSC method so that the tedious time- and chemical-consuming standard chemical method can be avoided in future routine testing.

Food authentication is the process by which a food is verified as complying with its label description (Dennis, 1998). Appropriate techniques for authenticating oils and fats ensure that correctly described products remain available to the consumer and that consumer confidence is maintained, which, in turn, ensures a place in the market for these products. DSC is being increasingly applied to authentication of oils and fats. Detection of adulteration of palm-based oils with a series of lard products (e.g., genuine, randomized, and interesterified forms) using DSC has been carried out by Marikkar et al. (2001, 2002, 2003). Encouraging results have been obtained for detecting lard in palm oil and palm olein using DSC. The authors' results showed that the detection limit of 1% lard could be clearly seen in the crystallization curves.

8.3.5   LIPID OXIDATION AND QUALITY CHANGES IN PALM-BASED OILS BY DSC

Lipid oxidation of edible oils is a problem for consumer acceptance of many products. It is one of the major deteriorative reactions in edible oils, and often results in a significant loss of quality. Therefore, fast and reliable instrumental methods for monitoring or testing both the immediate quality and the expected oxidation stage of edible oils are in great demand. In light of this knowledge, studies have been carried out on heated palm olein by calorimetric investigations ranging from $-100°C$ to $100°C$. This work is centered on the changes in thermal profiles of various vegetable oils during heating at $180°C$ and microwave heating.

Effects of microwave heating on the changes in DSC cooling or melting profiles, or both, of three palm oleins are reported by Tan et al. (2002). The influence of microwave power (low-, medium-, and high-power settings) and heating time on lipid deterioration produced during the microwave heating of palm olein is evaluated. The DSC method is based on the cooling or melting curve, or both, of oils at a scanning rate of $5°C/min$. The DSC results are explained on the basis of the endothermic or exothermic peak temperatures. In this study, the chemical analyses of the oils was based on peroxide value (PV), anisidine value (AnV), FFA content, IV, and C18:2/C16:0 ratio. A statistical comparative study was carried out on the DSC and chemical parameters. In general, there are good correlations between these parameters. Likewise, the experimental data showed that, for a given microwave-power setting, a good correlation between DSC curve parameters and heating periods was found. This is an indication that DSC can be used as an objective nonchemical and instrumental technique to monitor lipid oxidation in microwave-heated oils.

The oxidative stability of edible oils can be established by DSC. In the DSC method a pressure DSC cell is charged with oxygen under pressure after loading with sample. The instrument is then run at elevated temperature with isothermal conditions, and the time required for deviation from the baseline to occur is measured. This indicates onset of oxidation, and, by comparing times for different oil samples, allows comparison of their oxidative stability (Hassel, 1976). Normal-pressure DSC can be used, but running times can be several hours rather than 1 or 2 h (Cross, 1970). Also, an evaluation of the efficacy of antioxidants in palm olein by DSC was conducted by Tan et al. (2001a).

A separate comparative study was conducted to determine the oxidative stability of palm olein, palm kernel oil, and coconut by DSC and the oxidative stability index (OSI) instrument (Tan et al., 2001b). The DSC cell was set at four different isothermal temperatures: 110°C, 120°C, 130°C, and 140°C. A dramatic increase in evolved heat was observed, with the appearance of a sharp exothermic curve during initiation of the oxidation reaction. The oxidative induction time (T_o) was automatically determined by extrapolation of the downward portion of the DSC oxidation curve to the time axis. The high correlation found between DSC T_o values and OSI measurements implies that DSC can be recommended as an appropriate objective method for assessing the oxidative stability of palm-based and coconut oils.

DSC has been used to measure antioxidant activity in palm olein (Tan et al., 2001a). The oxidation temperature was 150°C and the oxygen flow rate was 50 ml/min. In this method, the thermal changes occurring during oxidation of the oil are recorded. Generally, the results show that antioxidants act mainly by increasing the induction period of lipid oxidation. The increase in induction time with increasing antioxidant concentrations is best fitted by linear or polynomial equations. These calorimetric results indicate that DSC is a valuable technique in the development and optimization of an antioxidant system for palm olein. This work could contribute to the selection of an appropriate antioxidant (or combination of antioxidants) at the optimum level in palm olein.

8.4   CONCLUSION

DSC thermal profiles of palm-based and coconut oils constitute an interesting area of research. DSC can be applied as a common analytical tool, capable of detecting many different physicochemical properties of these oils. Due to the dual physical states (liquid and solid) of palm-based oils and coconut oil, the versatility of DSC allows us to gain deeper knowledge at a molecular level of the mechanisms underlying the macroscopic processes in these oils. This alone has significant importance. But, furthermore, DSC could also become a unique tool for quality control, since its sensitivity to structural changes can provide alternative insights for analysis. Thermal changes between similar components are fine tools useful for quality control and authentication in the palm oil and coconut oil industries.

REFERENCES

Abdul Azis, A., Mohamud, Y., Roselina, K., Boo, H. C., Nyuk, L. C., and Che Man, Y. B. Rheological, chemical and DSC thermal characteristics of different types of palm oil/palm stearin-based shortenings. *International Food Research Journal* 18(1) (2011): 189–200.

Adhikari, P. and Hu, P. Enzymatic and chemical interesterification of rice bran oil, sheaolein, and palm stearin and comparative study of their physicochemical properties. *Journal of Food Science* 77(12) (2012): C1285–C1292.

Anihouvi, P. P., Blecker, C., Dombree, A., and Danthine, S. Comparative study of thermal and structural behavior of four industrial lauric fats. *Food and Bioprocess Technology* 6(12) (2013): 3381–3391.

Basiron, Y. Palm oil. In Y. H. Hui (ed.), *Bailey's Industrial Oil and Fat Products*, 5th edn., vol. 2, Edible Oils and Oilseeds, pp. 271–376. New York: Wiley (1996).

Berger, K. G. Palm oil products—Why and how to use them. *Food Technology* 40(3) (1986): 72–79.

Berger, K. G. and Akehurst, E. E. Some applications of differential thermal analysis to oils and fats. *Journal of Food Technology* 1 (1966): 237–247.

Booksh, K. S. and Kowalski, B. R. Theory of analytical chemistry. *Analytical Chemistry* 66(15) (1994): 782A–791A.

Braipson-Danthine, S. and Gibon, V. Comparative analysis of triacylglycerol composition, melting properties and polymorphic behavior of palm oil and fractions. *European Journal of Lipid Science and Technology* 109(4) (2007): 359–372.

Busfield, W. K. and Proschogo, P. N. Hydrogenation of palm stearine: Changes in chemical composition and thermal properties. *Journal of the American Oil Chemists' Society* 67(3) (1990): 176–181.

Canapi, E. C., Agustin, Y. T. V., Moro, E. A., Pedrosa Jr., E., and Bendaño, M. L. J. Coconut oil. In Y. H. Hui (ed.), *Bailey's Industrial Oil and Fat Products*, 5th edn., vol. 2, Edible Oils and Oilseeds, pp. 97–124. New York: Wiley (1996).

Che Man, Y. B., Haryati, T., Ghazali, H. M., and Asbi, B. A. Composition and thermal profile of crude palm oil and its products. *Journal of the American Oil Chemists' Society* 76(2) (1999): 237–242.

Che Man, Y. B. and Swe, P. Z. Thermal analysis of failed-batch palm oil by differential scanning calorimetry. *Journal of the American Oil Chemists' Society* 72(12) (1995): 1529–1532.

Chu, B. S., Tan, C. P., Ghazali, H. M., and Lai, O. M. Determination of iodine value of palm olein mixtures using differential scanning calorimetry. *European Journal of Lipid Science and Technology* 104(8) (2002): 472–482.

Coni, M. D. P., Coppolelli, P., and Bocca, A. Detection of animal fats in butter by differential scanning calorimetry: A pilot study. *Journal of the American Oil Chemists' Society* 71(8) (1994): 807–810.

Cross, C. K. Oil stability: A DSC alternative for the active oxygen method. *Journal of the American Oil Chemists' Society* 47(6) (1970): 229–230.

Danthine, S. and Deroanne, C. Blending of hydrogenated low-erucic acid rapeseed oil, low-erucic acid rapeseed oil, and hydrogenated palm oil or palm oil in the preparation of shortenings. *Journal of the American Oil Chemists' Society* 80(11) (2003): 1069–1075.

Dennis, M. J. Recent developments in food authentication. *Analyst* 123(9) (1998): 151R–156R.

Dimick, P. S., Reddy, S. Y., and Ziegler, G. R. Chemical and thermal characteristics of milk-fat fractions isolated by a melt crystallization. *Journal of the American Oil Chemists' Society* 73(12) (1996): 1647–1652.

D'Souza, V., deMan, L., and deMan, J. M. Chemical and physical properties of the high melting glyceride fractions of commercial margarines. *Journal of the American Oil Chemists' Society* 68 (1991): 153–162.

Dyszel, S. M. and Baish, S. K. Characterization of tropical oils by DSC. *Thermochimica Acta* 212 (1992): 39–49.

Gascon, J. P., Noiret, J. M., and Meunier, J. Oil palm. In G. Robbelen, R. K. Downey, and A. Ashri (eds), *Oil Crops of the World*, pp. 475–495. New York: McGraw-Hill (1989).

Graalmann, M. Laurics (Coconut/Palm kernel). In D. R. Erickson (ed.), *Edible Fats and Oils Processing: Basic Principles and Modern Practices: World Conference Proceedings*, pp. 270–274. Illinois: AOCS Press (1990).

Grimaldi, R., Gonçalves, L. A. G., Gioielli, L. A., and Simões, I. S. Interactions in interesterified palm and palm kernel oils mixtures. II—Microscopy and differential scanning calorimetry. *Grasas y Aceites* 52(6) (2001): 363–368.

Gunstone, F. D. Production of vegetable oils—forecasts for 2010/11. *Lipid Technology* 22(8) (2010): 192.

Haryati, T. Development and application of differential scanning calorimetric methods for physical and chemical analysis of palm oil. PhD Dissertation, Universiti Putra Malaysia, Serdang, Malaysia (1999).

Haryati, T., Che Man, Y. B., Asbi, A., Ghazali, H. M., and Buana, L. Determination of iodine value of palm oil by differential scanning calorimetry. *Journal of the American Oil Chemists' Society* 74(8) (1997): 939–942.

Hassel, R. L. Thermal analysis: An alternative method of measuring oil stability. *Journal of the American Oil Chemists' Society* 53(5) (1976): 179–181.

Herrera, M. L. and Añón, M. C. Crystalline fractionation of hydrogenated sunflowerseed oil. II. Differential scanning calorimetry (DSC). *Journal of the American Oil Chemists' Society* 68(11) (1991): 799–803.

Jacobsberg, B. and Ho, O. C. Studies in palm oil crystallization. *Journal of the American Oil Chemists' Society* 53(10) (1976): 609–617.

Jahurul, M. H. A., Zaidul, I. S. M., Nik Norulaini, N. A., Sahena, F., Abedin, M. Z., Mohamed, A., and Mohd Omar, A. K. Hard cocoa butter replacers from mango seed fat and palm stearin. *Food Chemistry* 154 (2014): 323–329.

Kawamura, K. The DSC thermal analysis of crystallization behavior in palm oil. *Journal of the American Oil Chemists' Society* 56(8) (1979): 753–758.

Kawamura, K. The DSC thermal analysis of crystallization behavior in palm oil, II. *Journal of the American Oil Chemists' Society* 57(1) (1980): 48–52.

Kellens, M., Gibon, V., Hendrix, M., and De Greyt, W. Palm oil fractionation. *European Journal of Food Lipid Science and Technology* 109(4) (2007): 336–349.

Krishnamurthy, R. and Kellens, M. Fractionation and winterization. In Y. H. Hui (ed.), *Bailey's Industrial Oil and Fat Products*, 5th edn., vol. 4, Edible Oil and Fat Products: Processing Technology, pp. 301–338. New York: Wiley (1996).

Lambelet, P. and Ganguli, N. C. Detection of pig and buffalo body fat in cow and buffalo ghees by differential scanning calorimetry. *Journal of the American Oil Chemists' Society* 60(5) (1983): 1005–1008.

Langstraat, A. and Jurgens, B. V. Characteristics and composition of vegetable oil bearing materials. *Journal of the American Oil Chemists' Society* 53(6) (1976): 241–244.

Latip, R. A., Lee, Y. Y., Tang, T. K., Phuah, E. T., Tan, C. P., and Lai, O. M. Physicochemical properties and crystallisation behaviour of bakery shortening produced from stearin fraction of palm-based diacyglycerol blended with various vegetable oils. *Food Chemistry* 141(4) (2013): 3938–3946.

Law, K. and Thiagarajan, T. Palm oil—Edible oil of tomorrow. In D. R. Erickson (ed.), *Edible Fats and Oils Processing: Basic Principles and Modern Practices: World Conference Proceedings*, pp. 260–269. Urbana, IL: AOCS Press (1990).

Marikkar, J. M. N., Ghazali, H. M., Che Man, Y. B., and Lai, O. M. Differential scanning calorimetric analysis for determination of some animal fats as adulterants in palm olein. *Journal of Food Lipids* 10(1) (2003): 63–79.

Marikkar, J. M. N., Lai, O. M., Ghazali, H. M., and Che Man, Y. B. Detection of lard and randomized lard as adulterants in refined-bleached-deodorized palm oil by differential scanning calorimetry. *Journal of the American Oil Chemists' Society* 78(11) (2001): 1113–1119.

Marikkar, J. M. N., Lai, O. M., Ghazali, H. M., and Che Man, Y. B. Compositional and thermal analysis of RBD palm oil adulterated with lipase-catalyzed interesterified lard. *Food Chemistry* 76(2) (2002): 249–258.

Md.Ali, A. R. and Dimick, P. S. Thermal analysis of palm mid-fraction, cocoa butter and milk fat blends by differential scanning calorimetry. *Journal of the American Oil Chemists' Society* 71(3) (1994): 299–302.

Morgan, N. R. and Sanford, S. Oilseeds and product trading. In Y. H. Hui (ed.), *Bailey's Industrial Oil and Fat Products*, 5th edn., vol. 1, Edible Oil and Fat Products: General Applications, pp. 45–82. New York: Wiley (1996).

Ng, S. P., Lai, O. M., Faridah, A., Lim, H. K., Beh, B. K., Ling, T. C., and Tan, C. P. Compositional and thermal characteristics of palm olein-based diacylglycerol in blends with palm super olein. *Food Research International* 55 (2014): 62–69.

Ng, W. L. A study of the kinetics of nucleation in a palm oil melt. *Journal of the American Oil Chemists' Society* 67(11) (1990): 879–882.

Ng, W. L. and Oh, C. H. A kinetic study on isothermal crystallization of palm oil by solid fat content measurements. *Journal of the American Oil Chemists' Society* 71(10) (1994): 1135–1139.

Noor Lida, H. M. D., Sundram, K., and Idris, N. A. DSC study on the melting properties of palm oil, sunflower oil, and palm kernel olein blends before and after chemical interesterification. *Journal of the American Oil Chemists' Society* 83(8) (2006): 739–745.

NorAini, I., Razzali, I., Habi, N., Miskandar, M. D., Miskadar, M. S., and Radauan, J. FTN2: Blending of palm products with commercial oil and fats for food applications. In: *2001 PIPOC International Palm Oil Congress Food Technology and Nutrition Conference*, pp. 13–22. 20–22 August, Malaysia (2001).

Nor Hayati, I., Che Man, Y. B., Tan, C. P., and Nor Aini, I. Stability and rheology of concentrated O/W emulsions based on soybean oil/palm kernel olein blends. *Food Research International* 40(8) (2007): 1051–1061.

Nor Hayati, I., Che Man, Y. B., Tan, C. P., and Nor Aini, I. Physicochemical characteristics of soybean oil, palm kernel olein, and their binary blends. *International Journal of Food Science and Technology* 44(1) (2009): 152–161.

O'Brien, R. O. Raw materials. In *Fats and Oils: Formulating and Processing for Applications*, pp. 1–46. Lancester: Technomic Publishing (1998).

Oh, F. C. H. Thermal analysis of palm oil and other oils. *PORIM Bulletin* 11 (1985): 24–33.

Orthoefer, F. T. Vegetable oils. In Y. H. Hui (ed.), *Bailey's Industrial Oil and Fat Products*, 5th edn., vol. 1, Edible Oil and Fat Products: General Applications, pp. 19–44. New York: Wiley (1996).

Padley, F. B. Major vegetable fats. In F. D. Gunstone, F. B. Padley, and J. L. Harwood (eds), *The Lipid Handbook*, 2nd edn., pp. 53–146. Glasgow: Chapman and Hall (1994).

Petrauskaite, V., De Greyt, W., Kellens, M., and Huyghebaert, A. Physical and chemical properties of trans-free fats produced by chemical interesterification of vegetable oil blends. *Journal of American Oil Chemists' Society* 75(4) (1998): 489–493.

Ramli, N., Said, M., Mizan, A. B. A., Tan, Y. N., and Ayob, M. K. Physicochemical properties of blends of palm mid-fraction, palm stearin and olive oil. *Journal of Food Quality* 37(1) (2014): 57–62.

Rossell, J. B. Differential scanning calorimetry of palm kernel oil products. *Journal of the American Oil Chemists' Society* 52(12) (1975): 505–511.

Saberi, A. H., Lai, O.-M., and Miskandar, M. S. Melting and solidification properties of palm-based diacylglycerol, palm kernel olein, and sunflower oil in the preparation of palm-based diacylglycerol-enriched soft tub margarine. *Food and Bioprocess Technology* 5(5) (2012): 1674–1685.

Saberi, A. H., Tan, C. P., and Lai, O.-M. Phase behavior of palm oil in blends with palm-based diacylglycerol. *Journal of the American Oil Chemists' Society* 88(12) (2011): 1857–1865.

Salunkhe, D. K., Chavan, J. K., Adsule, R. N., and Kadam, S. S. Oil palm. In *World Oilseeds: Chemistry, Technology, and Utilization*, pp. 217–248. New York: Van Nostrand Reinhold (1992a).

Salunkhe, D. K., Chavan, J. K., Adsule, R. N., and Kadam, S. S. Coconut. In *World Oilseeds: Chemistry, Technology, and Utilization*, pp. 280–325. New York: Van Nostrand Reinhold (1992b).

Siddique, B. M., Ahmad, A., Ibrahim, M. H., Hena, S., Rafatullahb, M., and Mohd, O. A. K. Physico-chemical properties of blends of palm olein with other vegetable oils. *Grasas y Aceites* 61(4) (2010): 423–429.

Siew, W. L. and Ng, W. L. Crystallisation behaviour of palm oleins. *Elaeis* 8 (1996): 75–82.

Smith, P. R., Cebula, D. J., and Povey, M. J. W. The effect of lauric-based molecules on trilaurin crystallization. *Journal of the American Oil Chemists' Society* 71(12) (1994): 1367–1372.

Swe, P. Z., Che Man, Y. B., and Ghazali, H. M. Composition of crystals of palm olein formed at room temperature. *Journal of the American Oil Chemists' Society* 72(3) (1995): 343–347.

Talbot, G. Fat eutectics and crystallization. In S. T. Beckett (ed.), *Physicochemical Aspects of Food Processing*, pp. 142–166. London: Chapman and Hall (1995).

Tan, C. P. Application of differential scanning calorimetric method for assessing and monitoring various physical and oxidative properties of vegetable oils. PhD Dissertation, University Putra Malaysia, Serdang, Malaysia (2001).

Tan, C. P. and Che Man, Y. B. Quantitative differential scanning calorimetric analysis for determining total polar compounds in heated oils. *Journal of the American Oil Chemists' Society* 76(9) (1999): 1047–1057.

Tan, C. P. and Che Man, Y. B. Differential scanning calorimetric analysis of edible fats and oils: Comparison of thermal properties and chemical composition. *Journal of the American Oil Chemists' Society* 77(2) (2000): 143–155.

Tan, C. P. and Che Man, Y. B. Differential scanning calorimetric analysis of palm oil, palm based products and coconut oil: Effects of scanning rate variation. *Food Chemistry* 76(1) (2002): 89–102.

Tan, C. P., Che Man, Y. B., Jinap, S., and Yusoff, M. S. A. Effects of microwave heating on changes in thermal properties of RBD palm olein by differential scanning calorimetry. *Innovative Food Science and Emerging Technologies* 3(2) (2002): 157–163.

Tan, C. P., Che Man, Y. B., Selamat, J., and Yusoff, M. S. A. Efficacy of natural and synthetic antioxidants in RBD palm olein by differential scanning calorimetry. In K. Nesaretnam and L. Packer (eds), *Micronutrients and Health: Molecular Biological Mechanisms*, pp. 108–118. Champaign: AOCS Press (2001a).

Tan, C. P., Che Man, Y. B., Selamat, J., and Yusoff, M. S. A. Application of Arrhenius kinetics to evaluate oxidative stability of vegetable oils by isothermal different scanning calorimetry. *Journal of the American Oil Chemists' Society* 78(11) (2001b): 1133–1138.

Tang, T. S. and Teoh, P. K. Palm kernel oil extraction—The Malaysian experience. *Journal of the American Oil Chemists' Society* 62(2) (1985): 254–258.

Traitler, H. and Dieffenbacher, A. Palm oil and palm kernel oil in food products. *Journal of the American Oil Chemists' Society* 62(2) (1985): 417–421.

Undurraga, D., Markovits, A., and Erazo, S. Cocoa butter equivalent through enzymic interesterification of palm oil midfraction. *Process Biochemistry* 36(10) (2001): 933–939.

Yap, P. H., de Man, J. M., and de Man, L. Polymorphism of palm oil and palm oil products. *Journal of the American Oil Chemists' Society* 66(5) (1989): 693–697.

Young, F. V. K. Palm kernel and coconut oils: Analytical characteristics, process technology and uses. *Journal of the American Oil Chemists' Society* 60(2) (1983): 374–379.

9 DSC Application to Lipid Modification Processes

Glazieli Marangoni de Oliveira, Monise Helen Masuchi, Rodrigo Corrêa Basso, Valter Luís Zuliani Stroppa, Ana Paula Badan Ribeiro, and Theo Guenter Kieckbusch

CONTENTS

9.1 INTRODUCTION

Chemically, all fats and oils are mixtures of glycerol esters, formed by attachment of fatty acids to the hydroxyl groups of glycerol. Nevertheless, the fatty acid components are distinguishable by three features: chain length, number and position of the double bonds, and their position within the glyceride molecule. The chemical and physical properties of fats and oils are largely determined by their fatty acid compositions and the arrangement of these components within the triacylglycerol (TAG) molecule. Variations in these characteristics are responsible for diversity in fats and oils (O'Brien 2008).

The use of fats and oils in products in the chemical, pharmaceutical, and food industries depends on their physical and chemical properties. However, the majority of natural fats and oils have limited applications in their unaltered state. The most significant physical characteristics for application in the food industry are consistency, plasticity, crystallization rate, microstructural configuration, and polymorphic stability (Ribeiro et al. 2010).

Some of the main modifications that can be performed in fats and oils to meet technical, economic, and nutritional specifications are interesterification, fractionation, and the addition of crystallization agents (seeding) or structural compounds to the lipid systems. These alterations allow fat bases with the desired characteristics to be produced.

Interesterification reactions can be performed using chemical catalysts or enzymes. In the chemical interesterification process, the fatty acid groups undergo either intramolecular or intermolecular rearrangement, that is, between the fatty acids of the same TAG or between the fatty acid groups of different TAG molecules, respectively. The free energy differences between the various combinations of TAGs are insignificant and therefore do not lead to selectivity of the fatty acids. The random distribution of the fatty acids in the product can be altered by crystallization, separating the saturated products with higher melting points by carrying out the reaction at temperatures below the crystallization temperature of the component with the highest melting point. In this case, chemical interesterification is classified as conducted or selective. The principle of the enzymatic reaction is the use of lipases as a means of transferring fatty acids among the TAG molecules. These enzymes act in both breakage and formation of ester bonds and can be selective for specific bonds (Ribeiro et al. 2010; Marangoni and Rousseau 1995; Liu 2004; Xu et al. 2006; Kodali and List 2005).

Fractionation is based on the controlled crystallization of TAGs, resulting in the separation of a solid phase and a liquid phase, which, in turn, can be further fractionated. If a melt of fat is cooled slowly, the TAGs with higher melting points than the tempering temperature will eventually form crystalline material, which can be filtered off from the liquid bulk. This process can be performed using only the fat system (dry fractionation) or by solubilizing it in an appropriated solvent (solvent fractionation) (Van Duijin et al. 2006; Wassell and Young 2007).

The most common components used to modify the crystallization properties and structure of fats and oils are saturated TAGs, partial acylglycerols, fatty acids, fatty alcohols, waxes, sorbitan alkylates, lecithin, and their mixtures. Generally, these additives change the thermodynamic equilibrium between the liquid and solid phases, modify the crystallization kinetics, and affect the nucleation process by providing heterogeneous nucleation sites, stabilizing the developing nuclei, and affecting the nucleation driving force by changing the components' solubility. In addition, these substances can form structured networks dispersed in the triacylglycerol mixtures, providing mechanical resistance and acting as a barrier to oil migration (Smith et al. 2011; Pernetti et al. 2007).

The modified fat and oil-based products require to be characterized in relation to their physical properties in order to determine their applications. Thermal analysis is a valuable tool to observe phase transitions, and, especially for complex systems such as lipids, the thermal events can be correlated with a variety of structural rearrangements during the heating or cooling process. Other techniques, such as Fourier transform infrared spectroscopy or x-ray diffraction, need to be used in combination to identify the structural pattern (Kalentunç 2009).

Differential scanning calorimetry (DSC) is the thermoanalytical technique used most commonly in the study of fats and oils. The various thermal events related to these materials are established by monitoring enthalpy changes and phase transitions in the diverse TAG blends (Che Man et al. 2003). Evaluation using DSC provides a direct measurement of the energy involved in the melting and crystallization processes of fats and oils. Oil crystallization causes a contraction in volume, which is associated with exothermic effects. In contrast, the melting of fats results in volume expansion, characterizing an endothermic situation (Tan and Che Man 2002a).

9.2 DSC PARAMETERS AND THEIR APPLICATION IN FAT AND OIL ANALYSIS

DSC can be defined as the technique based on the measurement of the difference between the heat flow rate to the sample and to a reference sample while both are subjected to the same controlled temperature program. DSC analysis allows quantification and correlation of thermally induced conformational transitions as well as phase transitions as a function of temperature. During temperature scanning, depending on the complexity of the material, peaks or inflection points reflecting the thermally induced transitions can be observed. The direction of the peak corresponds to the nature of the transition, either endothermic or exothermic. Values for thermal and thermodynamic changes in free energy (ΔG), enthalpy (ΔH), entropy (ΔS), and heat capacity (ΔC_P) of various transitions, in addition to determination of the bulk heat capacity of the material, can be obtained from calorimetric data (Kalentunç 2009).

Since oils and fats are composed of mixtures of different TAGs, they do not have specific melting and crystallization temperatures, but undergo melting or crystallization profiles (Tan and Che Man 2000). The extremely complex thermal behavior presented by fats and oils is highly dependent on the chemical composition and the DSC experimental protocol. Comparative studies of cooling and melting profiles of different raw materials under the same temperature-scanning rates showed that each lipid system has a specific thermal behavior, presenting different energy release in thermal events, determined by the enthalpy difference per peak, and distinct transition temperatures (Tan and Che Man 2002b).

When temperatures between −100°C and +80°C are investigated, several different thermal behaviors can be observed as a reflection of the TAG profile of the sample (Tan and Che Man 2002a). Consequently, it is difficult to standardize the DSC operational conditions, but the American Oil Chemists Society (AOCS) method Cj 1-94—DSC Melting Properties of Fats and Oils—recommends the following temperature program for a standardized evaluation of the crystallization or melting curves of fats or oils: 10 min (80°C); 80°C to −40°C (10°C/min); 30 min at −40°C; −40°C to 80°C (5°C/min) (AOCS 2009). However, the majority of studies report modified analytical conditions with respect to this temperature protocol (Ribeiro et al. 2009).

The heating and cooling rates (scanning) have a primary effect on the DSC output, and a correlation between scanning rate and response is critical for qualitative analyses of fats and oils (Timms 2003). On crystallization with high cooling rates, the TAG molecules usually crystallize in metastable polymorphic forms, which subsequently transform themselves into more stable polymorphs. On the other hand, under low cooling rates, TAGs with similar chain lengths associate themselves directly into more stable geometrical arrangements, resulting in the formation of a more stable polymorphic form. In the same way, the results obtained for melting curves are strongly associated with the heating rate used. At low rates (1°C/min or less), the heat transfer allows the occurrence of molecular rearrangements. Some thermal events, such as the melting of existing polymorphic forms, polymorphic transformations, and remelting of the polymorphs just formed, can be observed when a sample is melted slowly, but are not perceived when high heating rates (1°C/min or more) are applied. This results in melting of the sample before the structural alterations can establish themselves. The presence of more than one peak in a melting curve, if not related to the detection of polymorphs, may result from the existence of TAG groups that are dissimilar with respect to saturation and to the size of the TAG chain. These groups present distinct temperature ranges with respect to their behavior during the stages of crystallization and melting (Tan and Che Man 2002a; Ribeiro et al. 2009).

The polymorphic characteristics of the TAGs make the study of their thermal and structural properties complex. The different polymorphic forms have different melting points. The three main polymorphs (α, β', and β) are correlated with the subcell structure: hexagonal, orthorhombic-perpendicular, and triclinic-parallel, respectively. Of these three, the α-form has the lowest stability and easily transforms into either the β' or the β form, depending on the composition and thermal treatment (Sato and Ueno 2005).

The structural properties of fats have been studied by DSC, which is widely used for investigation of the transitions among crystalline forms during melting processes of fats (Szydlowska-Czerniak et al. 2005).

When a fat is heated it may exhibit multiple melting phases, and each recrystallization step represents the transition of a less stable polymorphic form to a more stable one. The peak transition temperature may be used as an indicator of the polymorphic form of a crystal, since the most stable crystalline form has a higher melting point (Ribeiro et al. 2009). However, DSC melting curves of fats are complex, and their interpretation is not straightforward. Therefore, the DSC crystallization curves, which are influenced by the chemical composition of the sample, are more reproducible and simpler than the melting curves (Szydlowska-Czerniak et al. 2005).

In a pioneering study, Loisel et al. (1998a) used the DSC technique to evaluate the phase transitions and polymorphism of cocoa butter, confirming the existence of six polymorphic forms for this fat. Numerous studies have used DSC as an important tool for the evaluation of polymorphic forms in fats, especially when they are subjected to chemical, physical, or enzymatic processes to obtain novel lipid fractions (Litwinenko et al. 2002; MacNaughtan et al. 2006; Himawan et al. 2007; Sato et al. 2009; Silva et al. 2009).

9.3   DSC AND LIPID MODIFICATION PROCESSES

9.3.1   EFFECTS OF ADDITIVES IN FAT SYSTEMS

Generally, two different mechanisms can be considered in the structuring of organic phases: dispersion of a foreign phase and specific molecular mechanisms, such as self-assembling. These mechanisms provide the elements that act as building blocks for the three-dimensional networks required in structuration of oil and fat phases. The size and shape of the structuring elements and their interactions will determine the final framework of the product and consequently its properties (Pernetti et al. 2007).

The addition of minor components to lipid systems can affect crystal nucleation, growth, or both. The effects may be distinct. Some components stimulate nucleation, while retarding growth; others, in contrast, stimulate or retard both. The effects measured with respect to nucleation are: alteration in the nucleation time, shift of the nucleation temperature, and change in the number and the nature of the nuclei formed (Smith et al. 2011).

The use of ingredients with limited solubility, which precipitate during gel formation, is another way to structure fat blends. Macromolecules such as polymers or proteins and low-molecular weight compounds such as fatty acids, fatty alcohols, or TAGs can act like structuring agents. They can interact in different ways: through covalent bonds, electrostatic forces, hydrogen bonds, van der Waals forces, or by steric entanglements. Generally, only a small number of building blocks are required to create a stable network (Pernetti et al. 2007).

9.3.1.1   DSC in the Analysis of Mixtures and Additives in Fat Systems

Addition or mixture of different components into fats and oils is used in the food industry to modify their physical behavior. DSC is an interesting technique for the evaluation of the several changes related to melting and crystallization properties caused by the presence of these components.

A study of the solubility behavior of tripalmitin (PPP), cocoa butter–stearin (CB-S), and palm oil–stearin (PO-S) in tricaprylin (CaCaCa), canola oil (CO), sunflower oil (SO), lard olein (LO), and palm oil–olein (PO-O) was performed by Zhou and Hartel (2006). In this study, the melting point and ΔH obtained using DSC were compared with the values calculated by the compositional method based on solubility lines derived from the composition of the liquid phase at equilibrium state. For systems that presented ideal solid–liquid equilibrium behavior, the values using DSC and composition analysis were, respectively, 63.4°C and 150.5 kJ/mol and 63.5°C and 161 kJ/mol for PPP; 35°C and 130 kJ/mol and 36.5°C and 119.4 kJ/mol for CB-S; and 58°C and 81 kJ/mol and 57.1°C and 73.9 kJ/mol for PO-S. The authors considered, in cases where ideal solutions were formed, that the values obtained by the two methods were similar, and suggest that the calorimetric approach may be sufficient for identifying equilibrium conditions. However, for nonideal systems the deviations between the two approaches were significant. The systems that presented nonideal behavior were the mixtures between CB-S and LO with PO-O, and PO-S with LO, and all the systems containing CaCaCa.

An indirect method in DSC isothermal analyses named the stop-and-return method, to be used when the crystallization process starts during cooling at isothermal temperature or during the equilibration period, was applied by Vereecken et al. (2009) in the study of the crystallization of different fat blends with palm midfraction oil, shea stearin, palm stearin, hydrogenated high-oleic soybean oil, and high-oleic sunflower oil. As previously reported by Foubert et al. (2008), this method is based on stopping the progression of the crystallization at different moments during isothermal crystallization and raising the sample temperature. The results indicated that PPP formed mixed crystals with POP, leading to a more efficient seeding of the crystallization process than when the mixed forms of SSS and SOS were used. Some blends showed two-step crystallization attributable to polymorphism. For blends with significant amounts of trisaturated TAGs, the stop-and-return method suggested an initial crystallization in an unstable polymorph followed by polymorphic transition during crystallization.

The effects of 1,2- and 1,3-dipalmitoylglycerol (PP) isomers on the crystallization of purified palm oil were studied by Siew and Ng (1999) using DSC at a cooling rate of 10°C/min. There was no difference in the crystallization behavior of samples containing 1% of both 1,2- and 1,3-PP and those containing 5% 1,2-PP, and the onset temperature was 20°C. In contrast, for the sample containing 5% 1,3-PP, the onset temperature of 30°C was reported, indicating earlier crystallization of the higher-melting-point TAGs of the palm oil. The peak at high-temperature range for the sample containing 5% 1,3-PP was broader than the one obtained for the sample containing 5% 1,2-PP, indicating higher intersolubility of the diacylglycerol in the purified palm oil.

The effect of the addition of tripalmitin (PPP) and commercial monopalmitin (MP) and monobehenin (MB) on palm oil crystallization was studied by Basso et al. (2010) using DSC at a cooling rate of 10°C/min. For the samples containing PPP, MP and MB, a displacement of the onset and peak temperatures for the stearin thermal events was observed in the higher-temperature region, indicating an acceleration of the crystallization process. Addition of monoacylglycerols showed lower thermal release during the crystallization process compared with PPP.

The thermal behavior of salad oil (canola:soybean 50:50) and rice bran wax (RBW) mixtures was characterized by DSC and the results compared with other waxes, concerning their organogel properties, by Dassanayake et al. (2009). Cooling and heating curves of samples containing 1%, 3%, 5%, 10%, 20%, 40%, 50%, and 100% of RBW were obtained using DSC at a temperature rate of 2°C/min. In order of increasing RBW concentration, the melting temperatures found were 54.3°C, 57.8°C, 60.8°C, 65.2°C, 68.7°C, 72.6°C, 73.4°C, and 78.2°C, and the crystallization temperatures were 48.1°C, 56.6°C, 57.9°C, 60.3°C, 66.3°C, 68.8°C, 70.1°C, and 70.1°C. From a graph of temperature versus RBW concentration, the authors concluded that RBW–salad oil mixtures tend, at RBW concentrations below 20%, to form a gel structure immediately after the temperature of the mixture reaches its melting point.

9.3.2 INTERESTERIFICATION OF FATS AND OILS

Interesterification is considered a versatile alternative for modifying the physical properties of fats and oils, and is currently the main method used to prepare plastic fats with low or even zero *trans* isomer contents. In contrast to hydrogenation, interesterification does not promote isomerization of the double bonds of the fatty acids or affect their degree of saturation, but allows modification of the behavior of the fats and oils, offering an important contribution to optimization of their use in food products (Norizzah et al. 2004). The interesterification of fats and oils can be applied for a variety of reasons: to influence the melting behavior, so as to provide the desired consistency at room temperature or under refrigeration; to improve or modify the crystalline behavior in order to facilitate production processes; and to decrease the tendency to recrystallize during the product's shelf life (Marangoni and Rousseau 1995).

Changes in the melting and solidification properties of the interesterified fats and oils are due to the relative proportions of acylglycerolic components after rearrangement of the fatty acids. In the majority of cases, interesterification leads to an increase in the melting point of the product due to the introduction of saturated fatty acids at the *sn*-2 position of the glycerol and the resulting increase in the levels of disaturated and trisaturated TAGs. Thus, this random recombination allows the production of fat bases with a wide variety of characteristics, and, by the correct selection of the oils and fats or their blends in adequate proportions, it is possible to obtain compounds with potential characteristics for a great diversity of applications (O'Brien 2008).

There are two types of interesterification in use: chemical and enzymatic. In the enzymatic process, biocatalysts, such as microbial lipases, are used to promote acyl migration from the acylglycerol molecules. In the widely used chemical interesterification process, the most frequently used catalyst is sodium methoxide, although other bases, acids, and metals are available (Marangoni and Rousseau 1995).

Chemical interesterification is random and proceeds entropically until thermodynamic equilibrium is reached. This equilibrium is based on the composition of the initial materials and can be predicted by the laws of probability (Marangoni and Rousseau 1995). Figure 9.1 illustrates the chemical interesterification between two different hypothetical TAGs, ABA and CCC. The randomization of these species, governed by thermodynamic and statistical laws, would theoretically result in 18 isomers that could be detected in the mixture at the end of the reaction.

Enzymes can be nonspecific, hydrolyzing the three fatty acids of the acylglycerols equally; *sn*-1,3 specific, hydrolyzing only the fatty acids at the *sn*-1 and *sn*-3

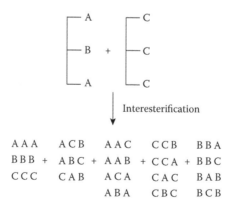

FIGURE 9.1 Schematic illustration of randomization promoted by chemical interesterification, exemplified by the triacylglycerols ABA and CCC.

positions on the glycerol; or specific in relation to determined fatty acids, hydrolyzing fatty acids with double bonds in defined positions. Independently of the specificity presented by the lipase, acyl migration can be observed during the reaction. This is a strictly chemical phenomenon leading to the spontaneous exchange of fatty acids from one ester site to another, leading to a random redistribution of the fatty acids on the acylglycerols, resulting in structures that cannot be predicted from the lipase applied (Carvalho et al. 2003).

Directing the industrial application of an interesterified fat requires a complete understanding of its physicochemical, functional, and technological properties, and also its stability during and after processing. A complete characterization of fats produced by chemical interesterification can be obtained by the use and correct interpretation of the results obtained from different analytical tools (Ribeiro et al. 2009).

9.3.2.1 DSC Analysis of Interesterified Fats and Oils

DSC is considered to be a sensitive technique for the characterization of interesterified products. When the TAG composition of a fat or oil is subjected to change by interesterification, concurrent changes in the thermal profiles can be observed. The melting and crystallization curves represent a valuable tool to check the alterations caused by randomization. DSC is widely used to determine the final point of interesterification of lard and of lauric fats, since the crystallization curves are significantly representative of the final TAG composition desired (Ribeiro et al. 2009).

Several parameters can be calculated to describe the thermal behavior of a sample and evaluate the modifications occurring as a function of interesterification: peak (maximum) crystallization (T_c) and melting (T_m) temperatures; onset crystallization (T_{oc}) and melting (T_{om}) temperatures; end of thermal phenomenon temperature (T_{end}) for crystallization and melting; and peak area corresponding to the enthalpies of crystallization (ΔH_c) and melting (ΔH_m). The value for ΔH_m is strongly related to the intermolecular arrangement of the TAG species and is usually modified by randomization (Campos 2005).

Figure 9.2 shows the crystallization curves for a blend of 60% soybean oil (SO) and 40% fully hydrogenated soybean oil (FHSBO) (w/w) before and after randomization. The subdivision of a crystallization curve of a fat and oil into different exothermic regions corresponds to the different melting profile TAGs (Campos 2005). The noninteresterified blend presents a prominent peak at the crystallization event (peak 1) of the trisaturated fraction of FHSBO with higher-melting-point TAG components. Interesterification causes the appearance of a second peak (peak 2) on the crystallization curve, characterizing the formation of TAGs with intermediary melting points (disaturated–monounsaturated and monosaturated–diunsaturated) promoted by the randomization process. At the same time, the first peak decreases in intensity and becomes wider, and the initial (T_{oc}) and the final (T_{end}) crystallization temperatures are reduced, allowing the interesterified blend to crystallize at temperatures lower than the original blend. Concomitantly, the value for ΔH_{c1} decreases as a result of the randomization, due to the decrease in concentration of trisaturated TAGs in the blends resulting from the rearrangement.

With respect to the melting curves, Figure 9.3 shows that two characteristic peaks were detected in the original blend, relative to the unsaturated fraction of the SO,

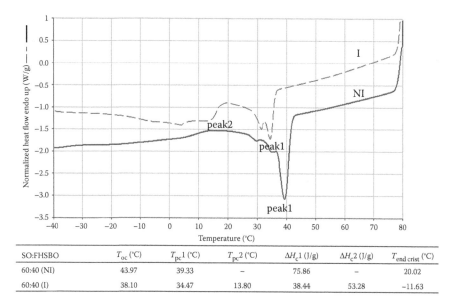

FIGURE 9.2 DSC crystallization parameters of 60% soybean oil (SO) and 40% fully hydrogenated soybean oil (FHSBO) (w/w) blend: crystallization onset temperature (T_{oc}), crystallization peak temperature (T_{pc}), crystallization enthalpy (ΔH_c), and crystallization final temperature ($T_{end\ crist}$). NI, not interesterified blend; I, interesterified blend.

SO:FHSBO	T_{oc} (°C)	$T_{pc}1$ (°C)	$T_{pc}2$ (°C)	ΔH_c1 (J/g)	ΔH_c2 (J/g)	$T_{end\ crist}$ (°C)
60:40 (NI)	43.97	39.33	–	75.86	–	20.02
60:40 (I)	38.10	34.47	13.80	38.44	53.28	−11.63

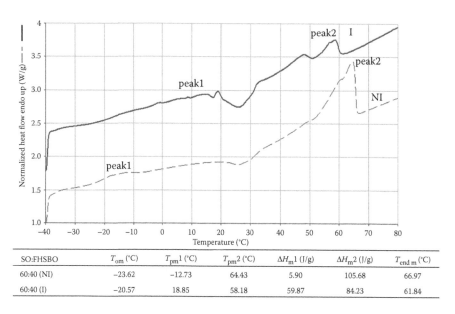

FIGURE 9.3 DSC melting parameters of 60% soybean oil (SO) and 40% fully hydrogenated soybean oil (FHSBO) (w/w) blend. Melting onset temperature (T_{om}), melting peak temperatures (T_{pm}), melting enthalpies (ΔH_m), and melting final temperature ($T_{end\ m}$). NI, not interesterified blend; I, interesterified blend.

SO:FHSBO	T_{om} (°C)	$T_{pm}1$ (°C)	$T_{pm}2$ (°C)	ΔH_m1 (J/g)	ΔH_m2 (J/g)	$T_{end\ m}$ (°C)
60:40 (NI)	−23.62	−12.73	64.43	5.90	105.68	66.97
60:40 (I)	−20.57	18.85	58.18	59.87	84.23	61.84

with lower-melting-point TAGs (peak 1) and the trisaturated fraction corresponding to the FHSBO (peak 2). After randomization, the first peak of the curve was displaced to the right, with respective increases in the values for T_{om} and melting peak1 temperature (T_{pm1}), related to the significant increase in TAGs with intermediate melting points (disaturated–monounsaturated and monosaturated–diunsaturated). Also, the values of ΔH_{m1} increased after the reaction, indicating an increase in the proportion of these TAG species in the interesterified blend. In parallel, interesterification caused a decline of the values for melting peak2 temperature (T_{pm2}), T_{end} (melting), and ΔH_{m2} as a result of the decreased proportion of trisaturated TAGs in the blend.

Numerous studies have considered DSC as a standard technique for determining the changes promoted by the randomization of TAGs. In a landmark study, Rossell (1975) studied the effects of randomization on palm kernel oil. The shapes of the melting curves were not modified by the process, but the temperatures of the peaks decreased, indicating the greater variety of TAG species present. Marangoni and Rousseau (1998) evaluated the extent of interesterification of palm oil/soybean oil and lard/canola oil based on the crystallization parameters T_{oc}, T_c, T_{end}, and ΔH_c. Kok et al. (1999) observed that the interesterification of highly saturated soybean oil promoted an increase in T_{oc} and T_{om} values. DSC was used by Rodríguez et al. (2001) as a tool to optimize the interesterification of lard/sunflower oil. Khatoon and Reddy (2005) reported a study on the interesterification of palm stearin/*Mangifera indica* oil and of palm stearin/*Madhuca latifolia* oil, evaluating the peak melting temperatures (T_m) and the melting enthalpy (ΔH_m). The initial blends were characterized by three distinct melting endotherm events, confirming the heterogeneous nature of the TAGs present. After randomization, the proportion of peaks related to higher-melting-point TAGs was reduced, while the peaks referring to unsaturated TAGs increased significantly. Norizzah et al. (2004) found the same behavior for the interesterification of palm stearin and palm kernel olein. Chiu et al. (2008) evaluated the properties of blends of abdominal chicken fat, chicken fat stearin, and medium-chain TAGs using DSC, before and after chemical interesterification. Ramli et al. (2005) found lower and higher values, respectively, for the T_m and ΔH_m of interesterified blends of fully hydrogenated palm kernel oil and goat's milk fat, as compared with noninteresterified blends.

9.3.3 THERMAL FRACTIONATION

The essential objective of the fractionation of oils and fats is to obtain lipid fractions for specific applications. The method consists of the crystallization of TAGs using one or more stages, which are usually characterized by the difference in solubility of the solid crystals within the liquid phase, the molecular weight and degree unsaturation degree of the molecules (Kellens et al. 2007; Timms 1997). The fractionation is a physical and reversible process. The solid phase, named stearin, is separated from the liquid phase, known as olein, by different techniques. For industrial-scale production, the fat crystals can be obtained by three different technologies: detergent, solvent, and dry fractionation (Kellens et al. 2007).

The analytical techniques generally applied for evaluating both fractions—stearin and olein—obtained from fractionation include: iodine value, usually by titration (Wijs method); TAG and fatty acid composition by chromatography; solid fat content by nuclear magnetic resonance; and thermal behavior by DSC. In conjunction, these techniques allow characterization of the fractions and, therefore, evaluation of the performance of different fractionation processes and conditions. According to Kellens et al. (2007), DSC displays direct information about the thermal behavior—both crystallization and melting—of fats and oils, and, therefore, it should be considered an important tool for assessing fraction quality and fractionation performance.

Detergent fractionation is performed with crude palm oil, and involves the steps of cooling, crystallization, and separation of crystals using a solution of sodium lauryl sulfate and an electrolyte to aid in the separation (Gupta 2008).

The method of solvent fractionation is very similar to the detergent method; the distinction between them is the use of acetone or hexane. Typically, solvent fractionation employs a ratio of oil to solvent from 1:3 to 1:4 (Hashimoto et al. 2001). There are many advantages to this process, including the more rapid nucleation and growth of crystals, allowing the use of high cooling rates and low temperatures, which favors the achievement of a lower crystallization time. Furthermore, the dilution of fat in the solvent facilitates heat transfer and decreased viscosity, and reduces the amount of liquid oil in the solid matrix, simplifying the filtration of the product (Timms 1997).

The solvent process is frequently used to obtain a high yield of the middle fractions of palm oil, especially those with TAGs containing long-chain fatty acids; when subjected to low temperatures, the solution becomes extremely viscous (Gupta 2008). Furthermore, these products contain TAGs with very similar melting points, making them difficult to separate.

The choice of solvent can affect the selectivity in the separation of lipid classes: that is, more polar lipids, such as diacylglycerols and free fatty acids, can be separated from TAGs with the use of a polar solvent, such as acetone. According to Hashimoto et al. (2001), fractionation using acetone tends to be more selective in the crystallization of symmetrical 1,3-disaturated TAGs (SUS) compared with nonsymmetrical 1,2- or 2,3-disaturated (SSU or USS) TAGs.

TAG composition techniques are very useful for evaluating the performance of the fractionating process and the quality of the separated fractions obtained from solvent. Besides this determination, thermal behavior can be another important tool to evaluate the olein and stearin obtained, as observed by Yanty et al. (2013). The authors evaluated the fractions obtained through acetone fractionation of avocado butter at three different processing temperatures by TAG composition and thermal analyses. The DSC temperature program started at 70°C and then cooled down to −90°C, at a rate of 5°C/min. After assessment of the crystallization history, the samples were heated to 70°C at the same rate. For the avocado butter before fractionation, the cooling curves showed three major exothermic transition peaks, characterizing the crystallization of TAG classes with three different melting-point ranges. The first peak, representing a group of higher-melting-point TAGs, was separated from the other two peaks, which corresponded to the lower-melting TAG groups, indicating that the fractionation of this raw material could easily proceed with good separation into high- and low-melting fractions. A shift in the onset of the crystallization temperature was verified for the stearin and olein samples. The stearin showed an onset of crystallization at 41.6°C and the olein at −10.2°C, confirming the good performance of the fractionation process.

Dry fractionation is regarded as economic and natural, requires no additional substances, and is considered a safe process. It is also based on the control of crystallization, in which an oil or melted fat is cooled under controlled conditions to achieve the desired crystal structure, in order to obtain effective separation of the solid phase and the liquid phase (Kellens et al. 2007).

In some specific dry fractionation processes in fats and oils, seeding techniques can be used in order to facilitate TAG crystallization and separation. Bootello et al. (2011) evaluated the dry fractionation of high-oleic high-stearic sunflower oil by adding a high-melting-point stearin powder. Different stearin fractions rich in disaturated TAGs were obtained and evaluated by DSC in terms of melting profile as well as solid fat content. Each sample was quickly cooled down from 20°C to −40°C, after which the temperature was increased to 90°C at a rate of 1°C/min. Solid fat content was determined by continuous integration of the DSC melting curves. Stearin samples containing higher amounts of disaturated TAGs presented a higher solids content, and the addition of 0.05%–0.25% of high-melting-point stearin powder as seeding material was considered important to improve the separation, thereby avoiding the use of solvent in the fractionation.

Knowledge of the phase behavior of mixtures of TAGs is fundamental to optimizing the fractionation performance. This concept of the phase behavior profile refers to the melting and crystallization of the mixture with different TAGs and their polymorphic nature in both liquid and solid state (Calliauw et al. 2007). Kellens et al. (2007) established binary (temperature–composition) phase diagrams by mixing and melting pure TAGs such as PPP/PSP and PPP/POO, based on powder x-ray diffraction and DSC data. When performing both techniques, the previous melted blends were quenched at −40°C and heated at a constant rate of 5°C/min. The peaks for crystallization and melting of the samples were obtained by DSC, while the polymorphic behavior was obtained by powder x-ray diffraction. The binary phase diagrams revealed different molecular structures (solid solution and eutectic interaction) of liquid blends and polymorphic forms depending on the temperature and the mixture ratios. These diagrams are important to finding out the crystal miscibility in the solid state, which affects the efficiency of fractional crystallization.

Mamat et al. (2005) evaluated, by DSC, the effect of the crystallization temperature on the physicochemical properties of the fractions obtained in the fractionation of mixtures of palm oil and sunflower oil. The authors observed that at low temperatures (15°C) it was possible to achieve an olein fraction with a higher iodine value: that is, there was a decrease in the content of monounsaturated TAGs and an increase in the content of di- and polyunsaturated TAGs.

The fractionation of palm oil produces various profiles of olein intended for cooking and frying oils, as well as several fractions of stearin, used in formulations of shortenings, margarine, and chocolate (Gupta 2008). These wide ranges of lipid fractions are achieved by multistage fractionation and add commercial value to the product. Zaliha et al. (2004) investigated the dry fractionation of palm oil at different temperatures: 18°C, 15°C, 11°C, and 9°C. The fractions obtained in a single-step process were evaluated in terms of cloud point, iodine value, fatty acid composition, solid fat content, crystal morphology, and thermal behavior. For the crystallization profile by DSC, the stearin and olein fractions were first heated to 70°C and held at this temperature for 10 min to totally destroy any fat nuclei, and then cooled to −40°C at a cooling rate of 5°C/min. The inverse thermal procedure was also performed. The crystallization curve of the palm oil before fractionation presented two main exothermic peaks: the first at 28.5°C, corresponding to the higher-melting TAGs, and the second at 12.4°C, corresponding to the lower-melting TAGs. The crystallization behavior of the stearin fraction at different temperatures was also evaluated, and showed different profiles: from the fractionation at 18°C, a harder stearin was obtained, characterized by an onset of crystallization at 29.9°C, while the process at 9°C produced a softer stearin, which started to crystallize at a temperature of 19.2°C. By performing the dry fractionation of palm oil at different temperatures and thermally characterizing the separated stearin and olein samples, different applications could be indicated for each fraction, according to the production and product requirements.

9.3.4 Cocoa Butter (CB), Cocoa Butter Replacers (CBR), Cocoa Butter Substitutes (CBS), and Cocoa Butter Equivalents (CBE)

9.3.4.1 Definition of CB, CBR, CBS, and CBE

CB is the most important component used in the manufacture of chocolate and confectionery products, and is responsible for several quality attributes of the final product, such as hardness and snap at room temperature, complete melting at mouth temperature, brightness, and release of aroma and flavor in tasting (Ribeiro et al. 2012, 2013). It contains more than 75% symmetrical TAGs of the SUS type, with oleic acid in the *sn*-2 position. These characteristics are essential to the functionality of CB. The main TAGs present in CB are POP (1,3-dipalmitoyl-2-oleoylglycerol), SOS (1,3-distearoyl-2-oleoylglycerol), and POS (1-palmitoyl-3-stearoyl-2-oleoylglycerol). The composition of fatty acids and TAGs in CB can vary depending on its area of origin (Ribeiro et al. 2012; Hernandez et al. 1991), according to Table 9.1. The TAG composition is also responsible for the melting point and sharp melting profile of CB (Beckett 2008; Awad and Marangoni 2005; Shukla 2005; Yamada et al. 2005).

TABLE 9.1

Most Common Triacylglycerol Components (%) Present in Cocoa Butter from Different Origins and the Total of Monounsaturated Symmetrical Triacylglycerols

TAGs	Brazil	Ecuador	Ivory Coast	Malaysia	Nigeria	Indonesia
POP	19.1±3.4	21.3±4.9	18.6±4.9	18.5±5.9	19.5±5.1	15.7±0.3
POS	37.3±3.2	37.0±0.6	37.4±0.7	38.5±2.2	38.0±1.0	39.2±1.2
SOS	25.5±2.5	29.1±3.5	28.5±2.5	32.2±1.7	29.8±3.8	28.4±1.7
SUS	81.9±3.4	87.3±9.0	84.5±7.4	89.2±5.4	87.3±7.9	83.2±2.6

Source: Hernandez, B., Castellote, A.I., Permanyer, J.J., *Food Chem.*, 41(3), 269, 1991; Marty, S., Marangoni, A.G., *Crystal Growth Design*, 9(10), 4415, 2009; Ribeiro, A.P.B., Basso, R.C., Gonçalves, L.A.G., Gioielli, L.A., Santos, A.O., Cardoso, L.P., Kieckbusch, T.G., *Grasas y Aceites*, 63(1), 89, 2012. With permission. SUS, monounsaturated symmetrical triacylglycerols.

The crystallization of the CB is determined by the composition of this fat, and the characteristics of this process determine the quality of the end products. For this reason, CB plays an important role in chocolates, and was very expensive before the 2000s. Since then, the national regulations of several countries have permitted the industry to use distinct fats, known as cocoa butter alternatives (CBA), in chocolate formulation. Continued research in the field of confectionery science has resulted in the development of fats with characteristics resembling CB, produced mainly by processes such as hydrogenation, interesterification, fractionation, and blending (Timms 2003). Generally these fats are classified as: CBE, having higher levels of the TAGs POP and SOS than the original CB; CBR, composed of nonlauric vegetable fats with physical, but not chemical, characteristics similar to CB; and CBS, produced from lauric vegetable fats under hydrogenation or fractionation, containing mainly lauric and myristic acids (Shukla 2005; Yamada et al. 2005).

According to Yamada et al. (2005), these fats can be classified as follows:

CBE: Fat that is chemically and physically similar to CB and can be used generally in any formulation (chocolates, coatings, and compounds). CBE can be produced by several methods (such as fractionation and interesterification) and from several oil sources (especially palm oil, illipe fat, and shea). Specific fats of fat with a characteristic TAG composition can be blended with other natural fats to result in a product with characteristics very similar to CB.

CBR: Nonlauric fat that is physically, but not chemically, similar to CB and can be used to replace part of the CB in a recipe, particularly coatings. It can be produced from a nonlauric source (palm oil or soybean oil) by hydrogenation, fractionation, or both. Similarly to CB, stearic, palmitic, and oleic acids are common in these fats, but the arrangement of these fatty acids in the glycerol molecules is more random, giving a different melting pattern. In addition, the presence of high levels of elaidic acid (*trans*) can limit the compatibility with CB (by maximum 25% of the total fat).

CBS: Lauric fat that is physically, but not chemically, similar to CB and can be used to replace all the CB in a recipe, particularly coatings. It is fully modified (by fractionation and hydrogenation) from palm kernel, coconut, or both. As a consequence of the presence of the lauric acid, this fat has a completely different TAG composition, resulting in incompatibility with CB. Due to the high chance of softening and bloom formation, the use of CB is limited to 5% in a CBS formulation.

As expected, CBA must have a similar chemical composition (fatty acids and TAGs), melting range, polymorphic behavior, and compatibility to CB. In addition, this fat must not require changes in the CB-based products and processing. Other characteristics such as good flavor, stability, and miscibility to CB are also desirable (Yamada et al. 2005).

CB and its analogous fats are applied in several industrialized products, mainly in the food and pharmaceutical industries. Chocolates, pralines, spreads, lipsticks, body lotions and other body care products, suppositories, ointments, some pharmaceutical drugs, and some emulsions are examples of products that have CB in their formulation.

9.3.4.2 Crystallization and Polymorphism of Cocoa Butter

Crystallization behavior is the main characteristic of fats and oils responsible for achieving desirable texture, structure, and quality during processing and storage of manufactured food products. Different fat crystal networks can be structured, depending on the TAG composition of the lipid system, other constituents present in the formulation, and processing conditions (shearing and temperature, for instance) (Dhonsi and Stapley, 2006).

According to several crystallographic studies, CB presents six different polymorphs. Each crystal form exhibits a specific melting point, thermodynamic stability, and latent heat of melting. The six crystal forms are termed, in increasing order of thermodynamic stability and melting point, I, II, III, IV, V, and VI or γ, α, β'_2, β'_1, β_2, and β_1. In chocolate products, form V is the desirable one, since it melts around 32°C—close to mouth temperature—and presents desirable properties of color, gloss, and snap (Schenk and Peschar 2004; Norberg 2006). Besides these properties, form V is associated with a specific chocolate structure able to trap liquid oil inside its crystal network, avoiding the migration of melted TAG to the product surface (Norberg 2006). In contrast, the most stable crystal polymorph of CB, form VI, is associated with the *fat bloom* formation, This phenomenon is characterized by phenomenon characterized by the presence of a grayish film, white spots, or both on the chocolate surface, mainly due to temperature oscillation during storage and transport or poorly tempered products. Polymorphism of fats and oils can be identified by x-diffraction and also by DSC (Metin and Hartel 2005; Sato and Ueno 2005).

9.3.4.3 Chocolate

Chocolate, the most important product of cocoa, is a special food commonly appreciated everywhere in the world. It is a dispersion of solid particles in a continuous fat phase. The solids come from cocoa powder, milk powder, and sugar. The fat phase is composed of CB, CBAs, milk fat, emulsifiers (soy lecithin and polyglycerol polyricinoleate [PGPR]), and flavors, or a combination of these. These ingredients can be combined to produce milk, dark, or white chocolate. In several countries, regional laws that regulate the ingredients have been created since the 2000s to standardize chocolate production (FDA 2013; European Parliament and the Council of 23 June 2000; Codex Alimentarius 2003).

The most critical step in chocolate production is tempering. This consists of the induction of the formation of a stable polymorph during chocolate crystallization by cooling the chocolate mass under shear to the tempering temperature of $28°C \pm 1°C$. It yields the Form V crystal in the final product, promoting contraction during the cooling step, which facilitates demolding.

The characteristics of chocolate are very dependent on the polymorphism of the fat phase. This is influenced by the fat phase composition and the way the crystallization is carried out (as in the tempering procedure). The presence of CB from different origins is enough to interfere in the crystallization. Futhermore, the presence of CBAs, milk fat, and emulsifiers can cause a decrease in the crystallization temperature and a delay in tempering, which may require modifications in the tempering conditions (Norberg 2006).

9.3.4.4 DSC Methodologies and Applications for Cocoa Butter

DSC is an important technique for evaluating melting, crystallization, and polymorphic characteristics of CB and its products. Through this thermal analysis, the peak temperature, onset temperature, heat of melting, final melting temperature, and identification of polymorphic behavior can be obtained.

9.3.4.4.1 DSC Methodologies for Pure Cocoa Butter

For pure CB and its analogous fats, there are several methodologies for different characterizations, such as melting and crystallization behavior, polymorphism identification, and solid fat content determination (Biliaderis 1983).

The AOCS (2009) official method for determining thermal properties by DSC, Cj 1-94, allows fats and oils to be characterized. Figure 9.4 exhibits a typical crystallization curve of CB by DSC, obtained at a cooling rate of 2°C/min. Different peak separation of the samples can be achieved by varying the rate of cooling or heating in the DSC temperature program.

Kinta and Hartel (2010) carried out a DSC analysis in order to investigate the crystallization and melting behavior of CB containing different seed CB amounts. Dynamic crystallization of the pure CB and blends (CB and fat base seeds) was performed by DSC from 30°C to 10°C at a cooling rate of 0.25°C/min, and afterward the same samples were heated at 5°C/min to evaluate the crystal form from the melting profile. Without seed addition, only unstable β′ crystals were observed; however, for the CB with seeds added, the presence of the stable β polymorph was detected.

For polymorphic identification in CB, a very slow heating rate should be applied, as demonstrated by Loisel et al. (1998a). In this reference study of the processing

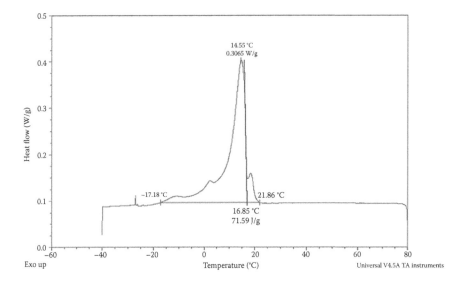

FIGURE 9.4 Crystallization curve of cocoa butter by DSC at a cooling rate of 2°C/min.

conditions for the determination of the polymorphic transition in CB, the sample was quenched at −40°C and subsequently heated at rates of 0.1, 1, and 2°C/min in order to evaluate polymorphic transitions.

Studies on CB polymorphism by DSC determined specific melting temperatures for each polymorphic form. However, there is great discrepancy among the data presented in the literature regarding the number of crystalline forms and their respective melting points. A survey indicates that the melting temperatures of polymorphs I, II, III, IV, V, and VI lie, respectively, within the following limits: 11.3°C–18.0°C; 15.1°C–24.2°C; 20.0°C–28.0°C; 21.1°C–33.0°C; 27.4°C–35.0°C; and 29.0°C–36.2°C (Ribeiro et al. 2012).

Figure 9.5 correlates the temperatures of the polymorphic transitions obtained for six different CB samples, corresponding to the peak temperatures in the melting curves. To induce the formation of crystals of type I, the samples were previously immersed in liquid nitrogen for 15 s and then transferred to the calorimeter, where they were kept at −45°C for 30 min. Heating was conducted in two steps: the temperature was raised to −10°C at a rate of 5°C/min and then increased to 40°C at a rate of 0.3°C/min. The melting peak temperatures, corresponding to the temperatures of polymorphic transitions, were determined. The polymorphic transition temperatures of the crystalline phases I, II, III + IV, and V were within 15.90°C – 16.80°C; 19.58°C – 20.16°C; 20.02°C – 21.94°C; and 28.72°C – 33.58°C, respectively. From this parameter, relationships can be established between the chemical composition and crystallization behavior of the samples, and they can be differentiated in terms of technological and industrial potential for use in tropical regions.

DSC is also considered as an alternative technique for the measurement of solid fat content (SFC) in CB and other fats. Walker and Bosin (1971) presented a comparative study on the use of SFI (solid fat index; dilatometry), DSC, and nuclear magnetic resonance (NMR) for determining solid–liquid ratios in fats. The results

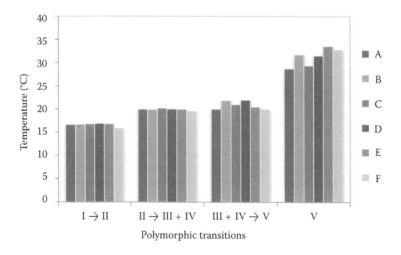

FIGURE 9.5 Polymorphic transition temperatures of cocoa butter samples.

confirmed that both NMR and DSC are adequate techniques for characterization of SFC of fats and oils. Recently, Márquez et al. (2012) proposed a correction and prior treatment for the estimation of SFC by DSC.

9.3.4.4.2 DSC Methodologies for Cocoa Butter Products

Chocolate after the tempering and molding stages can be evaluated by different DSC methodologies. Stapley et al. (1999) evaluated the effects of shearing and temperature history in chocolate by DSC, immediately after tempering using a concentric cylinder shearing apparatus. The conditions applied for data acquisition by DSC were divided into two steps: a cooling step for the crystallization of the tempered chocolate and a rewarming step for melting the structured chocolate. Before sample insertion, the scanning start temperature was set as the final tempering temperature, and then a 2°C/min or 5°C/min cooling rate was applied until the temperature reached −10°C (crystallization step). The sample was then rewarmed at a rate of 5°C/min, increasing the temperature to 50°C. Following this methodology, the thermal behavior of chocolate during solidification and subsequent melting was evaluated for different conditions of tempering, varying shear and temperature. Thermal analysis of chocolate was also performed by Loisel et al. (1998b) to investigate the crystallization behavior using a lab-scale scraped surface heat exchanger. After crystallization, the sample was placed in a preheated aluminum capsule at the same temperature as final tempered chocolate and heated at a rate of 10°C/min from 21°C to 50°C inside the DSC. The results indicated that chocolate crystallization during a dynamic process is a two-step process for crystallization temperatures higher than 26.2°C and a one-step process at temperatures below 26.2°C. Two peaks were found: the larger peak corresponded to the presence of monounsaturated TAGs, while the smaller peak was related to the trisaturated TAGs content of chocolate.

Besides evaluating tempering conditions for chocolate processing, DSC is also an important tool to monitor fat bloom formation. Walter and Cornillon (2001) studied fat bloom formation with a different thermal history and with the addition of sucrose, milk powder, and 1,3-dibehenoyl, 2-oleoylglycerol (BOB) in dark unsweetened chocolate. DSC was applied in order to determine the melting point of the samples. About 10 mg of chocolate stabilized between −5°C and 5°C was heated to 60°C at 10°C/min. The melting points were determined using the specific software of the equipment and the values ranged from 33°C to 40°C, depending on the thermal conditions and the additives added.

REFERENCES

AOCS (American Oil Chemists Society). *Official Methods and Recommended Practices of the American Oil Chemists Society.* Champaign, IL: AOCS Press (2009).

Awad, T. S. and Marangoni, A. G. Ingredient interactions affecting texture and microstructure of confectionery chocolate. In A. McPherson and A. G. Gaonkar (eds), *Ingredient Interactions: Effects on Food Quality*, pp. 423–476. New York: CRC Press (2005).

Basso, R. C., Ribeiro, A. P. B., Masuchi, M. H., et al. Tripalmitin and monoacylglycerols as modifiers in the crystallisation of palm oil. *Food Chemistry* 122(4) (2010): 1185–1192.

Beckett, S. (ed.). Crystallizing the fat in chocolate. In *The Science of Chocolate*, 2nd edn., pp. 103–124. New York: RSC Publishing (2008).

Biliaderis, C. S. Differential scanning calorimetry in food research: A review. *Food Chemistry* 10 (1983): 236–265.

Bootello, M. A., Gracés, R., Martínez-Force, E., and Salas, J. J. Dry fractionation and crystallization kinetics of high-oleic stearic sunflower oil. *Journal of the American Oil Chemists' Society* 88 (2011): 1511–1519.

Calliauw, G., Gibon, V., Greyt, W., Plees, L., Foubert, I., and Dewettinck, K. Phase composition during palm olein fractionation and its effect on soft PMF and superolein quality. *Journal of the American Oil Chemists' Society* 84(9) (2007): 885–891.

Campos, R. Experimental methodology. In A. G. Marangoni (ed.), *Fat Crystal Networks*, pp. 267–348. New York: Marcel Dekker (2005).

Carvalho, P., Renata, P., Campos, B., et al. Aplicação de lipases microbianas na obtenção de concentrados de ácidos graxos poliinsaturados. *Química Nova* 26(1) (2003): 75–80.

Che Man, Y. B., Shamsi, K., Yusoff, M. S. A., and Jinap, S. A study on the crystal structure of palm oil-based whipping cream. *Journal of the American Oil Chemists' Society* 80(5) (2003): 409–415.

Chiu, C. M., Gioielli, L. A., and Grimaldi, R. Lipídios estruturados obtidos a partir da mistura de gordura de frango, sua estearina e triacilgliceróis de cadeia média. II—Pontos de amolecimento e fusão. *Química Nova* 31(2) (2008): 238–243.

Codex Alimentarius. Codex standard for chocolate and chocolate products. STAN 87-1981, Rev. 1–2003.

Dassanayake, L. S. K., Kodali, D. R., Ueno, S., and Sato, K. Physical properties of rice bran wax in bulk and organogels. *Journal of the American Oil Chemists' Society* 86(12) (2009): 1163–1173.

Dhonsi, D. and Stapley, A. G. F. The effect of shear rate, temperature, sugar and emulsifier on the tempering of cocoa butter. *Journal of Food Engineering* 77(2006): 936–942.

European Parliament and the Council of 23 June 2000. Directive 2000/36/EC. *Official Journal of the European Communities* 197(2000): 19–25.

FDA (Food and Drug Administration). *Code of Federal Regulations Title 21 Part 163 Cacao Products*. United States (2013). http://www.accessdata.fda.gov/scripts/cdrh/cfdocs/cfcfr/CFRSearch.cfm?CFRPart=163&showFR=1.

Foubert, I., Fredrick, E., Vereecken, J., Sichien, M., and Dewettinck, K. Stop-and-return DSC method to study fat crystallization. *Thermochimica Acta* 471(1–2) (2008): 7–13.

Gupta, M. K. *Practical Guide to Vegetable Oil Processing*. Illinois: AOCS Press (2008).

Hashimoto, S., Nezu, T., Arakawa, H., Ito, T., and Maruzeni, S. Preparation of sharp-melting hard palm midfraction and its use as hard butter in chocolate. *Journal of the American Oil Chemists' Society* 78(5) (2001): 455–460.

Hernandez, B., Castellote, A. I., and Permanyer, J. J. Triglyceride analysis of cocoa beans from different geographical origins. *Food Chemistry* 41(3) (1991): 269–276.

Himawan, C., MacNaughtan, W., Farhat, I. A., and Stapley, A. G. F. Polymorphic occurrence and crystallization rates of tristearin/tripalmitin mixtures under non-isothermal conditions. *European Journal of Lipid Science and Technology* 109 (2007): 49–60.

Kalentunç, G. Calorimetric methods as applied to food: An overview. In *Calorimetry in Food Processing: Analysis and Design of Food Systems*, pp. 5–14. Iowa: Wiley and IFT Press (2009).

Kellens, M., Gibon, V., Hendrix, M., and De Greyt, W. Palm oil fractionation. *European Journal of Lipid Science and Technology* 109(4) (2007): 336–349.

Khatoon, S. and Reddy, S. R. Y. Plastic fats with zerotrans fatty acids by interesterification of mango, mahua and palm oils. *European Journal of Lipid Science and Technology* 107(11) (2005): 786–791.

Kinta, Y. and R. W. Hartel. Bloom formation on poorly-tempered chocolate and effects of seed addition. *Journal of the American Oil Chemists' Society* 87(1) (2010): 19–27.

Kodali, D. R. and List, G. R. *Trans Fats Alternatives*. Champaign: AOCS Press (2005).

Kok, L. L., Fehr, W. R., Hammond, E. G., and White, P. J. Trans-free margarine from highly saturated soybean oil. *Journal of American Oil Chemists' Society* 76(10) (1999): 1175–1181.

Litwinenko, J. W., Rojas, A. M., Gerschenson, L. N., and Marangoni, A. G. Relationship between crystallization behavior, microstructure, and mechanical properties in a palm oil-based shortening. *Journal of the American Oil Chemists' Society* 79 (2002): 647–654.

Liu, L. How is chemical interesterification initiated: Nucleophilic substitution or α-proton abstraction? *Journal of American Oil Chemists' Society* 81(4) (2004): 331–337.

Loisel, C., Keller, G., Lecq, G., Bourgaux, C., and Ollivon, M. Phase transitions and polymorphism of cocoa butter. *Journal of the American Oil Chemists' Society* 75(4) (1998a): 425–439.

Loisel, C., Lecq, G., Keller, G., and Ollivon, M. Dynamic crystallization of dark chocolate as affected by temperature and lipid additives. *Journal of Food Science* 63(1) (1998b): 73–79.

MacNaughtan, W., Farhat, I. A., Himawan, C., Starov, V. M., and Stapley, A. G. F. A differential scanning calorimetry study of the crystallization kinetics of tristearin-tripalmitin mixtures. *Journal of the American Oil Chemists' Society* 83 (2006): 1–9.

Mamat, H., Aini, I. N., Said, M., and Jamaludin, R. Physicochemical characteristics of palm oil and sunflower oil blends fractionated at different temperatures. *Food Chemistry* 91(4) (2005): 731–736.

Marangoni, A. G. and Rousseau, D. Engineering triacylglycerols: The role of interesterification. *Trends in Food Science and Technology* 6 (1995): 329–335.

Marangoni, A. G. and Rousseau, D. The influence of chemical interesterification on physicochemical properties of complex fat systems. 1. Melting and crystallization. *Journal of American Oil Chemists' Society* 75 (1998): 1265–1271.

Márquez, A. L., Pérez, M. P., and Wagner, J. R. Solid fat content estimation by differential scanning calorimetry: Prior treatment and proposed correction. *Journal of the American Oil Chemists' Society* 90(4) (2012): 467–473.

Marty, S. and Marangoni, A. G. Effects of cocoa butter origin, tempering procedure, and structure on oil migration kinetics. *Crystal Growth Design* 9(10) (2009): 4415–4423.

Metin, S. and Hartel, R. W. Crystallization of fats and oils. In F. Shahidi (ed.), *Bailey's Industrial Oil and Fat Products*, 6th edn., pp. 45–76. New Jersey: Wiley-Interscience (2005).

Norberg, S. Chocolate and confectionery fats. In F. D. Gunstone (ed.), *Modifying Lipids for Use in Food*, pp. 488–514. Cambridge: CRC Press (2006).

Norizzah, A. R., Chong, C. L., Cheow, C. S., and Zaliha, O. Effects of chemical interesterification on physicochemical properties of palm stearin and palm kernel olein blends. *Food Chemistry* 86(2) (2004): 229–235.

O'Brien, R. D. *Fats and Oils: Formulating and Processing for Applications*, 3rd edn. Boca Raton: CRC Press (2008).

Pernetti, M., van Malssen, K. F., Flöter, E., and Bot, A. Structuring of edible oils by alternatives to crystalline fat. *Current Opinion in Colloid and Interface Science* 12(4–5) (2007): 221–231.

Ramli, N., Said, M., and Loon, N. T. Physicochemical characteristics of binary mixtures of hydrogenated palm kernel oil and goat milk fat. *Journal of Food Lipids* 12(3) (2005): 243–260.

Ribeiro, A. P. B., Basso, R. C., Gonçalves, L. A. G., Gioielli, L. A., Santos, A. O., Cardoso, L. P., and Kieckbusch, T. G. Physico-chemical properties of Brazilian cocoa butter and industrial blends. Part II—Microstructure, polymorphic behavior and crystallization characteristics. *Grasas y Aceites* 63(1) (2012): 89–99.

Ribeiro, A. P. B., Basso, R. C., Grimaldi, R., Gioielli, L. A., and Gonçalves, L. A. G. Instrumental methods for the evaluation of interesterified fats. *Food Analytical Methods* 2(4) (2009): 282–302.

Ribeiro, A. P. B., Basso, R. C., and Kieckbusch, T. G. Effect of the addition of hardfats on the physical properties of cocoa butter. *European Journal of Lipid Science and Technology* 115(3) (2013): 301–312.

Ribeiro, A. P. B., Gonçalves, L. A. G., Gioielli, L. A., and Basso, R. C. Interesterification: Alternative for obtaining zero trans fat bases for food applications. In A. K. Haghi (ed.), *Advances in Food Science and Technology*, pp. 113–182. New York: Nova Science Publishers (2010).

Rodríguez, A., Castro, E., Salinas, M. C., López, R., and Miranda, M. Interesterification of tallow and sunflower oil. *Journal of American Oil Chemists' Society* 78(4) (2001): 431–436.

Rossell, J. B. Differential scanning calorimetry of palm kernel oil products. *Journal of the American Oil Chemists' Society* 52 (1975): 505–511.

Sato, K., Kigawa, T., Ueno, S., Gotoh, N., and Wada, S. Polymorphic behavior of structured fats including stearic acid and ω-3 polyunsaturated fatty acids. *Journal of American Oil Chemists' Society* 86 (2009): 297–300.

Sato, K. and Ueno, S. Polymorphism in fats and oils. In F. Shahidi (ed.), *Bailey's Industrial Oil and Fat Products*, 6th edn., vol. 1, pp. 77–120. Hoboken, New Jersey: Wiley-Interscience (2005). http://onlinelibrary.wiley.com/doi/10.1002/047167849X.bio020/full.

Schenk, H. and Peschar, R. Understanding the structure of chocolate. *Radiation Physics and Chemistry* 71(3–4) (2004): 829–835.

Shukla, V. K. S. Confectionery lipids. In F. Shahidi (ed.), *Bailey's Industrial Oil and Fat Products*, 6th edn., pp. 159–173. Hoboken, New Jersey: Wiley-Interscience (2005).

Siew, W. L. and Ng, W. L. Influence of diglycerides on crystallisation of palm oil. *Journal of the Science of Food and Agriculture* 79(5) (1999): 722–726.

Silva, J. C., Plivelic, S., Herrera, M. L., Ruscheinsky, N., Kieckbusch, T. G., Luccas, V., and Torriani, I. Polymorphic phases of natural fat from cupuassu (*Theobroma grandiflorum*) beans: A WAXS/SAXS/DSC study. *Crystal Growth and Design* 9(12) (2009): 5155–5163.

Smith, K. W., Bhaggan, K., Talbot, G., and Malssen, K. F. Crystallization of fats: Influence of minor components and additives. *Journal of the American Oil Chemists' Society* 88(8) (2011): 1085–1101.

Stapley, A. G. F., Tewkesbury, H., and Fryer, P. J. The effects of shear and temperature history on the crystallization of chocolate. *Journal of the American Oil Chemists' Society* 76(6) (1999): 677–685.

Szydlowska-Czerniak, A., Karlovits, G., Lach, M., and Szlyk, E. X-ray diffraction and differential scanning calorimetry studies of β′→β transitions in fat mixtures. *Food Chemistry* 92 (2005): 133–141.

Tan, C. P. and Che Man, Y. B. Differential scanning calorimetric analysis of edible oils: Comparison of thermal properties and chemical composition. *Journal of the American Oil Chemists' Society* 77(2) (2000): 143–155.

Tan, C. P. and Che Man, Y. B. Comparative differential scanning calorimetric analysis of vegetable oils: I. Effects of heating rate variation. *Phytochemical Analysis* 13(3) (2002a): 129–141.

Tan, C. P. and Che Man, Y. B. Differential scanning calorimetric analysis of palm oil, palm oil based products and coconut oil: Effects of scanning rate variation. *Food Chemistry* 76(1) (2002b): 89–102.

Timms, R. E. Fractionation. In *Lipid Technologies and Applications*, pp. 199–222. New York: Marcel Dekker (1997).

Timms, R. E. *Confectionery Products Handbook*. Bridgwater: The Oily Press (2003).

Van Duijin, G., Dumelin, E. E., and Trautwein, E. A. Virtually trans free oils and modified fats. In C. William and J. Buttriss (eds), *Improving the Fat Content of Foods*, pp. 490–507. Cambridge: Woodhead Publishing (2006).

Vereecken, J., Foubert, I., Smith, K. W., and Dewettinck, K. Effect of SatSatSat and SatOSat on crystallization of model fat blends. *European Journal of Lipid Science and Technology* 111(3) (2009): 243–258.

Walker, R. C. and Bosin, W. A. Comparison of SFI, DSC and NMR methods for determining solid-liquid ratios in fats. *Journal of the American Oil Chemists' Society* 48 (1971): 50–53.

Walter, P. and Cornillon, P. Influence of thermal conditions and presence of additives on fat bloom in chocolate. *Journal of the American Oil Chemists' Society* 78(9) (2001): 927–932.

Wassell, P. and Young, N. W. G. Food applications of trans fatty acid substitutes. *International Journal of Food Science and Technology* 42(5) (2007): 503–517.

Xu, X., Guo, Z., Zhang, H., Vikbjerg, A. F., and Damstrup, M. L. Chemical and enzymatic interesterification of lipids for use in food. In F. Gunstone (ed.), *Modifying Lipids for Use in Food*, pp. 234–272. Cambridge: Woodhead Publishing (2006).

Yamada, K., Ibuki, M., and Mcbrayer, T. Cocoa butter, cocoa butter equivalents, and cocoa butter replacers. In O. M. Lai and C. C. Akoh (eds), *Healthful Lipids*. Champaign: AOCS Press (2005).

Yanty, N. A. M., Marikkar, J. M. N., and Che Man, Y. B. Effect of fractional crystallization on composition and thermal characteristics of avocado (*Persea americana*) butter. *Journal of Thermal Analysis and Calorimetry* 111(3) (2013): 2203–2209.

Zaliha, O., Chong, C. L., Cheow, C. S., Norizzah, A. R., and Kellens, M. J. Crystallization properties of palm oil by dry fractionation. *Food Chemistry* 86(2) (2004): 245–250.

Zhou, Y. and Hartel, R. W. Phase behavior of model lipid systems: Solubility of high-melting fats in low-melting fats. *Journal of the American Oil Chemists' Society* 83(6) (2006): 505–511.

10 DSC Application to Characterizing Food Emulsions

Song Miao and Like Mao

CONTENTS

10.1 INTRODUCTION

A large majority of foods, such as milk, butter, and orange juice, exist, either partly or wholly, as emulsions. Emulsions can also act as delivery systems for bioactive compounds, for example, vitamins, carotenoids, and fish oil, to protect them from degradation, improve solubility, and control release (McClements et al., 2007). Emulsions are thermodynamically unstable, and destabilized emulsions have impaired functionality. Thermal treatments, including heating, cooling, and freezing (and subsequent thawing), are common processes during food manufacturing, transportation, storage, and consumption. Thermal history influences emulsion properties and, in turn, the organoleptic profile (Euston et al., 2001). When bioactives are incorporated in the dispersed phase, thermal stress also has a big effect on the physicochemical properties of the bioactives. Therefore, knowledge about the thermal behavior of emulsion ingredients and emulsion properties is essential for food scientists to develop food products with desired properties. Conventionally, emulsions are thermally treated and properties are then evaluated by, for example, particle size analysis, creaming layer measurement, and detection of oxidized products. In some cases, such as concentrated emulsions, conventional approaches do not always work, as dilution is required before analysis.

In this chapter, we introduce differential scanning calorimetry (DSC) to characterize food emulsions. DSC is a technique based on heat exchange from phase transition of the ingredients in an emulsion, or from polymorphic transition of crystals, as a function of temperature and time (Höhne, 1997). The area of the thermogram peak represents the enthalpic change, and the direction of the peak indicates whether the thermal event is exothermic or endothermic. DSC is particularly useful to track the solidification (crystallization) and melting processes of ingredients in emulsions. As thermal behaviors of emulsions are connected with emulsion properties, the thermograms obtained can be used to evaluate emulsion properties. A great advantage of DSC analysis is that it can differentiate the thermal behavior of different ingredients (oils, proteins, etc.) in one thermal cycle without extraction, which is difficult to accomplish through conventional methods. The objective of this chapter is to review the application of DSC to evaluate the thermal behavior of oils, water, and emulsifiers in emulsions; to track mass transfer of oils and water within emulsions; and then to use the information obtained to characterize emulsions. Examples are mainly focused on oil-in-water (O/W) emulsions, but water-in-oil (W/O) emulsions and water-in-oil-in-water (W/O/W) and oil-in-water-in-oil (O/W/O) emulsions are discussed as well.

10.2   FUNDAMENTAL OF FOOD EMULSIONS

10.2.1   Food Emulsions

Emulsions consist of two immiscible phases, one (the dispersed phase) of which is dispersed in the other (the continuous phase) as small droplets. The continuous phase constitutes the majority of the emulsion (McClements, 2005). Although a gas phase can be part of an emulsion (as in whipped topping), most food emulsions are liquid/liquid emulsions. Basically, there are two types of emulsions: O/W emulsion and W/W emulsion. In O/W emulsions, oil droplets are dispersed in a continuous water phase. Typical O/W food emulsions include milk, mayonnaise, some sauces, and so on. On the other hand, W/O emulsions are dispersions of water droplets in an oil phase. Butter and spreads are two well-known examples of W/O emulsions. In some food emulsions, the dispersed phase itself is a dispersion, which means that the dispersed droplets contain even smaller droplets. These emulsions are called multiple emulsions (or double emulsions), and they can be divided into W/O/W and O/W/O emulsions (Garti and Bisperink, 1998). Food emulsions are complex systems, and other ingredients may be present in either of the two immiscible phases. For example, polysaccharides, salts, and acids can be incorporated in the water phase, while vitamin A or D, and many volatile flavor compounds, can be dissolved in the oil phase. In order to maintain the stability of food emulsions (discussed later), amphiphilic ingredients such as proteins, phospholipids, and monoglycerides will be present in the systems as well, mainly located at the droplet surface (McClements, 2005).

10.2.2 Emulsion Formation and Stability

As indicated earlier, the dispersed phase is present as small droplets. Mechanical forces are generally applied to disrupt the bulk phase into droplets, and this process is called homogenization. In the food industry, the most widely used homogenizing equipment is high-speed blenders, colloidal mills, and high-pressure valve homogenizers (Schultz et al., 2004). During the homogenization process, a premixed dispersion (also called coarse emulsion) of oil and water phases is forced to pass through the narrow slit (between homogenizing valves or rotor and stator) of the equipment, and experiences collision, cavitation, shearing, friction, and other forces. In the meantime, amphiphilic ingredients are adsorbed onto the droplet surface, creating a stabilizing interfacial layer, which prevents the favorable aggregation of the newly produced droplets and finally leads to a fine emulsion (Schultz et al., 2004). To produce multiple emulsions, two steps of homogenization are generally required. Emulsions can also be made under mild conditions, through membrane emulsification or microchannel emulsification; these methods are mostly limited to lab scale (Charcosset et al., 2004; Sugiura et al., 2002).

Emulsions are thermodynamically unstable, since the free energy of mixing is exceeded due to the large interfacial area of the aqueous and oil phases. Theoretically, only the separated system is "stable." Emulsions with smaller droplets are prone to be more unstable, due to the larger interfacial area. However, smaller droplets experience stronger Brownian motion, which can counteract the separation of the two immiscible phases to a certain extent. Therefore, emulsions can be kinetically stable. An unstable emulsion results in flocculation, coalescence, creaming or sedimentation, Ostwald ripening, or phase inversion (Figure 10.1). In most situations, two or more types of the instability coexist or take place sequentially (McClements, 2005). Flocculation and coalescence can result in an increase in mean droplet size, and will normally lead to creaming (or sedimentation). However, flocculation and coalescence are caused by different mechanisms. Flocculation happens when two droplets are associated with each other, but without losing their individual identity. In other words, the droplets do not merge into one droplet and they have separate interfacial films. Therefore, flocculated droplets can be redispersed by blending or diluting. Flocculation is the result of van de Waals attraction, and the force increases dramatically when the distance between droplets decreases. Depending on the polymer types present in the emulsion, depletion flocculation or bridging flocculation can occur. Flocculation can be inhibited when there is a repulsive force between droplets, such as electrostatic repulsion or hindrance repulsion. In the case of coalescence, two or more droplets merge to form a bigger droplet, and the contents of each droplet are mixed. Severe coalescence is the beginning of creaming (take an O/W emulsion as example), as bigger droplets are more likely to move to the top of the system. Coalescence is the immediate result of the rupture of interfacial films (Ivanov et al.,

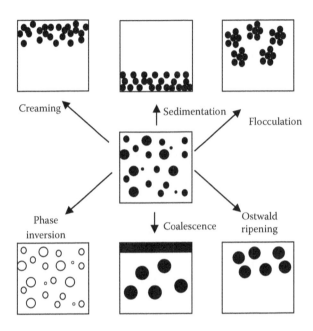

FIGURE 10.1   Schematic diagram of the destabilization of emulsion.

1999; Borwankar et al., 1992). Therefore, methods of strengthening the mechanical properties of the film can inhibit coalescence, for example, adsorption of polymers at the interface (Dickinson, 2011).

Due to the difference in density, the oil phase and aqueous phase inherently separate after a certain storage time, appearing as creaming or sedimentation (Robins, 2000). Separation is driven by gravitational force (or centrifugal force), and it happens when the gravitational force overcomes Brownian motion. Stokes' law predicts the separation rate of the droplets:

$$V_0 = \frac{2R^2 \Delta \rho g}{9\eta} \tag{10.1}$$

where:

V_0 = the separation rate (creaming or sedimentation)
R = the radius of droplets
$\Delta \rho$ = density difference between the oil and water phases
g = the acceleration force due to gravity or centrifuge
η = the system viscosity (Hunter, 1986)

From Equation 10.1 it can be found that increase in viscosity and reduction in droplet size can reduce the separation rate, and this can be achieved by addition of thickening agents and increased input of mechanical force. The density difference can be reduced by adding weighting agents to the oil phase to match the density of the water phase (Chanamai and McClements, 2000). It should be noted that, when droplet size is reduced to <0.1 μm, Stokes' law may not apply, as the Brownian motion can retard phase separation. When droplet size is <10 nm, the separation could be completely inhibited, which is actually the case for thermodynamically stable microemulsions (Russel, 1981).

Phase inversion is defined as the change of emulsion type from an O/W emulsion to a W/O emulsion, and vice versa. Phase inversion mostly happens when the properties of the emulsifier change. For example, in O/W emulsions stabilized with polyoxyethylene-type nonionic surfactants (e.g., Tween 20, 80), increased temperature can make the emulsifiers more hydrophobic, and the oil phase can become the continuous phase, forming a W/O emulsion (Scherman and Parkinson, 1978).

10.2.3 CONVENTIONAL APPROACHES TO CHARACTERIZE FOOD EMULSIONS

The functional properties of food emulsions, for example, color, shelf life, and texture, mostly rely on particle size distribution and emulsion stability, and these two parameters are widely adopted to characterize food emulsions. For an ideal emulsion, all the droplets are of the same size, and the emulsions are monodispersed. In reality, droplets in emulsions are of different sizes, and the emulsions are polydispersed. In this case, emulsions are generally characterized by mean particle size, accompanied by a polydisperse index. Droplets in food emulsions are generally in the micrometer and submicrometer ranges. As the naked human eye can only resolve particles with size >100 μm, specific techniques or equipment are required to observe size distribution in emulsions. Widely used methods of particle size evaluation include microscopy (scanning electronic microscopy, transmission electronic microscopy, confocal laser scanning microscopy, etc.), static/dynamic light scattering, and sedimentation techniques. The principles, applications, and operational techniques of these methods have been well reviewed (Barth and Sun, 1993; McClements, 2005). It should be noted that most of the above techniques require emulsion dilution, and the end result may not represent the original information. Dilution can break droplet association, interfere with droplet interaction, or change the environmental conditions (pH, ion strength, or concentration of other ingredients), and thus the diluted samples have properties different from the original samples. In this regard, it is difficult to correctly characterize droplet distribution of concentrated emulsions.

Emulsion stability collapses when severe droplet aggregation (flocculation, coalescence) occurs, which may finally lead to emulsion separation (creaming or sedimentation). Various factors can influence emulsion stability, for example, dispersed phase volume, droplet crystallization, emulsifier concentration, or environmental stresses (pH, salt, temperature, etc.) (McClements, 2005). Emulsion stability can be described as the change of particle size (flocculation, coalescence), the level of phase separation (thickness of creaming layer), amount of free oil (ratio of emulsified oil and destabilized oil), or change in turbidity (change in light signal). When chemical stability is concerned, emulsion stability can be expanded to oxidative stability of oil and chemical stability of incorporated bioactives (Berton et al., 2012; Mao et al., 2009; McClements et al., 2007). In recent years, the physicochemical stability of food emulsions during digestion (in the oral cavity, in the GI tract) has attracted much attention, and many *in vitro* and *in vivo* studies have been carried out (Sarkar et al., 2009; Hur et al., 2009).

10.3 CRYSTALLIZATION IN EMULSIONS

10.3.1 Fat Crystallization

Fat crystallization greatly influences the physicochemical properties of both O/W emulsions and W/O emulsions. In some foods, fat crystallization determines the rheology, stability, and appearance of the food. During the production of ice cream (an O/W emulsion), fat in the emulsion is partially crystallized when the temperature is lowered. The crystallized fat promotes droplet aggregation (also called partial coalescence) to form a two-dimensional network with air bubbles trapped, and also a three-dimensional network in the continuous phase that contributes to the stability and texture of the ice cream. With an increase of fat crystallization, severe droplet aggregation could happen, and the O/W emulsion would transform to a W/O emulsion. This inversion is the main step in the production of butter, margarines, and spreads (W/O emulsions). In these foods, emulsion stability and rheological properties are mainly determined by the presence of a fat crystal network in the continuous phase. Formation of the fat crystal network increases emulsion viscosity and prevents the sedimentation of water droplets, as they are trapped in the network. Level of fat crystallization also has a pronounced effect on the firmness and spreadability of the products.

In an O/W emulsion, oil crystallizes through a mechanism different from its crystallization in bulk oil, which has been widely confirmed using DSC. When oil is disrupted into small droplets, it generally crystallizes at a much lower temperature than it does in the bulk state. This phenomenon is called supercooling (or undercooling). In an emulsion containing hydrogenated palm kernel oil (HPKO), Cornacchia and Roos found that bulk HPKO crystallized at 22.5°C (peak temperature), while the emulsified oil HPKO had a crystallizing peak at around 5°C. When both HPKO and emulsified HPKO were present in a system, two crystallizing peaks were observed (Figure 10.2) (Cornacchia and Roos, 2011).

Bulk oil generally crystallizes through heterogeneous homogenization, as naturally occurring lipids contain a certain amount of impurities, which can act as crystal templates. When oil is emulsified into small droplets, the concentration of impurities in each droplet is dramatically reduced, and many droplets are devoid of impurities. As a result, a much lower temperature is required to promote oil droplet crystallization in emulsion systems. In this case, the crystallization is mainly through a homogeneous mechanism. In an emulsion system, each dispersed particle can be regarded as an independent sample. As the particles can be as small as several nanometers, a much lower temperature is required to promote nucleation and crystallization. If the interfacial film is continuous (or partially continuous) (e.g., concentrated emulsion system, coalescence emulsion), fast crystallization will happen once the first nucleus forms, as the interface propagates the crystallization. Furthermore, if there are fat crystals in some droplets, droplet collision will accelerate the crystallization

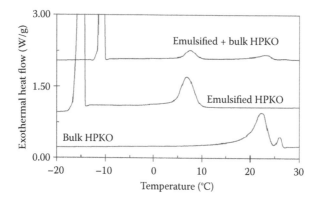

FIGURE 10.2 DSC cooling scans of HPKO showing its crystallization peaks in bulk, 40 wt% O/W emulsion, and mixture of both. (Reprinted from *Food Hydrocolloids*, 25, Cornacchia, L. and Roos, Y.H., Lipid and water crystallization in protein-stabilized oil-in-water emulsions, 1726–1736, Copyright (2011), with permission from Elsevier.)

(McClements et al., 1993a). Due to the progressive development of crystallization in emulsions, the crystallizing peak of the emulsified fat is wider than that of a bulk fat (Figure 10.3) (Clausse et al., 2005).

Palanuwech and Coupland (2003) made confectionery coating fat emulsions with different droplet sizes (0.76–7.44 μm), and they found that the oil crystallized via heterogeneous mechanisms when the droplet size was >5 μm, and via homogeneous mechanisms when the droplet size was <1 μm. Therefore, oil crystallization in emulsions is dependent on droplet size, and emulsions with smaller droplet size will have a much lower crystallizing temperature (a higher level of supercooling).

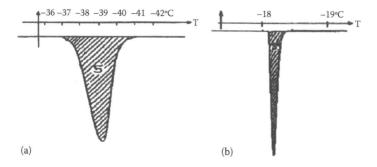

FIGURE 10.3 DSC cooling curves for water dispersed in an emulsion (a) and for bulk water (b). (Reprinted from *Advances in Colloid and Interface Science*, 117, Clausse, D., Gomez, F., Pezron, I., Komunjer, L., and Dalmazzone, C., Morphology characterization of emulsions by differential scanning calorimetry, 59–74, Copyright (2005), with permission from Elsevier.)

In an emulsion system, emulsifiers affect not only droplet size but also interfacial structure, which then influence the crystallization behavior of the oil droplets. Several studies have reported that both hydrophilic emulsifiers (diacylglycerols) and hydrophilic emulsifiers (propylene glycerol monostearate, polyglycerol fatty acid esters) can accelerate the crystallization of fat in O/W emulsions (Awad et al., 2001; Sonoda et al., 2006; Awad and Sato, 2002). Emulsifiers at the interface can aggregate with coexistent emulsifiers (added during emulsification, e.g., Tween 20), the oil phase, or both. The aggregates can act as catalytic templates for the heterogeneous nucleation of the fat in the emulsions, as the hydrophobic tails of the aggregates solidify early on cooling. Furthermore, the self-assembled structures (micelles or reverse micelles) of the emulsifiers can catalyze the nucleation of the oil. Awad and Sato, using DSC, found that, in an emulsion containing palm kernel oil, the emulsified oil began to crystallize at around 0°C, and the crystallizing temperature shifted to about 9°C with the addition of stearic acid sucrose oligoester (S-170 in Figure 10.4). Emulsifiers have no effect on the crystallization temperature of the bulk oil. Reports have indicated that the nucleation rate is proportional to interfacial area but not volume (Skoda and Van den Tempel, 1963; McClements et al., 1993c).

X-ray diffraction (XRD) analysis revealed that the addition of S-170 was favorable for the formation of the β′ polymorph, while the α polymorph formed in

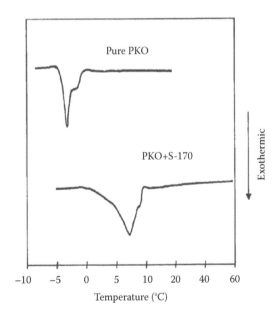

FIGURE 10.4 Exothermic peaks obtained by DSC during cooling of palm kernel oil (PKO)/water emulsion without (pure) and with the addition of 1 wt% S-170. (Reprinted from *Colloids and Surfaces B: Biointerfaces*, 25, Awad, T., Sato, K., Acceleration of crystallization of palm kernel oil in oil-in-water emulsion by hydrophobic emulsifier additives, 45–53, Copyright (2002), with permission from Elsevier.)

the system without S-170. It was reported that even-numbered n-alkanes (carbon number 16–22) can form unique crystal structures in O/W emulsions that were not found in the bulk oil system. Therefore, emulsions can be used to form new crystalline polymorphic structures (Ueno et al., 2003). Emulsifier concentration also contributes to the crystallizing behavior of oil in emulsions. Even with a low amount of emulsifiers added, where the nucleation sites were small, nucleation could also be accelerated to a certain degree. With a higher amount of emulsifier added, the acceleration of nucleation reached a maximum rate at a certain concentration.

For a specific emulsifier, the original location of the emulsifier, that is, in the water phase or in the oil phase, could lead to different crystallizing temperatures of the oil phase. In an emulsion stabilized by sodium caseinate (20% palm oil), palm oil had a lower crystallizing temperature when propylene glycol monostearate (PGMS) was added into the oil phase than that when PGMS was added into the water phase (Kalnin et al., 2004). Increase of PGMS content in the aqueous phase led to an increase in the crystallization temperature of the oil, but a decrease in the enthalpy of the phase transition (Figure 10.5). This finding was attributed to the polymorphism transition of the oil, as the presence of PGMS interferes with the crystallization of the oil.

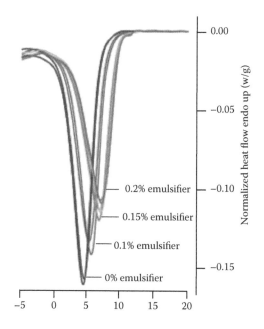

FIGURE 10.5 DSC recordings obtained as a function of PGMS concentration in the range of 0%–0.2% of the total weight of emulsion, showing a decrease of both crystallization temperature and enthalpy. (Reprinted with permission from Kalnin, D., Schafer, O., Amenitsch, H., and Ollivon, M. *Crystal Growth and Design*, 4, 1283–1293, Copyright (2004), American Chemical Society.)

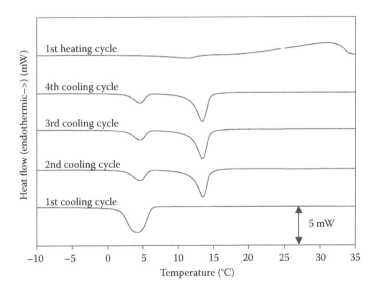

FIGURE 10.6 Successive cooling curves and a heating curve of 40 wt.% CCF emulsion stabilized with 2 wt.% Tween 20 measured at 1.5°C min⁻¹. Exotherms are shown downward. (Reprinted from *Colloids and Surfaces A: Physicochemical Engineering Aspects*, 223, Palanuwech, J. and Coupland, J.N., Effect of surfactant type on the stability of oil-in-water emulsions to dispersed phase crystallization, 251–262, Copyright (2003), with permission from Elsevier.)

Accelerated crystallization can also be achieved when there are solid particles in the oil droplets. The solid particles collide with each other and induce nucleation (McClements et al., 1994; Dickinson et al., 1993).

Due to the difference in the crystallizing temperature between emulsified oil and bulk oil, DSC can be used to quantify the destabilized oil (free oil) in an emulsion when partial (or complete) separation occurs under thermal treatment. Figure 10.6 shows that, in the first cooling cycle, only one crystallizing peak was detected, indicating that all the oil was in the emulsified state and the emulsion was stable. The emulsion was then heated and underwent a second cooling–heating cycle. In the second cooling stage, two crystallizing peaks appeared, indicating that part of the oil was destabilized (free oil). In the following cooling–heating cycles, the crystallizing peak of free oil became more significant while the crystallizing peak of emulsified oil diminished, suggesting more oil was released (Palanuwech and Coupland, 2003).

The ratio of the enthalpies at different crystallizing peaks can be used to calculate the proportion of destabilized oil in the emulsion (Equation 10.2).

$$\text{Percentage of fat destabilization} = A1/(A1 + A2) \times 100 \qquad (10.2)$$

where $A1$ and $A2$ are the areas under the crystallizing peaks of emulsified oil and destabilized oil, respectively. The measured free fat ratio based on Equation 10.2 correlated well with the ratio of free fat added (Palanuwech and Coupland,

2003). Compared with solvent extraction methods, the calorimetric method has evident advantages when applied to an emulsion system: (1) small amount of sample required; (2) no solvent required; (3) time- and cost-effective; and (4) fully automatic and software controlled.

10.3.2 WATER CRYSTALLIZATION

Water crystallization, that is, freezing, is a common process in the food industry. Freezing of the water phase in emulsions is essential for the production of some foods, for example, ice cream. Freezing is mainly used to improve the shelf life of food, as ice crystals can inhibit undesirable microbial growth, biochemical processes, and chemical reactions (Fellows, 2000). The crystallization of water as a bulk continuous phase in O/W emulsion follows the same behavior as normal water, and dissolved ingredients can have great effects on ice formation and growth. It should be noted that oil droplets constitute most of the ice nuclei, and water in an O/W emulsion crystallizes at a much higher temperature than pure water does (>10°C) (Donsì et al., 2011). The mechanism of water crystallization is well understood, and DSC is widely applied to elucidate the mechanism (Ghosh and Rousseau, 2009). The present chapter is not intended to focus on DSC application to evaluate water crystallization, but to emphasize the big effect of water crystallization on emulsion properties. Examples of DSC application to evaluate the effect of water crystallization on emulsion properties are discussed in Section 10.5.2.

In most situations, water crystallization in an emulsion can lead to droplet aggregation and phase separation, depending on the magnitude of freezing stress and emulsion compositions. When water crystals form, oil droplets are forced much closer together, and penetration of ice crystals into oil droplets leads to interfacial membrane rupture (Saito et al., 1999). Consequently, flocculation and coalescence take place. Moreover, water crystallization can result in protein denaturation and emulsifier oversaturation (Chang et al., 2000; Carvajal et al., 1999), which then promote phase separation. Other studies revealed that emulsifiers can adsorb onto the surface of ice crystals, thereby reducing the concentration of emulsifiers covering oil droplets (Hillgren et al., 2002). The freezing stability of emulsions can be improved when cryoprotectants are added, for example, sucrose, lactose, or maltodextrin, or when a stronger interfacial layer is formed, for example, an interface containing a protein and polysaccharide multilayer (Hartel, 2001).

In terms of W/O emulsions, in which water is dispersed as small droplets, supercooling is required to initiate water crystallization. However, a high degree of supercooling is seldom applied in the food industry. It was observed from DSC thermograms that water droplets in polyglycerol polyricinoleate (PGPR)-stabilized canola oil emulsion crystallize at −40°C (Figure 10.7), about 20°C–25°C lower than the crystallization temperature of bulk water (Ghosh and Rousseau, 2009).

10.3.3 EMULSIFIER CRYSTALLIZATION AND INTERFACIAL CRYSTALLIZATION

The emulsifiers being discussed here refer to polar lipids in the food industry, which are lipids in nature but have amphiphilic properties. Therefore, these types

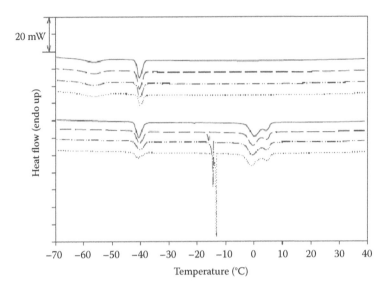

FIGURE 10.7 DSC thermograms of freezing/thawing cycles of PGPR-stabilized water in canola oil emulsion. Samples were temperature cycled from 40°C to −70°C at a rate of 5°C/min. (Reprinted from *Journal of Colloids and Interface Science*, 339, Ghosh, S. and Rousseau, D., Freeze-thaw stability of water-in-oil emulsions, 91–102, Copyright (2009), with permission from Elsevier.)

of emulsifiers can adsorb at the water–oil interface to stabilize emulsions, and form crystals when cooled. Monoglycerides or monoacylglycerides (MGs) and their organic acid derivatives are, by volume, the most commonly used polar lipids in the food industry, and their crystallization properties and phase behavior have a big effect on emulsion properties (Krog, 2001).

An emulsifying MG (hydrophilic–lipophilic balance value ~2–5) can be used to stabilize W/O emulsions, and also O/W emulsions with the coexistence of water-soluble emulsifiers (McClements, 2005; Hwang et al., 1991). When mixed with water or oil, MG can form self-assembled structures (liquid crystals). In an MG–oil–water gel, MG will develop into a highly hydrated crystalline lamellar phase (Lα), covering oil droplets, and form a mesomorphic gel with some fat-like characteristics (Batte et al., 2007; Calligaris et al., 2010), which can be used to produce fat-reduced food. In an emulsion system, MGs showed different crystallizing behavior from that in water gel and in powdered MG. As Figure 10.8 illustrates, MG in emulsion and MG powder had similar thermal behavior, while the peak temperatures varied (Mao et al., 2012), which is due to the emulsification process as well as the effect of the matrix (Povey, 2001). Different melting and crystallizing peaks in the DSC graphs showed the polymorphic properties of MG: in the first heating cycle, the sharp melting peak (peak temperature of 62.5°C) corresponded to the β polymorph of the MG, which is normally formed after a relatively long storage time and is rather stable (Krog, 2001). The following cooling cycle led to

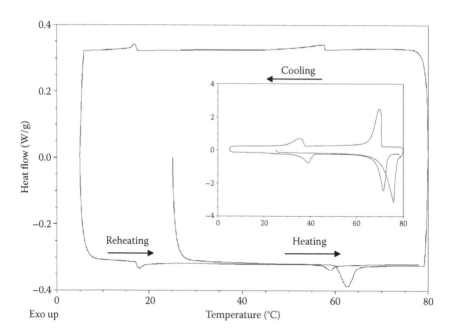

FIGURE 10.8 Thermal behavior of MG in TW (1% w/w) emulsion with 2% w/w MG and 20% w/w oil through a heating–cooling–reheating course on D4 (oil: 20% w/w). Inset DSC graph shows the thermal behavior of MG powder. Heating: from 25°C to 80°C at 5°C/min; cooling: from 80°C to 5°C at 5°C/min; reheating: from 5°C to 80°C, at 5°C/min. (Reprinted from *Food Research International*, 48, Mao, L., O'Kennedy, B.T., Roos, Y.H., Hannon, J.A., and Miao, S., Effect of monoglyceride self-assembled structure on emulsion properties and subsequent flavour release, 233–240, Copyright (2012), with permission from Elsevier.)

the development of two different crystalline forms, which were then melted in the reheating cycle. These two peaks probably corresponded to the crystallization of α and sub-α polymorphic forms, which are unstable and could eventually transform into the stable β form (Krog and Sparsø, 2004; Vereecken et al., 2009). In some studies, a third crystallization peak (named sub-α2) was also reported (Vereecken et al., 2009). According to the DSC data, the α (sub-α) form of the MG crystals has lower melting enthalpies and melting peak temperatures than the β form, which further confirms the existence of different MG polymorphic forms at different heating/cooling stages, as a more stable crystal generally has higher melting enthalpy and entropy (Himawan et al., 2006).

The formation of MG polymorphic crystals in the emulsions was time and temperature dependent (Figure 10.9) (Mao et al., 2014). On the day of Tween 20 (TW) emulsion preparation (D1, 25°C), MG only showed a weak melting peak (α-form crystal) at 57.39°C ± 0.04°C. After 3 days' storage (D4, 25°C), the peak transformed to a bigger peak (β-form crystal) at 62.02°C ± 0.28°C. Further storage (D7) did not

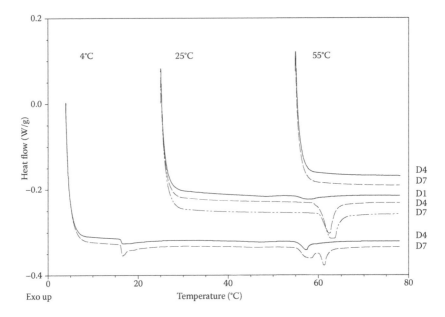

FIGURE 10.9 Effect of storage temperature on the melting behaviors of MG in emulsions stabilized by TW on day 1 (D1), day 4 (D4), and day 7 (D7) (TW 1%, oil 20%, MG 2%, w/w). All the emulsions were originally prepared at 25°C. (Reprinted from *Food Research International*, 58, Mao, L., Calligaris, S., Barba, L., and Miao, S., Monoglyceride self-assembled structure in O/W emulsion: Formation, characterization and its effect on emulsion properties, 81–88, Copyright (2014), with permission from Elsevier.)

result in significant change in peak temperature and melting enthalpy, indicating stable crystals were formed and maximum crystallinity was reached. When the emulsion was stored at 4°C, a third melting peak (sub-α-form crystal) was observed at 17.01°C ± 0.03°C (Vereecken et al., 2009; Chen and Terentjev, 2010). The hydration of the glycerol in the bilayer resulted in the formation of a thin layer inside the inverse lamellar bilayer, which disturbed the packing of MG molecules, leading to a decrease in the crystallization temperature of sub-α-form crystals (Chen and Terentjev, 2010). On D7, the three forms of MG crystals were present, indicating polymorphic transformation of sub-α- and α-form MG crystals to β-form crystals (Vereecken et al., 2009; Chen and Terentjev, 2010). It was also shown that MG was crystallizing at a much lower rate at 4°C, and would take a longer time to reach the highest crystallinity. When the emulsion was stored at 55°C, no melting peak was detected throughout the study. This was because the α-form crystals melted at 55°C, which inhibited the growth of the crystals and further transformation to β-form crystals.

Emulsifiers have different effects on the crystallizing behavior of MGs in emulsions. In whey protein isolate (WPI)-stabilized emulsion, the DSC profile shows that MG crystals formed on D1 and remained stable thereafter (Figure 10.10) (Mao et al., 2014). Emulsions stored at 4°C and 25°C had similar MG thermal behavior,

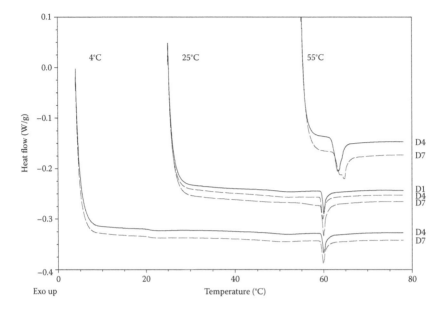

FIGURE 10.10   Effect of storage temperature on the melting behaviors of MG in emulsions stabilized by WPI on day 1 (D1), day 4 (D4), and day 7 (D7) (WPI 1%, oil 20%, MG 2%, w/w). All the emulsions were originally prepared at 25°C. (Reprinted from *Food Research International*, 58, Mao, L., Calligaris, S., Barba, L., and Miao, S., Monoglyceride self-assembled structure in O/W emulsion: Formation, characterization and its effect on emulsion properties, 81–88, Copyright (2014), with permission from Elsevier.)

with melting peaks at ~59.6°C. The melting peak shifted to ~62.8°C when WPI emulsion was stored at 55°C. By calculating the melting enthalpy of MG crystals during storage, it was found that most crystal formed on D1, and following storage the crystallinity only increased slightly. The lower melting enthalpy here suggested that only a small amount of MG crystal was formed in WPI emulsion compared with that in TW emulsion. In WPI emulsion, MG could present at the interface by replacing the adsorbed protein, because MG is more surface active than protein (Pelan et al., 1997; Pugnaloni et al., 2004). The lower crystallinity of MG in WPI emulsion was due to the interaction between MG and WPI (Leenhouts et al., 1997; Boots et al., 1999), and it also modified the thermal behavior of MG (Anker et al., 2002; Boots et al., 1999).

Similarly to the crystallizing behavior of common lipids in emulsions, MG has a reduced crystallizing rate in emulsions compared with that in the powdered state. This results mostly from homogenization and emulsification. Although a large inter-facial area is believed to promote crystallization due to surface crystallization and droplet collision (Povey, 2001), the presence of an interfacial film composed of TW or WPI in the current emulsion systems may inhibit this process.

The polymorphic behavior of MG crystals was affected by the size of the oil droplets in the emulsions. Figure 10.11 illustrates the crystallizing behavior of MG in three TW emulsions with different droplet sizes (Mao et al., 2014). Emulsions

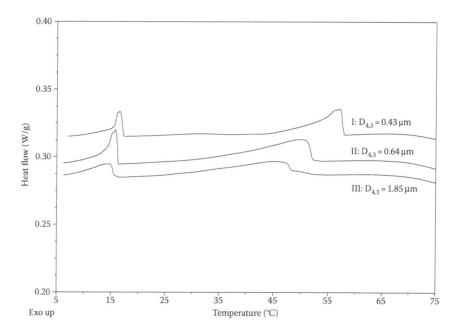

FIGURE 10.11 Effect of droplet size on the crystallizing behaviors of MG in TW emulsions (TW 1%, oil 20%, MG 2%, w/w). (Reprinted from *Food Research International*, 58, Mao, L., Calligaris, S., Barba, L., and Miao, S., Monoglyceride self-assembled structure in O/W emulsion: Formation, characterization and its effect on emulsion properties, 81–88, Copyright (2014), with permission from Elsevier.)

prepared with a microfluidizer (I), a lab homogenizer (II), and an ULTRA-TURRAX (III) had droplet sizes ($D_{4,3}$) of 0.43, 0.64, and 1.85 μm, respectively. It was found that in emulsion I MG tended to form α-form crystal at much higher temperatures (with an onset temperature of 57.68°C), whereas in emulsion II and emulsion III the formation of the crystal was considerably delayed (onset temperature of 52.12°C and 48.64°C, respectively). However, crystallizing enthalpies of the MG crystals in the three emulsions were quite close to each other, although the crystallizing peaks in emulsions II and III were tailed. On the other hand, the formation of sub-α-type crystals had a similar onset temperature at ~15°C in the three emulsions.

The formation of an MG crystalline structure results in increased viscosity of the emulsion and also makes the emulsion more sensitive to environmental stresses (e.g., pH, salt). Several studies have shown that an MG crystalline structure in emulsions could provide accommodation for bioactive compounds, and controlled release could be achieved (Sagalowicz et al., 2006; Mao et al., 2013).

10.4 MASS TRANSFER IN EMULSIONS

Oil exchange happens in mixed emulsions (different emulsions mixed together) or multiple emulsions (W/O/W emulsions or O/W/O emulsions). Generally, these emulsions contain two or more types of oils. As different oils have different thermal behavior, DSC can be used to track the mass transfer of the oils in emulsions.

Several studies reported oil exchange in mixed emulsions containing two emulsions with different oils stabilized by either a nonionic surfactant or a protein. The mechanism is not well understood, but could be due to the solubilization and transportation of the oils by emulsifiers. DSC was applied to track the exchange process. McClements et al. (1993b) made emulsions containing *n*-hexadecane or octadecane, and the two emulsions were mixed before cooling in the DSC. They observed that, immediately after mixing, discrete crystallizing peaks of *n*-hexadecane and octadecane were observed. After a certain period of storage, the two crystallizing peaks moved closer toward each other due to the incorporation of another type of oil, indicating the occurrence of oil exchange between the droplets. After further storage, the two crystallizing peaks completely merged, and only one crystallizing peak in the middle of the original crystallizing peaks of the two oils was observed (Figure 10.12).

As the author proposed, the driving force for oil exchange is the free energy of mixing. It is more entropically favorable for oil to be evenly distributed among all the droplets than to be separately distributed. The exchange is achieved when oil is reversibly bound to protein molecules (in the aqueous phase) and transported across the water region to the neighboring oil droplets. It was hypothesized that the oils were in a dynamic equilibrium between those inside the droplets and those bound with emulsifier. When additional emulsifier is added to the mixed emulsion, more oil can be transported, and the rate of oil exchange increases linearly with emulsifier concentration (Figure 10.13). As the free energy of mixing is much higher than that

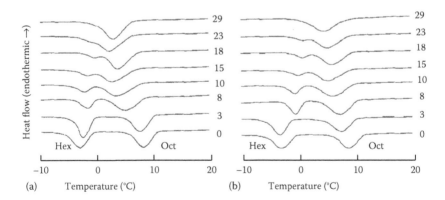

FIGURE 10.12 Variation in crystallization behavior over time of emulsions that initially contained a mixture of 10 wt% *n*-hexadecane droplets and 10 wt% octadecane droplets. The emulsions were stabilized with (a) WPI or (b) casein. Additional protein (2.5 wt%) was added to the aqueous phase after emulsification. The time is indicated in days at the right side of the thermograms. (Reprinted from *Journal of Colloid and Interface Science*, 156, McClements, D.J., Dungan, S.R., German, J.B., and Kinsella, J.E., Evidence of oil exchange between oil-in-water emulsion droplets stabilized by milk proteins, 425–429, Copyright (1993b), with permission from Elsevier.)

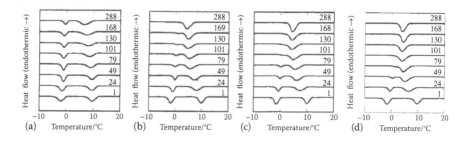

FIGURE 10.13 DSC measurements of the dependence of the crystallization behavior of emulsions on time, shown in hours with each trace. Emulsions initially contained a mixture of 10 wt% n-hexadecane droplets and 10 wt% octadecane droplets. Cooling rate = 5°C min^{-1}. Amount of additional Tween 20 added after emulsification: (a) 0%, (b) 1%, (c) 2%, and (d) 4%. (Reprinted with permission from McClements, D.J. and Dungan, S.R., *Journal of Physical Chemistry*, 97, 7304–7308, Copyright (1993), American Chemical Society.)

for Ostwald ripening, oil exchange does not result in large variance in the droplet size (McClements and Dungan, 1993).

Droplet association due to Brownian motion, flocculation, and coalescence could accelerate the exchange. As different emulsifiers have different binding capacity for oil molecules, the exchange rate is also affected by emulsifier type (Elwell et al., 2004). Oil exchange may result in the loss of some nutrients. For example, when an ω-3 oil-enriched emulsion (with antioxidant) is added to milk, oil exchange happens between ω-3 oil and milk fat, and the redistribution of the oils can break the antioxidant system originally designed for the ω-3 oil emulsion. In this case, additional antioxidant is required when the emulsions are mixed.

Oil exchange also happens in a single emulsion that contains two oil phases, that is, an O/W/O multiple emulsion. Avendano-Gomez et al. (2005) made a tetradecane/water/hexadecane emulsion, and the crystallization thermograms of the oil phases were evaluated by DSC. The peaks in the first thermograms (1 min) represented the crystallizing of the three different phases in the emulsion, that is, water phase (the far left peak), internal oil phase (tetradecane) (the middle peak), and the bulk external oil phase (hexadecane) (the far right peak). Similarly to the oil exchange in mixed emulsions, with increased storage time the peaks of the two oils moved closer toward each other. The last thermogram illustrated that, after 26 days of storage, the crystallizing peak of tetradecane almost disappeared, and a broader peak with intermediate crystallization temperature was observed, indicating the end of the oil exchange (Figure 10.14). The findings also revealed that oil exchange was dominated by the movement of tetradecane to hexadecane, and the tetradecane phase finally disappeared. Therefore, the O/W/O emulsion finally transformed into a W/O emulsion.

According to the mass balance law, the ratio of the oils in the final emulsion can be calculated. To validate this calculation, DSC evaluation was carried out on a mixture of bulk tetradecane and hexadecane in the same ratio as in the original O/W/O emulsion (Avendano-Gomez et al., 2005). The thermograms in Figure 10.15 show that the thermal behavior of the oils in the two systems fitted well, indicating that the oil composition of the emulsion was the same as that of the bulk oil mixture.

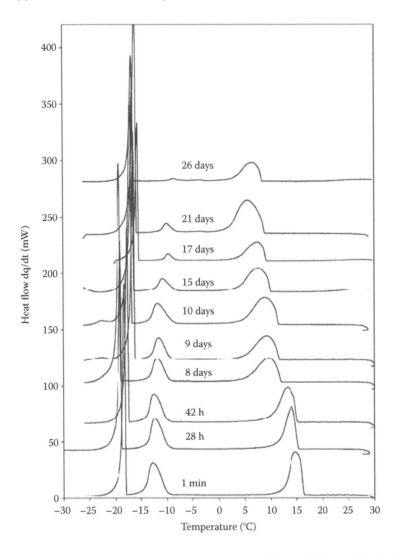

FIGURE 10.14 Evolution of crystallization thermograms of the multiple emulsion $O_1/W/O_2$ tetradecane/water/hexadecane containing 2 wt% Tween 20 surfactant in the aqueous membrane. (Reprinted from *Journal of Colloid and Interface Science*, 290, Avendano-Gomez, J.R., Grossiord, J.L., and Clausse, D., Study of mass transfer in oil-in-water multiple emulsions by differential scanning calorimetry, 533–545, Copyright (2005), with permission from Elsevier.)

It was hypothesized that the progressive mixing of the two oils was achieved through mass transfer of tetradecane across the water phase, driven by osmotic pressure. Similarly to the mechanism of oil exchange in a mixed emulsion, the emulsifier in the water phase played the role as a carrier to transport the internal oil phase into the external oil phase. Although some tetradecane molecules can move to the external oil phase without binding to emulsifier, the movement was much slower than the

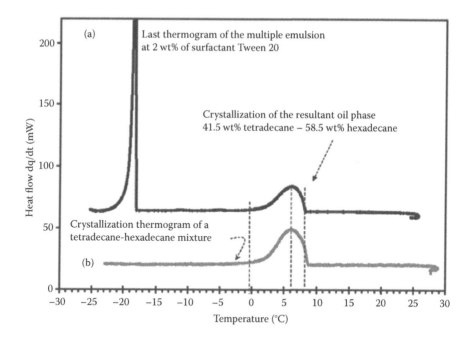

FIGURE 10.15   The superposition of (a) the last thermogram from the evolution of the multiple emulsion and (b) a thermogram of a bulk mixture composed of tetradecane and hexadecane at the composition expected for the mass balance once the transfer reaches the end. (Reprinted from *Journal of Colloid and Interface Science*, 290, Avendano-Gomez, J.R., Grossiord, J.L., and Clausse, D., Study of mass transfer in oil-in-water multiple emulsions by differential scanning calorimetry, 533–545, Copyright (2005), with permission from Elsevier.)

transportation with emulsifier. As a result, increased emulsifier concentration could accelerate the rate of oil exchange in multiple emulsions.

In order to confirm that oil exchange in an O/W/O emulsion was a result of the osmotic pressure gradient, a multiple emulsion containing the same oil (i.e., hexadecane) in the internal and external oil phases was calorimetrically evaluated (Avendano-Gomez et al., 2005). The thermogram in Figure 10.16 shows that the two crystallizing peaks of the two oil phases remained unchanged during the storage test, suggesting that no oil exchange occurred. In this emulsion, the osmotic pressure difference through the water phase remained constant, as there was no compositional difference between the two sides.

Similarly, DSC was also used to track the mass transfer of water in mixed W/O emulsions and W/O/W emulsions (Clausse et al., 1995).

10.5   EVALUATION OF EMULSION PROPERTIES USING DSC

10.5.1   EMULSION TYPE

Based on the knowledge that food ingredients have different thermal behavior in the bulk state and the emulsified state, DSC can be applied to determine the types of emulsions, for example, O/W, W/O, O/W/O, or W/O/W emulsions. This was determined by comparing the thermograms of the bulk water or oil and their corresponding

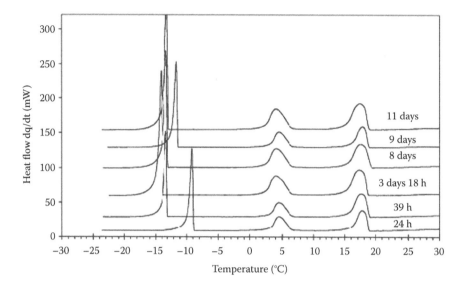

FIGURE 10.16 Sequence of thermograms of an isoosmotic $O_2/W/O_2$ multiple emulsion composed of hexadecane-in-water-in-hexadecane. The hexadecane constitutes the external and the internal phase. (Reprinted from *Journal of Colloid and Interface Science*, 290, Avendano-Gomez, J.R., Grossiord, J.L., and Clausse, D., Study of mass transfer in oil-in-water multiple emulsions by differential scanning calorimetry, 533–545, Copyright (2005), with permission from Elsevier.)

emulsions. For example, in a W/O emulsion, as the water is divided into small droplets, a much lower temperature is required to initiate crystallization (−39°C in Figure 10.17b). However, for a bulk water phase, ice crystals are formed at a higher temperature (−24°C in Figure 10.17a) (Dalmazzone et al., 2009). It should be noted that melting of the ice crystals takes place at 0°C whether or not the water is emulsified. When an O/W emulsion is concerned, the thermogram (if in the temperature range) should contain a water crystallizing peak similar to the one for bulk water in Figure 10.17a.

DSC can be applied for identifying some complex emulsions, for example, W/O/W emulsions. As Figure 10.17c illustrates, the thermogram of a W/O/W emulsion contains

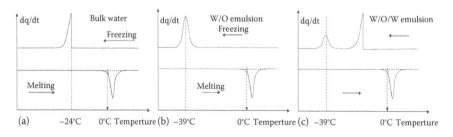

FIGURE 10.17 Cooling and heating thermograms of bulk water (a), water in W/O emulsion (b), and water in W/O/W emulsion (c). (Reprinted from *Oil and Gas Science and Technology*, 5, Dalmazzone, C., Noïk, C., and Clausse, D., Application of DSC for emulsified system characterization, 543–555, Copyright (2009), with permission from FP Energies nouvelles.)

two crystallizing peaks of water, which correspond to the internal water phase (at −39°C) and the external water phase (at −24°C), respectively (Dalmazzone et al., 2009). Similar identification can also be made by comparing the thermograms of the oil phase in the bulk state and the emulsified state. In complicated food systems, one would expect temperature deviation of the crystallizing peak due to the influence of other ingredients, droplet size, and so on, but the difference in crystallizing temperature of the bulk phase and emulsified phase would dominate. As discussed earlier, DSC is also able to track the transition of emulsion types, for example, from O/W/O emulsion to W/O emulsion.

10.5.2 EMULSION STABILITY AND PARTIAL COALESCENCE

It is now clear that DSC can be used to detect the occurrence of destabilized fat (or free fat) in an O/W emulsion. As thermal treatment is essential during DSC analysis and thermal force can trigger emulsion instability, DSC is particularly useful to evaluate the freeze–thaw stability of emulsions. For this specific application, freeze–thaw cycles can be mimicked in a DSC pan under a temperature program. In an O/W emulsion, crystallization of the oil phase or the water phase, and the order of their crystallization, affects emulsion stability.

In HPKO emulsions stabilized by WPI, cooling resulted in the sequential crystallization of HPKO and water. As Figure 10.18a illustrates, HPKO destabilization occurred in the second cooling cycle, and most HPKO was released from the oil droplets after four cycles. Cornacchia and Roos (2011) found that addition of sucrose could inhibit the destabilization. Sucrose modified the thermal behavior of the water phase during freezing–thawing, and the sucrose syrup protected the oil particle from interface damage caused by ice formation and growth. In the case of sunflower oil emulsions, oil crystallization took place well after water crystallization, and four consecutive freezing–thawing cycles did not show any shift in the crystallization peak, indicating that the emulsions were resistant to thermal treatment. Water crystallization did not induce fat destabilization in the emulsion, probably because the mechanical force from ice formation was tolerated by the flexible liquid oil particles.

In most cases, fat crystallization does not result in immediate emulsion separation, but leads to partial coalescence. Partial coalescence occurs in partially

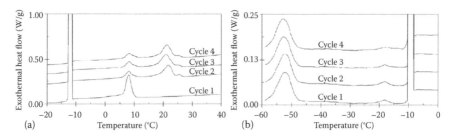

FIGURE 10.18 Thermograms of 40 wt% HPKO emulsions (a) and SO emulsions (b) stabilized by WPI, showing lipid and water crystallization upon cooling during four successive freeze–thaw cycles from 50°C to −40°C. (Reprinted from *Food Hydrocolloids*, 25, Cornacchia, L. and Roos, Y.H., Lipid and water crystallization in protein-stabilized oil-in-water emulsions, 1726–1736, Copyright (2011), with permission from Elsevier.)

crystalline droplets, where the crystals in one droplet penetrate the interface of a second droplet. The liquid oil then flows out and strengthens the association between the two droplets. As the mechanical strength of the crystal network is thought to be sufficient to overcome the Laplace pressure, the droplets maintain their individual shape (McClements, 2005; Vanapalli et al., 2002). The crystal network collapses when the fat is melted and the two droplets aggregate (i.e., true coalescence). In some dairy products, for example, ice creams and whipped toppings, partial coalescence is essential for the characteristic texture and mouthfeel.

10.5.3 Particle Size Distribution

As discussed earlier, the crystallizing temperature of oil droplets is dependent on droplet size. Therefore, qualitative information regarding the size of oil droplets could be obtained from different thermograms (Clausse et al., 2005).

Theoretically, quantitative information on droplet size may also be deduced from thermograms. The theory is based on the fact that formation of ice nuclei is required to induce icing, and there is a quantitative relation between droplet radius R and freezing temperature T:

$$R^3 = 0.69 \frac{3T}{4\pi \int_{T_f}^{T} A.\exp\left[-\frac{16\pi\gamma^3 V_s}{3L_f^2 \ln^2\left(T/T_f\right)}\frac{1}{kT}\right]dT} \qquad (10.3)$$

where:

T (K s^{-1})   is the scanning rate
A (s^{-1} m^{-3})   is the preexponential factor in the expression of the nucleation rate
k (N m K^{-1})   is the Boltzmann constant
γ (N m^{-1})   is the interfacial tension between water and the ice germ
V_s (m^3 mole^{-1}) is the molar volume of the germ
L_f (N m mole^{-1}) is the molar melting heat
T_f (K)   is the melting temperature (Dalmazzone et al., 2009)

10.5.4 Other Applications

Some other physicochemical properties of emulsions, including specific heat capacity (Jamil et al., 2011), equilibrium liquid temperature (Kousksou et al., 2007; Jamil et al., 2011), and dispersed phase volume fraction (Dalmazzone et al., 2009), can also be determined by DSC. As they are not widely used to describe emulsions except for specific applications, method details are not discussed in this chapter.

10.6 THERMAL BEHAVIOR OF PROTEINS IN EMULSIONS

Proteins, especially milk proteins, are present in many food emulsions, and mainly play the role as emulsifiers due to their unique amphiphilic properties

(Hoffmann and Reger, 2014). The emulsifying capacity of proteins is highly dependent on protein structure, which can be modified when the proteins are heat-treated. When protein is heated at temperatures above 65°C, the proteins unfold and expose more hydrophobic groups. Furthermore, unfolded proteins are capable of interacting with themselves or other food ingredients. Depending on thermal condition (e.g., heating rate, heating time, and final temperature) and environmental conditions (e.g., pH, ion strength, and other ingredients), change of protein structure may lead to enhanced emulsifying properties and improved emulsion stability, and may result in protein aggregates and emulsion collapse (Raikos, 2010). Thermal information from DSC can be used to track the structural change and determine equilibrium thermodynamic stability and folding mechanism of the proteins, as well as the interactions between the proteins and other food ingredients (Johnson, 2013).

Protein denaturation is the process of unfolding of the globular structure, and is regarded as an endothermic process, as energy is required to break intramolecular bonds (noncovalent or disulfide). Figure 10.19a shows the thermogram of WPI (3.0 wt%) in distilled water. The endothermic peak at 75°C is due to the denaturation of β-lactoglobulin, while the shoulder peak at 62°C represents the denaturation of α-lactabumin. When 100 mM NaCl was added into the solution, an exothermic peak was detected at 87°C, while at lower NaCl concentration the thermograms were devoid of detectable exothermic peaks (Fitzsimons et al., 2007). The exothermic peak was attributed to the aggregation of the unfolded protein molecules, a process that involves the formation of new intermolecular bonds. However, the exothermic peak was obscured when the denaturation proceeded further by increasing NaCl content or protein concentration during heating. In fact, the aggregation process affects protein denaturation as well. The study showed that with increased aggregate size the enthalpy values of denaturation increased to a maximum value and

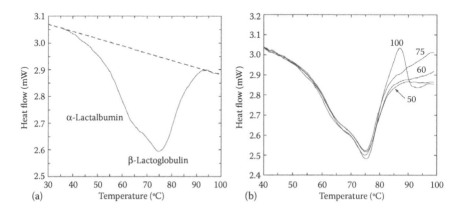

FIGURE 10.19   DSC heating scan (1.0°C min^{-1}) of 3.0 wt% WPI in water (a) and in solution containing 50, 60, 75, and 100 mM NaCl (b). (Reprinted from *Food Hydrocolloids*, 21, Fitzsimons, S.M., Mulvihill, D.M., and Morris, E.R., Denaturation and aggregation process in thermal gelation of whey proteins resolved by differential scanning calorimetry, 638–644, Copyright (2007), with permission from Elsevier.)

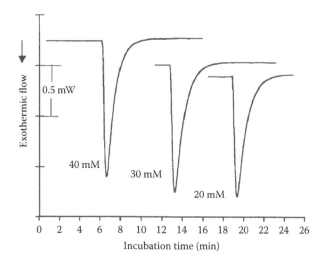

FIGURE 10.20 Differential scanning calorimetry exothermic curves obtained by incubation of 10 wt% WPI solution with 20, 30, or 45 mM $CaCl_2$ at 45°C for 1 h. (Reprinted from *Journal of Dairy Science*, 82, Ju, Z.Y., Hettiarachchy, N., and Kilara, A., Thermal properties of whey protein aggregates, 1882–1889, Copyright (1999), with permission from Elsevier.)

then decreased, but change of aggregate size affected the denaturation temperature. Another study revealed that addition of $CaCl_2$ resulted in more profound protein aggregation (Ju et al., 1999). Clear exothermic peaks (large enthalpy values) were detected when protein solutions were incubated at 45°C for 1 h in the presence of >20 mM $CaCl_2$, and the onset temperatures of the exothermic event were dependent on the concentration of $CaCl_2$ (Figure 10.20). When the incubation temperature was reduced, the occurrence of exothermic peaks was markedly delayed. This finding suggested that the aggregation of protein involves breakdown of hydrophobic bonds (Ma and Harwalkar, 1988).

In emulsion systems, polysaccharides are usually present to stabilize the emulsions by increasing the system viscosity. In fact, polysaccharide can interact with protein and then modify the emulsifying properties of the proteins (Dickinson, 2011). Figure 10.21 illustrates the thermograms of β-lactoglobulin and a mixture of β-lactoglobulin and pectin. As indicated, the presence of pectin modified not only the onset temperature of the exothermic peak, but also the enthalpy. When the protein–pectin complex was used to stabilize emulsions, the emulsions had improved stability against lower pH or higher salt concentration (Benjamin et al., 2012).

10.7 CONCLUSION

Food emulsions are complex systems including many ingredients with different structures and in different phases. Under thermal treatment, both properties of the ingredients and structures of the emulsions will change, and subsequently influence the functionalities of the emulsions. Thermal changes during processing reflect the

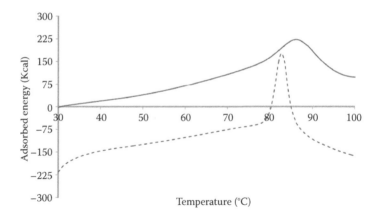

FIGURE 10.21 Differential scanning calorimetry thermogram of β-lactoglobulin (solid line) and β-lactoglobulin + pectin (dashed line) solution at pH 4.0. (Reprinted from *International Dairy Journal*, 26, Benjamin, O., Lassé, M., Silcock, P., and Everett, D.W., Effect of pectin adsorption on the hydrophobic binding sites of β-lactoglobulin in solution and emulsion systems, 36–40, Copyright (2012), with permission from Elsevier.)

nature of some food ingredients. DSC with a well-controlled temperature program can be used to simulate/evaluate the historical thermal process. Calorimetric information from DSC diagrams can be directly used to understand the thermal transitions that the food systems may undergo during processing or storage: for example, crystallization, melting, mass transfer, and protein denaturation. DSC is easy to operate and in most cases no special sample preparation is required. Many examples discussed in this chapter showed that DSC is effective in characterizing food emulsions, and in some cases it presented advantages over conventional approaches. It should be noted that the calorimetric information obtained is more useful for qualitative evaluation than for quantitative determination. When calibration curves are plotted based on the areas of the entropic peaks, the calorimetric results can be used to precisely calculate the degree of thermal transition. In this regard, the operation of DSC and analysis of the results should be carefully manipulated to ensure high reproducibility. Furthermore, the DSC technique can be used in combination with other techniques, for example, light scattering or x-ray diffraction, to better characterize food emulsions.

REFERENCES

Anker, M., Berntsen, J., Hermansson, A.M., and Stading, M. (2002). Improved water vapour barrier of whey protein films by addition of an acetylated monoglyceride. *Innovative Food Science and Emerging Technologies*, 3, 81–92.

Avendano-Gomez, J. R., Grossiord, J. L., and Clausse, D. (2005). Study of mass transfer in oil-in-water multiple emulsions by differential scanning calorimetry. *Journal of Colloid and Interface Science*, 290, 533–545.

Awad, T., Hamada, Y., and Sato, K. (2001). Effects of addition of diacylglycerols on fat crystallization in oil-in-water emulsion. *European Journal of Lipid Science and Technology*, 103, 735–741.

Awad, T. and Sato, K. (2002). Acceleration of crystallization of palm kernel oil in oil-in-water emulsion by hydrophobic emulsifier additives. *Colloids and Surfaces B: Biointerfaces*, 25, 45–53.

Barth, H. G. and Sun, S. T. (1993). Particle size analysis. *Analytical Chemistry*, 65, 55R–66R.

Batte, H. D., Wright, A. J., Rush, J. W., Idziak, S. H. J., and Marangoni, A. G. (2007). Phase behaviour, stability, and mesomorphism of monostearin-oil-water gels. *Food Biophysics*, 2(1), 29–37.

Benjamin, O., Lassé, M., Silcock, P., and Everett, D. W. (2012). Effect of pectin adsorption on the hydrophobic binding sites of β-lactoglobulin in solution and emulsion systems. *International Dairy Journal*, 26, 36–40.

Berton, C., Ropers, M. H., Bertrand, D., Viau, M., and Genot, C. (2012). Oxidative stability of oil-in-water emulsions stabilised with protein or surfactant emulsifiers in various oxidation conditions. *Food Chemistry*, 131, 1360–1369.

Boots, J. W. P., Chupin, V., Killian, J. A., Demel, R. A., and de Kruijffe, B. (1999). Interaction mode specific reorganization of gel phase monoglyceride bilayers by β-lactoglobulin. *Biochimica et Biophysica Acta*, 1420, 241–251.

Borwankar, R. P., Lobo, L. A., and Wasan, D. T. (1992). Emulsion stability—Kinetics of flocculation and coalescence. *Colloids and Surfaces*, 69, 135–146.

Calligaris, S., Pieve, S. D., Arrighetti, G., and Barba, L. (2010). Effect of the structure of monoglyceride–oil–water gels on aroma partition. *Food Research International*, 43(3), 671–677.

Carvajal, P. A., MacDonald, G. A., and Lanier, T. C. (1999). Cryostabilization mechanism of fish muscle proteins by maltodextrins. *Cryobiology*, 38, 16–26.

Chanamai, R. and McClements, D. J. (2000). Impact of weighting agents and sucrose on gravitational separation of beverage emulsions. *Journal of Agricultural and Food Chemistry*, 48, 5561–5565.

Chang, B. S., Kendrick, B. S., and Carpenter, J. F. (2000). Surface-induced denaturation of proteins during freezing and its inhibition by surfactants. *Journal of Pharmaceutical Sciences*, 85, 1325–1330.

Charcosset, C., Limayem, I., and Fessi, H. (2004). The membrane emulsification process—A review. *Journal of Chemical Technology and Biotechnology*, 79, 209–218.

Chen, C. H. and Terentjev, E. M. (2010). Effect of water on aggregation and stability of monoglycerides in hydrophobic solutions. *Langmuir*, 26, 3095–3105.

Clausse, D., Drelich, A., and Fouconnier, B. (2011). Mass transfer within emulsions studied by differential scanning calorimetry (DSC)— Application to composition ripening and solid ripening. In H. Nakajima (ed.), *Mass Transfer—Advanced Aspects*, pp. 743–778. Rijeka, Croatia: InTech.

Clausse, D., Gomez, F., Pezron, I., Komunjer, L., and Dalmazzone, C. (2005). Morphology characterization of emulsions by differential scanning calorimetry. *Advances in Colloid and Interface Science*, 117, 59–74.

Clausse, D., Pezron, I., and Gauthier, A. (1995). Water transfer in mixed water-in-oil emulsions studied by differential scanning calorimetry. *Fluid Phase Equilibria*, 110, 137–150.

Cornacchia, L. and Roos, Y. H. (2011). Lipid and water crystallization in protein-stabilized oil-in-water emulsions. *Food Hydrocolloids*, 25, 1726–1736.

Dalmazzone, C., Noïk, C., and Clausse, D. (2009). Application of DSC for emulsified system characterization. *Oil and Gas Science and Technology*, 5, 543–555.

Dickinson, E. (2011). Mixed biopolymers at interface: Competitive adsorption and multilayer structure. *Food Hydrocolloids*, 25, 1966–1983.

Dickinson, E., Kruizenga, F. J., Povey, M. J. W., and Vandermolen, M. (1993). Crystallization in oil-in-water emulsions containing liquid and solid droplets. *Colloids and Surfaces A: Physicochemical and Engineering Aspects*, 81, 273–279.

Donsì, F., Wang, Y., and Huang, Q. (2011). Freeze-thaw stability of lecithin and modified starch-based nanoemulsions. *Food Hydrocolloids*, 25, 1327–1336.

Elwell, M., Roberts, R. F., and Coupland, J. N. (2004). Effect of homogenization and surfactant type on the exchange of oil between emulsion droplets. *Food Hydrocolloids*, 18, 413–418.

Euston, S. R., Finnigan, S. R., and Hirst, R. L. (2001). Heat-induced destabilization of oil-in-water emulsions formed from hydrolysed whey protein. *Journal of Agricultural and Food Chemistry*, 49, 5576–5583.

Fellows, P. (2000). *Food Processing Technology: Principles and Practice*, 2nd edn. Chichester, UK: Ellis Horwood.

Fitzsimons, S. M., Mulvihill, D. M., and Morris, E. R. (2007). Denaturation and aggregation process in thermal gelation of whey proteins resolved by differential scanning calorimetry. *Food Hydrocolloids*, 21, 638–644.

Garti, N. and Bisperink, C. (1998). Double emulsions: Progress and applications. *Current Opinion in Colloid and Interface Science*, 3, 657–667.

Ghosh, S. and Rousseau, D. (2009). Freeze-thaw stability of water-in-oil emulsions. *Journal of Colloids and Interface Science*, 339, 91–102.

Hartel, R. W. (2001). *Crystallization in Foods*. Gaithersburg, MD: Aspen.

Hillgren, A., Lindgren, J., and Alden, M. (2002). Protection mechanism of Tween 80 during freeze-thawing of a model protein, LDH. *International Journal of Pharmaceutics*, 237, 57–69.

Himawan, C., Starov, V. M., and Stapley, A. G. F. (2006). Thermodynamic and kinetic aspects of fat crystallization. *Advances in Colloid and Interface Science*, 122, 3–33.

Hoffmann, H. and Reger, M. (2014). Emulsions with unique properties from proteins as emulsifiers. *Advances in Colloid and Interface Science*, 205, 94–104.

Höhne, G. W. H. (1997). Fundamentals of differential scanning calorimetry and differential thermal analysis. In V. B. F. Mathot (ed.), *Calorimetry and Thermal Analysis of Polymers*, pp. 47–91. Munich: Hanser.

Hunter, R. J. (1986). *Foundations of Colloid Science*. Oxford, UK: Oxford Science.

Hur, S. J., Decker, E. A., and McClements, D. J. (2009). Influence of initial emulsifier type on microstructural changes occurring in emulsified lipids during in vitro digestion. *Food Chemistry*, 114, 253–262.

Hwang, J. K., Kim, Y. S., and Pyun, Y. R. (1991). Comparison of the effect of soy protein isolate concentration on emulsion stability in the absence or presence of monoglyceride. *Food Hydrocolloids*, 5(3), 313–317.

Ivanov, I. B., Danov, K. D., and Kralchevsky, P. A. (1999). Flocculation and coalescence of micron-size emulsion droplets. *Colloids and Surfaces A: Physicochemical and Engineering Aspects*, 152, 161–182.

Jamil, A., Caubet, S., Grassl, B., Kousksou, T., El Omari, K., Zeraouli, Y., and Le Guer, Y. (2011). Thermal properties of non-crystallizable oil-in-water highly concentrated emulsions. *Colloids and Surfaces A: Physicochemical and Engineering Aspects*, 382, 266–273.

Johnson, C. M. (2013). Differential scanning calorimetry as a tool for protein folding and stability. *Archives of Biochemistry and Biophysics*, 531, 100–109.

Ju, Z. Y., Hettiarachchy, N., and Kilara, A. (1999). Thermal properties of whey protein aggregates. *Journal of Dairy Science*, 82, 1882–1889.

Kalnin, D., Schafer, O., Amenitsch, H., and Ollivon, M. (2004). Fat crystallization in emulsion: Influence of emulsifier concentration on triacylglycerol crystal growth and polymorphism. *Crystal Growth and Design*, 4, 1283–1293.

Kousksou, T., Jamil, A., Zeraouli, Y., and Dumas, J. P. (2007). Equilibrium liquidus temperatures of binary mixtures from differential scanning calorimetry. *Chemical Engineering Science*, 62, 6516–6523.

Krog, N. J. (2001). Crystallization properties and lyotropic phase behaviour of food emulsifiers. In N. Garti and K. Sato (eds), *Crystallization Processes in Fats and Lipid Systems*, pp. 505–526. New York: Marcel Dekker.

Krog, N. J. and Sparsø, F. V. (2004). Food emulsifiers: Their structures and properties. In S. E. Friberg, K. Larsson, and J. Sjöblom (eds), *Food Emulsions*, 4th edn., pp. 44–90. New York: Marcel Dekker.

Leenhouts, J. M., Demel, R. A., de Kruijff, B., and Boots, J. W. P. (1997). Charge-dependent insertion of β-lactoglobulin into monoglyceride monolayers. *Biochimica et Biophysica Acta*, 1330, 61–70.

Ma, C. Y. and Harwalkar, Z. M. (1988). Studies of thermal transitions of glycinin determined by differential scanning calorimetry. *Journal of Food Science*, 53, 531–534.

Mao, L., Calligaris, S., Barba, L., and Miao, S. (2014). Monoglyceride self-assembled structure in O/W emulsion: Formation, characterization and its effect on emulsion properties. *Food Research International*, 58, 81–88.

Mao, L., O'Kennedy, B. T., Roos, Y. H., Hannon, J. A., and Miao, S. (2012). Effect of monoglyceride self-assembled structure on emulsion properties and subsequent flavour release. *Food Research International*, 48, 233–240.

Mao, L., Roos, Y. H., and Miao, S. (2013). Volatile release from self-assembly structured emulsions: Effect of monoglyceride content, oil content, and oil type. *Journal of Agricultural and Food Chemistry*, 61, 1427–1434.

Mao, L., Xu, D., Yang, J., Yuan, F., Gao, Y., and Zhao, J. (2009). Effects of small and large molecule emulsifiers on the characteristics of β-carotene nanoemulsions prepared by high pressure homogenization. *Food Technology and Biotechnology*, 47, 336–342.

McClements, D. J. (2005). *Food emulsions: Principles, practices and techniques*, 2nd edn. Boca Raton, FL: CRC Press.

McClements, D. J., Decker, E. A., and Weiss, J. (2007). Emulsion-based delivery systems for lipophilic bioactive components. *Journal of Food Science*, 72, R109–R124.

McClements, D. J., Dickinson, E., Dungan, S. R., Kinsella, J. E., Ma, J. G., and Povey, M. J. W. (1993a). Effect of emulsifier type on the crystallization kinetics of oil-in-water emulsions containing a mixture of solid and liquid droplets. *Journal of Colloids and Interface Science*, 160, 293–297.

McClements, D. J. and Dungan, S. R. (1993). Factors that affect the rate of oil exchange between oil-in-water emulsion droplets stabilized by a non-ionic surfactant: Droplet size, surfactant concentration, and ionic strength. *Journal of Physical Chemistry*, 97, 7304–7308.

McClements, D. J., Dungan, S. R., German, J. B., and Kinsella, J. E. (1993b). Evidence of oil exchange between oil-in-water emulsion droplets stabilized by milk proteins. *Journal of Colloid and Interface Science*, 156, 425–429.

McClements, D. J., Dungan, S. R., German, J. B., Simoneau, C., and Kinsella, J. E. (1993c). Droplet size and emulsifier type affect crystallization and melting of hydrocarbon-in-water emulsions. *Journal of Food Science*, 58, 1148–1152, 117–118.

McClements, D. J., Han, S. W., and Dungan, S. R. (1994). Interdroplet heterogeneous nucleation of supercooled liquid droplets by solid droplets in oil-in-water emulsions. *Journal of American Oil Chemists' Society*, 71, 1385–1389.

Palanuwech, J. and Coupland, J. N. (2003). Effect of surfactant type on the stability of oil-in-water emulsions to dispersed phase crystallization. *Colloids and Surfaces A: Physicochemical Engineering Aspects*, 223, 251–262.

Pelan, B. M. C., Watts, K. M., Campbell, I. J., and Lips, A. (1997). The stability of aerated milk protein emulsions in the presence of small molecule surfactants. *Journal of Dairy Science*, 80, 2631–2638.

Povey, M. J. W. (2001). Crystallization in oil-in-water emulsion. In N. Garti and K. Sato (eds), *Crystallization Processes in Fats and Lipid Systems*, pp. 251–288. New York: Marcel Dekker.

Pugnaloni, L. A., Dickinson, E., Ettelaie, R., Mackie, A. R., and Wilde, P. J. (2004). Competitive adsorption of proteins and low-molecular-weight surfactants: Computer simulation and microscopic imaging. *Advances in Colloid and Interface Science*, 107, 27–49.

Raikos, V. (2010). Effect of heat treatment on milk protein functionality at emulsion interface. A review. *Food Hydrocolloids*, 24, 259–265.

Robins, M. M. (2000). Emulsions—creaming phenomena. *Current Opinion in Colloid and Interface Science*, 5, 265–272.

Russel, W. B. (1981). Brownian motion of small particles suspended in liquids. *Annual Review of Fluid Mechanics*, 13, 425–455.

Sagalowicz, L., Leser, M. E., Watzke, H. J., and Michel, M. (2006). Monoglyceride self-assembly structures as delivery vehicles. *Trends in Food Science and Technology*, 17, 204–214.

Saito, H., Kawagishi, A., Tanaka, M., Tanimoto, T., Okada, S., Komatsu, H., and Handa, T. (1999). Coalescence of lipid emulsions in floating and freeze-thawing processes: Examination of the coalescence transition state theory. *Journal of Colloid and Interface Science*, 219, 129–134.

Sarkar, A., Goh, K. K. T., and Singh, H. (2009). Colloidal stability and interactions of milk-protein-stabilized emulsions in an artificial saliva. *Food Hydrocolloids*, 23, 1270–1278.

Scherman, P. and Parkinson, C. (1978). Mechanism of temperature induced phase inversion in O/W emulsions stabilised by O/W and W/O emulsifier blends. *Progress in Colloid and Polymer Science*, 63, 10–14.

Schultz, S., Wagner, G., Urban, K., and Ulrich, J. (2004). High-pressure homogenization as a process for emulsion formation. *Chemical Engineering and Technology*, 27, 361–368.

Skoda, W. and Van den Tempel, M. (1963). Crystallization of emulsified triglycerides. *Journal of Colloid Science*, 18, 568–584.

Sonoda, T., Takata, Y., Ueno, S., and Sato, K. (2006). Effect of emulsifiers on crystallization behavior of lipid crystals in nanometers-size oil-in-water emulsion droplets. *Crystal Growth and Design*, 6, 306–312.

Sugiura, S., Nakajima, M., and Seki, M. (2002). Preparation of monodispersed emulsion with large droplets using microchannel emulsification. *Journal of the American Oil Chemists' Society*, 79, 515–519.

Ueno, S., Hamada, Y., and Sato, K. (2003). Controlling polymorphic crystallization of *n*-alkane crystals in emulsion droplets through interfacial heterogeneous nucleation. *Crystal Growth and Design*, 3, 935–939.

Vanapalli, S. A., Palanuwech, J., and Coupland, J. N. (2002). Influence of fat crystallization on the stability of flocculated emulsions. *Journal of Agricultural and Food Chemistry*, 50, 5224–5228.

Vereecken, J., Meeussen, W., Foubert, I., Lesaffer, A., Wouters, J., and Dewettinck, K. (2009). Comparing the crystallization and polymorphic behaviour of saturated and unsaturated monoglycerides. *Food Research International*, 42, 1415–1425.

Index